MW01612080

Ruwan's Civil PE Morning Breadth Exam Handbook

by Ruwan Rajapakse, PE, CCM, CCE, AVS

Printed in the United States of America

Library of Congress cataloging in Publication Data

Civil PE Morning Breadth Exam Handbook

ISBN-10:   1939493064
ISBN-13:   978-1-939493-06-4

 www.geotechweb.com

<u>Must Have for the Morning Session........</u>
<u>Four sample Exams for the Civil</u>
<u>PE Exam (Second Edition)</u>

All Modules Covered.....

Exam type problems and step by step solutions

- **WATER RESOURCES/ENVIRONMENTAL**
- **GEOTECHNICAL**
- **TRANSPORTATION**
- **CONSTRUCTION AND**
- **STRUCTURAL**

Ruwan Books

Four Sample Exams for
the Civil PE Exam
All Modules
Second Edition
With Step by Step Solutions

(WATER RESOURCES/ENVIRONMENTAL,
GEOTECHNICAL, TRANSPORTATION,
CONSTRUCTION AND STRUCTURAL)

Ruwan Rajapakse, PE, CCM, CCE, AVS

<u>FOUR SAMPLE EXAMS FOR THE CIVIL PE EXAM (ALL MODULES) (Second Edition)</u>

<u>As you know learning theory alone is not enough to pass the exam. Practicing problems is extremely important. This book is designed to provide the student to practice problems. There are four sample exams with step by step solutions provided.</u>

Four sample exams covering all five modules. (Water Resources, Transportation, Structural, Geotechnical, Construction);
* Illustrations and photographs
* All problems are similar to the real PE exam
321 pages

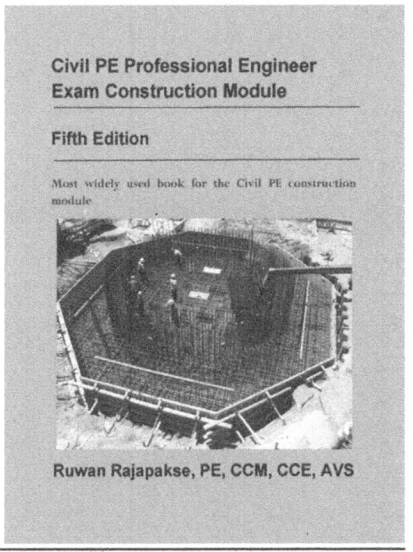

Civil PE Construction Module **Fifth Edition**

Most widely used book for the Construction Module. New Fifth Edition has been changed to match the highly evolving Construction module. New chapters has been added in

- Concrete construction, ACI testing methods, formwork design and construction (SP 4)
- Steel fabrication, shop drawing preparation and erection
- Scheduling, resource allocation and leveling, CPM methods

451 pages

CIVIL PE CONSTRUCTION MODULE PRACTICE PROBLEMS
Second Edition Now Available

Hundreds of Exam type practice problems -
Mass - Haul Diagrams (Free haul distance, waste, surplus)
- Borrow pit computations, Scheduling (CPM, Activity on arrow networks, Floats, Resource leveling)
- Concrete (Mixing, Placement, Reinforcements, Slump tests, Mass concrete, joints, pre-stressed concrete, post tensioned concrete, admixtures), Formwork (Formwork for slabs, columns, shoring, re-shoring), Steel and Timber, Masonry (Brick, concrete block, mortar mix, estimating)
- Equipment Production (Tower cranes, Mobile cranes, Backhoes, Graders, Dozers)
- Paperback: 301 pages, Publisher: Ruwan Rajapakse;

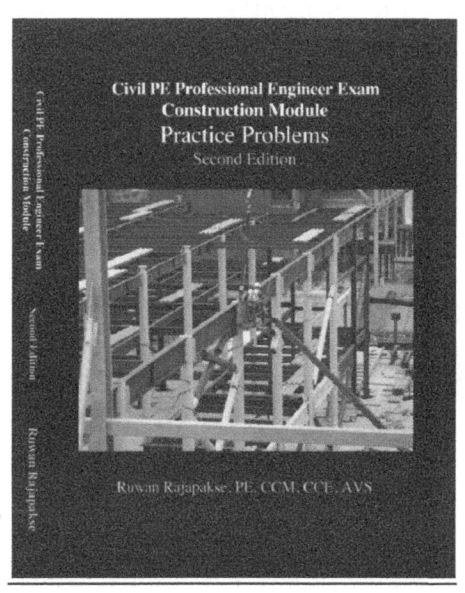

CIVIL PE GEOTECHNICAL MODULE

This book covers the depth portion of the Geotechnical module. Also this book will be great help for construction students as well. Geotechnical is the sub module of construction. All the subject matter needs to pass the geotechnical module is covered in this book.

Hundreds of Exam type practice problems;   Detailed solutions and Illustrations.

UNITS:

fps Units	SI Units
Length	
1 ft = 0.3048 m	1 m = 3.28084 ft
1 inch = 2.54 cm	
Pressure	
1 ksf = 1,000 psf	1 Pascal = 1 N/m^2
1 ksf = 0.04788 MPa	1 MPa = 20.88543 ksf
1 ksf = 47,880.26 Pascal	1 MPa = 145.0377 psi
1 ksf = 47.88 kPa	1 kPa = 0.020885 ksf
1 psi = 6,894.757 Pascal	1 kPa = 0.1450377 psi
1 psi = 6.894757 kPa	1 bar = 100 kPa
1 psi = 144 psf	
Area	
1 ft^2 = 0.092903 m^2	1 m^2 = 10.76387 ft^2
1 ft^2 = 144 in^2	1 Acre = 43,560 sq. ft
Volume	
1 ft^3 = 0.028317 m^3	1 m^3 = 35.314667 ft^3
1 gallon = 8.34 lbs	1 cu. ft = 7.48 gallons
Density	
1 lbs/ft^3 = 157.1081 N/m^3	1 kN/m^3 = 6.3658 lbs/ft^3
1 lbs/ft^3 = 0.1571081 kN/m^3	
Weight:	
1 kip = 1,000 lbs	1 kg = 9.80665 N
1 lb = 0.453592 kg	1 kg = 2.2046223 lbs
1 lb = 4.448222 N	1 N = 0.224809 lbs
1 ton (short) = 2,000 lbs	1 N = 0.101972 kg
1 ton = 2 kips	1 kN = 0.224809 kips

DENSITY OF WATER

1 g per cubic centimeter = 1,000 g per liter = 1,000 kg/m^3

    = 62.42 pounds per cu. feet

Table of Contents

I.0 Project Planning - 4 Questions

Projects need to be completed on time and on budget. Hence it is important to conduct a good cost estimate and a well thought out project schedule. This chapter is concerned of quantity takeoff, cost estimating and scheduling.

1.1 Quantity take-off methods:

The first step in cost estimating is to conduct a quantity take off using design drawings. You are required to obtain quantities using design drawings. During this step, estimator will find out how much concrete, rebars, steel and all other material needed to complete the project. Following examples will help the student to obtain quantities of formwork, concrete, steel and masonry.

1.1.1 Formwork and Concrete

Practice Problem: Find the following quantities using the design drawing given.
   a)   Volume of soil to be excavated
   b)   Concrete volume for footings
   c)   Area of formwork required for footings

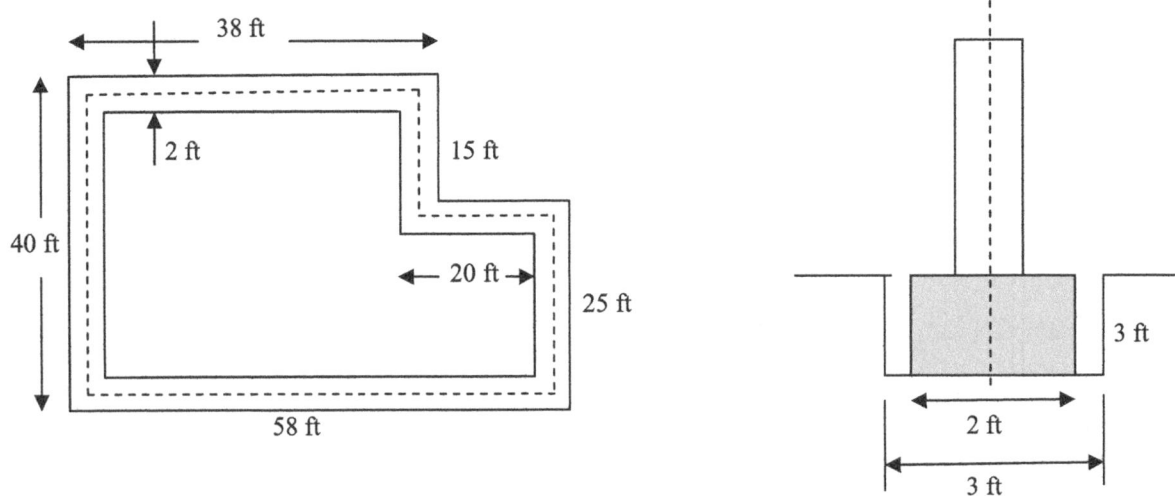

Plan of the Footing is shown above. Please note that the plan does not show the excavation or the wall. Plan view shows only the footing. Thickness of the footing is 2 ft.

Solution: Note: Length of footing needs to be measured along the centerline.
STEP 1)   Volume of soil to be excavated = (3 x 3) x length of the footing
Length of the footing along center line of the footing = (38 – 1 – 1) + 15 + 20 + (25 – 1– 1) + (58 - 1- 1) + (40 – 1 – 1)
= 188 ft
Volume of soil to be excavated = (3 x 3) x 188 = 1,692 cu. ft = 62.7 cu. yds   (1 cu. yd = 27 cu. ft)
STEP 2)   Concrete volume of footings = 2 x 3 x 188 = 1,128 cu. ft = 41.8 cu. yds
STEP 3)   Area of formwork required for footings:

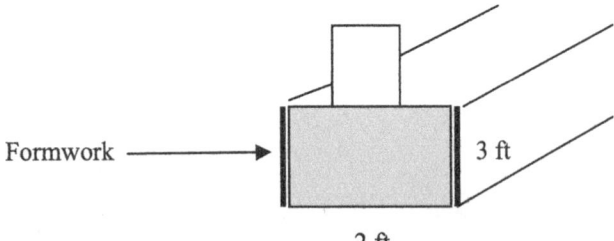

Formwork has to be erected on either side of the footing.

Formwork area for one side of the footing = (3 x 188) sq. ft = 564 sq. ft
Formwork area for both sides of the footing = 2 x 570 sq. ft = 1,128 sq. ft

Practice Problem: Find the quantities for following items for the 200 ft long retaining wall shown.
1)   Excavation volume
2)   Concrete volume
3)   Gravel volume
4)   Backfill volume

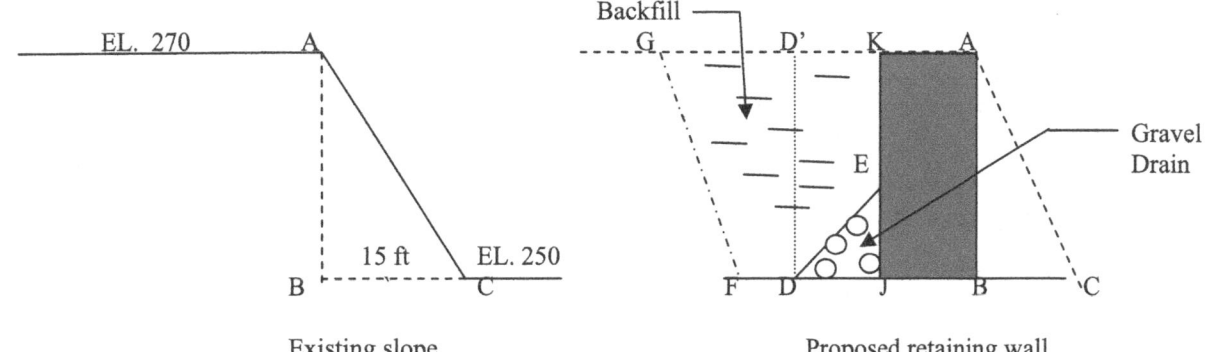

Existing slope   Proposed retaining wall

Following distances are given:
BC = 15 ft,   JB = 10 ft;   DJ = 3 ft;   EJ = 2.5 ft   FD = 3 ft;   GA = 20 ft

Construction Procedure:
STEP 1)   Excavate and remove soil volume GFCA
STEP 2)   Construct the retaining wall KAJB
STEP 3)   Construct the gravel drain EDJ
STEP 4)   Backfill GKEDF

Quantity Takeoff:
STEP 1)   Excavation volume = Area GFCA x 200   (200 ft is the length)
Area of GFCA = (GA + FC)/2 x Height KJ
GA is given to be 20 ft.
FC = BC + FB = BC + JB + DJ + FD = 15 + 10 + 3 + 3 = 31 ft
Area of GFCA = (GA + FC)/2 x Height KJ = (20 + 31)/2 x (EL 270 - El 250)
= (20 + 31)/2 x 20 = 510
Excavation volume = 510 x 200 cu. ft = 102,000 cu. ft = 3,778 cu. yds

STEP 2)   Concrete volume of the retaining wall = Area KAJB x 200 = (20 x 10) x 200
                    = 40,000 cu. ft = 1,481 cu.yds

STEP 3)   Volume of gravel required = Area EDJ x 200 = (3 x 2.5)/2 x 200 = 750 cu. ft = 28 cu. yds

STEP 4)   Backfill soil Volume = Area GKEDF x 200
Area GKEDF can be broken down to GD'DF and D'KED.
Area GD'DF = (FD + GD')/2 x D'D = (3 + 7)/2 x 20 = 100
Area D'KED = (D'D + EK)/2 x D'K = (20 + 17.5)/2 x 3 = 56.3
Area GKEDF = 100 + 56.3 = 156.3
Backfill volume = 156.3 x 200 = 31,260 cu. ft = 1,158 cu. yds

1.1.2   Masonry Quantity Takeoff

Quantity takeoff for masonry work is considered.

Practice Problem: Find the mortar volume required per brick in the wall shown. Bricks are 15" x 4" x 4".
Mortar thickness is 1 inch.

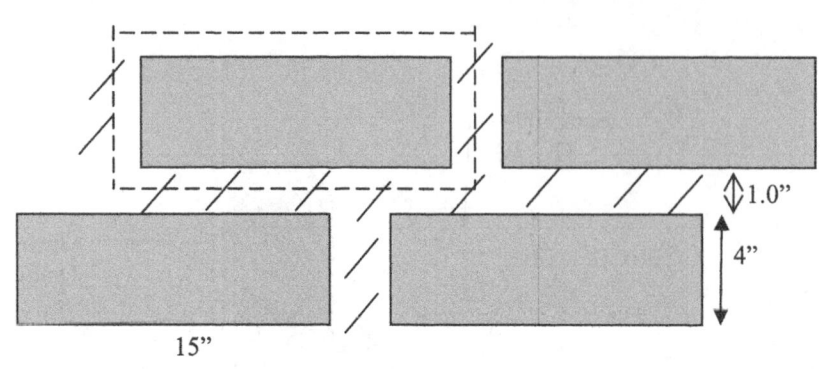

Elevation

Side View

Solution: Length of a brick with the mortar = 15 + 1/2 + 1/2 = 16 in
Height of a brick with the mortar =   4 + 1/2 + 1/2 = 5 in
Total volume of a brick with mortar = 16 x 5 x 4 = 320 cu. inches
Volume of the brick without mortar = 15 x 4 x 4 = 240 cu. inches
Volume of mortar per brick = 80 cu. inches

Practice Problem: 100 ft long, 4 in wide and 11 ft high wall is constructed using above shown brick (15" x 4" x 4") and 1.0" mortar. Find the following:
a)   Number of bricks required
b)   Volume of mortar required
Solution:
Total volume of the wall = (100 x 12) x (11 x 12) x 4 in = 633,600 cu. inches
Total volume of a brick with mortar = 16 x 5 x 4 = 320 cu. inches
(See the previous example)

Number of bricks required = Total volume of the wall/Volume of a brick with mortar = 633,600/320 = 1,980 bricks
Note: To obtain the number of bricks required for a wall, use the volume of a brick with the mortar. If you use bare volume of the brick (in this case 240 cu. in), you would get a wrong answer.

Volume of brick alone = 15 x 4 x 4 = 240 cu. in
Mortar volume in each brick = 320 - 240 = 80 cu. inches
Total mortar volume = Number of bricks x Mortar vol. per brick = 1980 x 80 = 158,400 cu. in
Your answers can be checked.
Total volume of the wall = Volume of bricks + volume of mortar
Volume of bricks only = 1980 x 240 = 475,200 cu. in
Volume of bricks + Volume of mortar = 475,200 + 158,400 = 633,600 cu. in
This is equal to the volume of the wall.

Standard Bricks: Bricks come in various sizes. Bricks known as standard bricks are 2 ¼ x 3 ¾ x 8 inches.

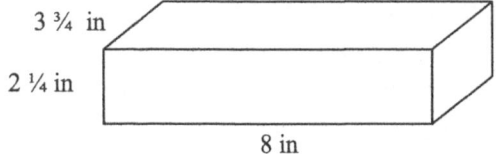

Practice Problem: A wall is built using standard bricks. The wall is one brick thick. How many bricks are required for 100 sq. ft of the wall? Assume mortar is ¼ in thick.

Solution:
Length of one brick with the mortar = 8 +1/8 + 1/8 = 8 ¼ in
Height of one brick with mortar = 2 ¼ + 1/8 + 1/8 = 2 ½ in
Area of one brick with mortar = (8 ¼) x (2 ½) = 8.25 x 2.5 = 20.625 sq. in
Bricks in 100 sq. ft of wall = (100 x144)/20.625 = 698

Practice Problem: Rectangular building has a length of 120 ft and width of 80 ft. Building walls are 9 ft high. There are total of 180 sq. ft of openings for windows and doors. Walls are built of standard bricks and are of one brick thick. Find the number of bricks and volume of mortar. Standard bricks are 8 in x 2 ¼ in x 3 ¾ in. Assume mortar thickness to be 0.5 in.

Solution:

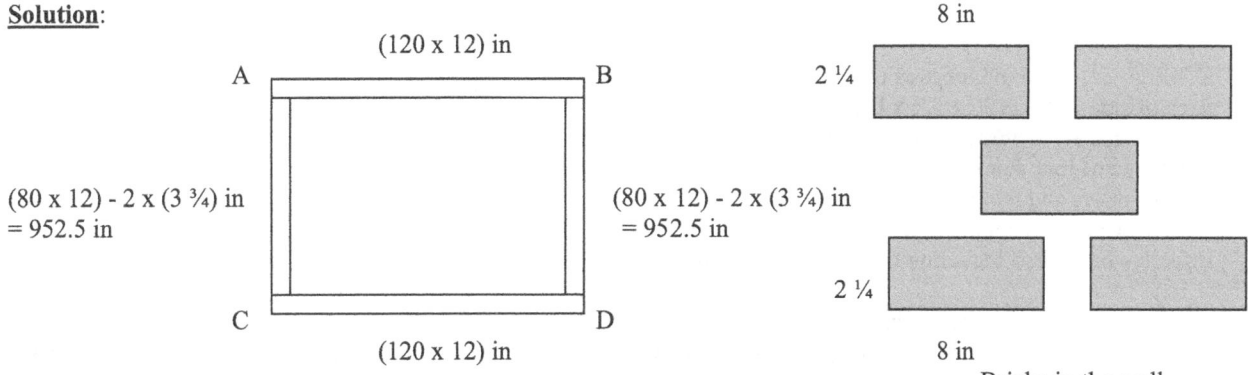

Bricks in the wall

Note: The brick at the edge need to be deducted to avoid double counting.

STEP 1:   Find the area of brick walls:
Area of side AB = (120 x 12) x height = (120 x12) x (9 x 12) = 155,520 sq. in
Area of side AC = width x height = 952.5 x (9 x 12) = 102,870 sq. in
Total area of four sides = 2 x (155,520 + 102,870) = 516,780 sq. in
Area of doors and windows = 180 sq. ft = 180 x 144 = 25,920 sq. in
Area of walls after reducing for openings = 516,780 - 25,920 = 490,860 sq. in
Effective brick area including the mortar thickness = (8 + 0.5) x (2.25 + 0.5) = 23.375 sq. in

STEP 2:   Number of bricks = 490,860/23.375 = 20,999 bricks
STEP 3:   Find the mortar volume:

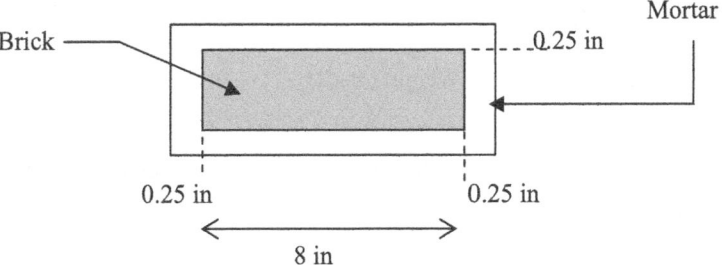

Area of mortar per brick = Total area with mortar - brick area = (8.5 x 2.75) - (8 x 2.25) = 5.375 sq. in
Wall is one brick thick. The thickness of the brick is 3.75 in.
Volume of mortar per brick = 5.375 x 3.75 = 20.156 cu. in
Number of bricks = 20,999   (See step 2)
Volume of mortar required = 20,999 x 20.156 cu. in = 423,255 cu. in = 245 cu. ft = 9.07 cu. yds

Note to candidates: Each question would have four answers. Depending upon the answers given in the exam you may opt to use a quick and dirty analysis. For an instance, area of the wall can be obtained as follows quickly, without paying attention to edges.

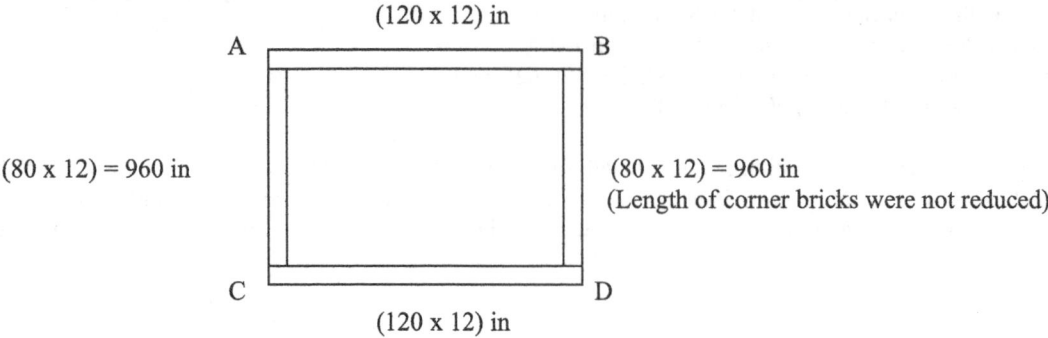

STEP 1:   Find the area of brick walls:
Area of side AB = (120 x 12) x height = (120 x12) x (9 x 12) = 155,520 sq. in
Area of side AC = 960 x (9 x 12) = 103,680 sq. in
Total area of four sides = 2 x (155,520 + 103,680)   = 518,400 sq. in
Area of doors and windows = 180 sq. ft = 25,920 sq. in
Area of walls after reducing for openings = 518,400 - 25,920 = 492,480 sq. in
Effective brick area including the mortar thickness = (8 + 0.5) x (2.25 + 0.5) = 23.375 sq. in

STEP 2:   Number of bricks = 492,480/23.375 = 21,069 bricks
As you could see, analysis that is more precise gave 20,999 bricks while quick and dirty analysis gave 21,069 bricks.

1.1.3 Quantity Takeoff (Steel)

Steel structures contain W - sections, S - sections, L - shapes, Channels, Angles, Hollow tubular sections and hollow rectangular sections.

W Shapes: (Wide Flange)

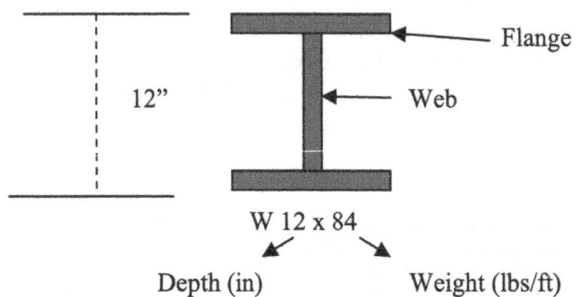

Above 12 indicates the approximate depth. Above 84 indicates weight of the section in lbs per foot.

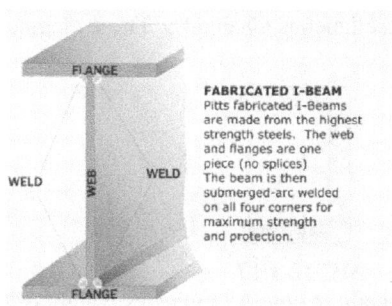

Channels: **(American Standard Channel)**

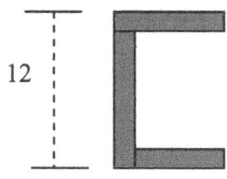

C 12 x 18.6

Above 12 indicates the approximate depth. Above 18.6 indicates weight of the channel section in lbs per foot.

S - Sections: S sections are similar to W - sections. They are known as American Standard Steel. W sections have a wider flange than S sections.

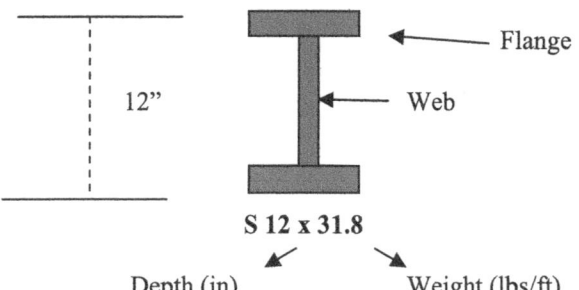

S 12 x 31.8

Depth (in)   Weight (lbs/ft)

M - Sections: M - Sections are known as "Miscellaneous Beam". These beams are manufactured by various steel manufacturers with varying thicknesses and depths.

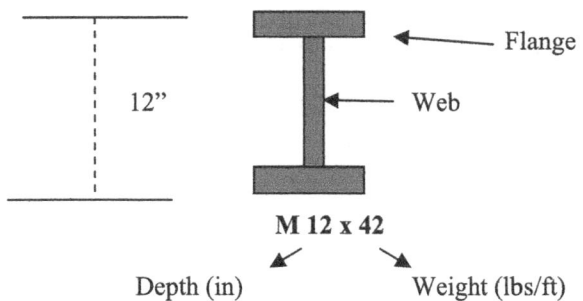

M 12 x 42

Depth (in)   Weight (lbs/ft)

MC Sections: MC stands for "Miscellaneous Channel". These channels are produced by various manufacturers with varying thicknesses and depths.

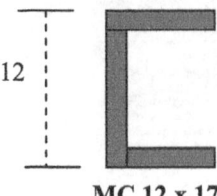

MC 12 x 17

Above 12 indicates the approximate depth. Above 17 indicates weight of the channel section in lbs per foot.

L - Angles: Angles are represented with "L". Angles may have equal or unequal legs.

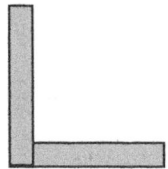

L 9 x 4 x 5/8

Above 9 indicates the length of one leg in inches and 4 indicates the length of the other leg in inches and 5/8 represents the thickness of the section. Weight of the L section has to be obtained from steel tables. Steel angles are shown below.

HP - Piles (Or simply H - Piles):   HP sections are used for piles.

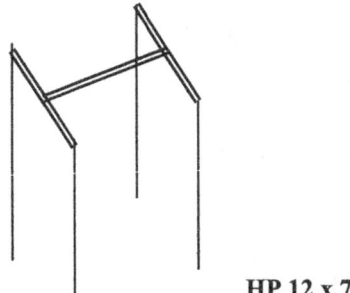

HP 12 x 72

(12 is the depth and 72 is the weight of the pile per linear foot given in lbs.

WT Sections: MT sections are cut from W (Wide flange) sections. Typically, a W section is obtained and one flange is cut off to obtain a T shape.

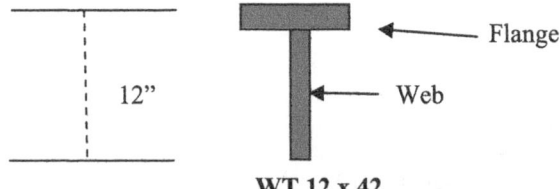

WT 12 x 42

(Above 12 is the depth and 42 is the weight per foot in lbs)

MT Sections: M section is obtained and one flange is removed to obtain a MT section. Similar to WT sections.

ST Sections: S section is obtained and one flange is removed to obtain a ST section. Similar to WT sections.

Practice Problem: Find the weight of steel in the building shown.

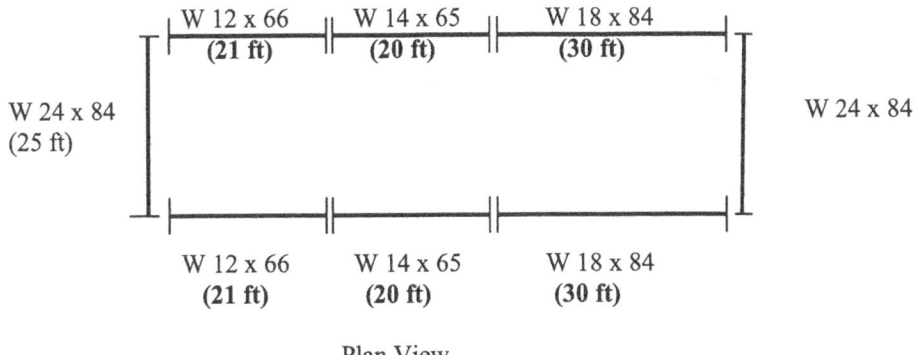

Plan View

STEP 1:   Find the weight of each beam:
W 12 x 66 (21 ft):   Weight = 21 x 66 lbs = 1,386 lbs
W 14 x 65 (20 ft)   Weight = 20 x 65 lbs = 1,300 lbs
W 18 x 84 (30 ft)   Weight = 30 x 84 lbs = 2,520 lbs
Subtotal = 5,206 lbs
Multiply by 2 to get the weight of other side. (This is possible since the structure is symmetrical).
Subtotal = 10,412 lbs

Two side beams:
W 24 x 84 (25 ft)   Weight = 25 x 84 lbs = 2,100 lbs
Subtotal = 4,200 lbs
Total weight of beams = 10,412 + 4,200 = 14,612 lbs = 7.306 tons

1.1.4 Quantity Takeoff - Reinforcement Bars (Rebars)

Concrete without rebars do not have any significant tensile strength. Tensile strength for concrete is provided by reinforcing bars or simply known as rebars. Below table provide rebar sizes and weight per foot.

Bar No:	Nominal diameter (in)		Nominal diameter (mm)	Nominal Weight (lb/ft)
3	0.375	(3/8)	9.5	0.376
4	0.500	(4/8)	12.7	0.668
5	0.625	(5/8)	15.9	1.043
6	0.750	(6/8)	19.1	1.502
7	0.875	(7/8)	22.2	2.044
8	1.000		25.4	2.670
9	1.128		28.7	3.400
10	1.270		32.3	4.303
11	1.410		35.8	5.313
14	1.693		43	7.650

Practice Problem: Construction site needs 2,000 ft of #3 rebars and 150 ft of #7 rebars. Find the total weight of rebars.

Solution: #3 rebars (0.376 lbs per foot) -   2000 x 0.376 = 752 lbs
                #7 rebars (2.044 lbs per foot) -   150 x 2.044 = 306.6
                Total weight = 1,058.6 lbs

Practice problem: 60 ft long 7.5 ft high wall needs rebars as shown. Find the total rebars needed in tons.

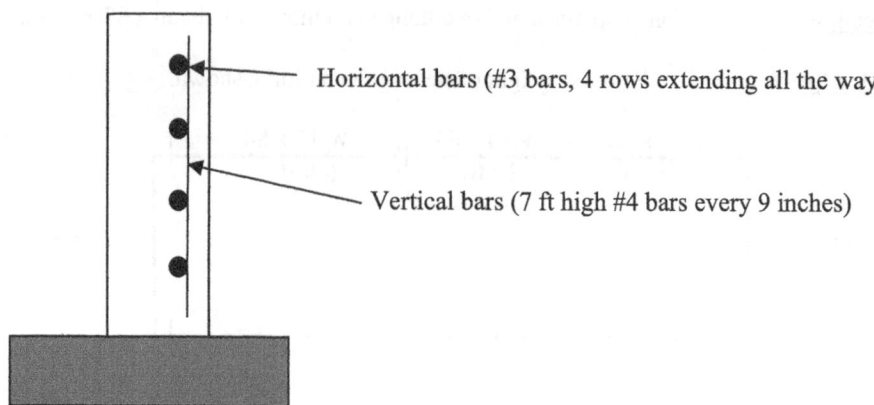

Horizontal bars (#3 bars, 4 rows extending all the way)

Vertical bars (7 ft high #4 bars every 9 inches)

Solution:

Horizontal bars: The wall is 60 ft long. Four rows of horizontal bars needed as shown in the drawing. Weight of #3 bars is 0.376 lbs per foot.

   4 x 60 x 0.376 = 90.24 lbs

Vertical bars: The wall is 60 ft long. That is 720 inches. Vertical bars are placed every 9 inches.
Number of vertical bars = 720/9 + 1= 81
Each vertical bar is 7 ft long.
Total length of vertical bars = 7 x 81 = 567 ft
Weight of 567 ft of #4 bars = 567 x 0.668 = 378.8 lbs
Total weight of rebars = 90.24 + 378.8 = 469 lbs = 0.234 tons

Practice Problem: Find the weight of steel rebars in the drilled shaft shown. Drilled shaft is 35 ft long and ties are attached every 5 ft. There are 8 ties. Diameter of the shaft is 2 ft.

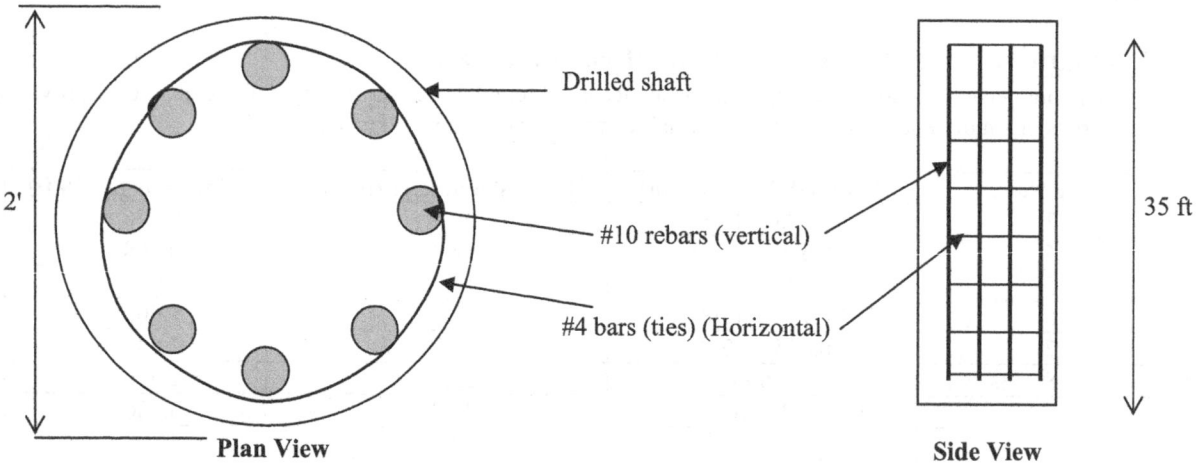

Drilled shaft

#10 rebars (vertical)

#4 bars (ties) (Horizontal)

2'

35 ft

Plan View

Side View

Solution:

STEP 1:   Length of #10 vertical bars = 35 ft
               Number of vertical bars = 8
               Total length of vertical bars = 35 x 8 = 280 ft
               Weight of #10 vertical bars = 4.303 lbs/ft
               Weight of vertical bars = 4.303 x 280 = 1,205 lbs

STEP 2:   Length of ties = Circumference of the drilled shaft (approximately)
               Length of ties = π x Diameter = π x 2 = 6.283 ft

Number of ties = 8
Total length of ties = 8 x 6.283 = 50.3 ft
Weight of ties (#4 bars) = 0.668 lbs/ft
Total weight of ties = 0.668 x 50.3 = 33.6 lbs
Total weight of rebars = 1,205 + 33.6 = 1,238.6 lbs

1.2   Cost Estimating

Cost estimating is a very important function in any project. Basic steps involved in cost estimating are;

- Quantity takeoff (Which was covered in the previous chapter)
- Assess the cost of labor, equipment and material
- Assess overhead and profits

The estimator should take into account of the location, time of the year and any other factors that could influence the cost. Cost of a project during winter could be higher than the summer. Construction project on top of a hill or middle of a major city could also be high due to efficiency, traffic and various other factors.

The estimator needs to pay attention to complexity of a project. For example, construction of a car dealership may be straight forward compared to a wastewater plant. Each company has their own specialty. A company that is specialized in building construction may not be able to construct a highway in a cost effective manner. In such situations, it may be profitable to have a subcontractor on board.

Average construction project always start with site clearing. Then the contractor may decide to build temporary roads for truck traffic and construction vehicles. Almost all construction projects will have some concreting. In today's world concrete is the most common construction material followed by steel. Following major subject areas can be seen in an average building project.

1)   Site clearing (Includes cutting down trees, removal of debris)
2)   Earthwork (Grading)
3)   Excavation for footings
4)   Pile driving (In some building projects, piles may not be necessary)
5)   Construction of footings
6)   Construction of walls and columns
7)   Construction of slabs
8)   Construction of upper floors
9)   Plumbing (sewer, water, drainage, fire standpipes)
10)   Mechanical and electrical work

Labor:
Crew hour rate is defined as cost per crew hour. Let us assume formwork crew consists of 3 carpenters plus 2 laborers. Let us say carpenters earn $50 per hour while laborers earn $30 per hour.
The crew rate would be (3 x 50 + 2 x 30) = $210 per hour. In other words if the crew cannot work for some delay, the contractor will lose $210 per hour.
Crew production rate is the output per hour by the crew. Output can be increased by increasing the crew. If the contractor adds more carpenters, he may be able to get more formwork done. On the other hand, cost of the crew will also increase.

Practice Problem: What crew is better?
Crew A: 3 carpenters + 3 laborers   (Production 100 sq. ft of formwork per hour)
Crew B: 5 carpenters + 3 laborers   (Production 125 sq. ft of formwork per hour)

Carpenter = $60 per hour   Laborer = $30 per hour

Solution:
Crew A - Cost per crew hour = (3 x 60) + (3 x 30) = $270
Crew A - Production = 100 sq. ft per crew hour
Cost per sq. ft = 270/100 = $2.7 per sq. ft of formwork

Crew B - Cost per crew hour = (5 x 60) + (3 x 30) = $390
Crew B - Production = 125 sq. ft per crew hour
Cost per sq. ft = 390/125 = $3.12 per sq. ft of formwork
The contractor has to pay $3.12 for each sq. ft of formwork if he uses crew B. Obviously, crew A is cheaper than crew B. On the other hand, crew B produces 125 sq. ft per hour compared to crew A, who produce only 100 sq. ft per hour. If the project has to be completed sooner, then crew B should be used even though that crew costs more.

Labor Hour (LH):   Examples given by NCEES guidebook uses labor hours (LH). Hence, it is important that you should be familiar with LH. Labor hour is obtained by dividing the crew hour by number of workers.

Practice Problem: Concreting crew consists of 3 masons and 2 laborers.
Mason = $70 per hour   Laborer = $30 per hour
Crew production = 10 cu. yds per hour
Find crew hour and labor hour. (LH)

Solution:
Crew hour = (3 x 70) + (2 x 30) = $270
Labor hour is obtained by dividing the cost of crew hour by total number of workers in the crew. In this case, there are total of 5 workers.

Hence cost of labor hour = $270/5 = $54
Crew production is given to be 10 cu. yds per hour.
Production per labor hour is obtained by dividing the production per crew hour by number of workers in that crew.
Production per labor hour = 10/5 = 2 cu. yds per labor hour (LH)

Labor Rates: Workers need to be paid various insurances beyond the base wages. It is important to understand benefits and fringes involved in payment to workers.

Base Pay: Base pay is the starting pay rate for a worker. Base wages are negotiated or agreed upon with the union.

Social Security Tax: All employers and employees have to pay social security tax by national law. Social security tax goes to a collective fund maintained by the federal government.

Unemployment Insurance: When workers are unemployed, they are eligible for unemployment benefits. Unemployment insurance will be maintained by the state.

Workers Compensation Insurance: Workers are entitled for payment in the case of an injury while at work. Workers compensation insurance covers injury to workers.

General Liability Insurance;   General liability insurance covers any harm or injury to third party due to the action of the worker. This insurance may cover property damage to public or any other entity.

Fringe Benefits: Fringe benefits covers workers health, vacation and pension plans.

Overtime Pay: Normal workday in USA is 8 hours a day. Typically any hours worked beyond 8 hours is paid at 50% extra known as time and a half. Usually workers get paid double time on Sundays and holidays. Some workers get double time on Saturdays while some others get time and a half. These rates are dependent upon union agreements in the locality.

Practice Problem: Mason works 11 hours per day from Monday to Friday. He works 7 hours on Saturday and 7 hours on Sunday. As per union agreement, worker is entitled to time and a half on any hours beyond 8 hours during weekdays. Worker is entitled to time and half on Saturdays and double time on Sundays.
a)   Worker is entitled to how many hours of pay?
Solution:
Monday to Friday (Hours per day) = (8 Standard time + 3 Overtime) = 8 + (3 x 1.5) = 12.5 hours

Total hours for Monday to Friday = 5 x 12.5 = 62.5 hours
Saturday = 7 x 1.5 = 10.5 hours
Sunday = 7 x 2 = 14 hours
Total hours = 62.5 + 10.5 + 14 = 87 hours

Practice Problem: Carpenter's base wage is $45 per hour. Unemployment insurance is 3% of his actual wages. Social security tax is 6% of actual wages. Worker's compensation insurance is 7% of base wages. General liability insurance is 4.5% of base wages. Fringe benefits are $5.30 per hour. Carpenter works 40 hours a week. (All standard hours).
a)   What is the weekly cost of the carpenter?
b)   What is the hourly rate of the carpenter?

Solution:
STEP 1:   Find the base wages per week;
Base wages per week = $45 x 40 = $1,800
STEP 2:   Add insurance and benefits;
Unemployment insurance 3% of actual wages = 3/100 x 1,800   = $54
Social security tax 6% of actual wages = 6/100 x 1,800   = $108
Worker's compensation insurance 7% of base wages = 7/100 x 1,800   = $126
General liability insurance 4.5% of base wages = 4.5/100 x 1,800   = $81
Fringe benefits are $5.30 per hour. Carpenter works 40 hours a week. Hence fringe benefits are 40 x 5.30 = $212
a)   Total cost per week to hire a carpenter = 1,800 + 54 + 108 + 126 + 81 + 212 = $2,381
b)   Hourly cost of the carpenter = $2,381/40 = $59.525
Base wage is computed without overtime hours. Below example makes this point clear.

Practice Problem: To meet schedule obligations, contractor is planning to have the carpenter in the above problem work additional 8 hours on Saturday. As per union agreement, contractor is required to pay time and a half for Saturdays.

Data Given: Carpenter's base wage is $45 per hour. Unemployment insurance is 3% of his actual wages. Social security tax is 6% of actual wages. Worker's compensation insurance is 7% of base wages. General liability insurance is 4.5% of base wages. Fringe benefits are $5.30 per hour. Carpenter works 40 hours a week from Monday to Friday and 8 hours on Saturday.
a)   What is the new weekly cost of the carpenter?
b)   What is the new hourly rate of the carpenter?

Solution:
STEP 1:   Find the base wages per week;
Carpenter works total 48 hours including the 8 hours on Saturday.
Base wages per week = $45 x 48 = $2,160
STEP 2:   Find the actual wages per week without insurance and benefits;
Carpenter works 40 hours during the week and 8 hours on Saturday. Saturday 8 hours has to be paid at time and half.
Hours carpenter should get paid = 40 + 1.5 x (8) = 52 hours
Actual wages without insurance and benefits = $45 x 52 = $2,340

STEP 3:   Add insurance and benefits;
Unemployment insurance 3% of actual wages = 3/100 x 2,340   = $ 70.2
Social security tax 6% of actual wages = 6/100 x 2,340   = $ 140.4
Worker's compensation insurance 7% of base wages = 7/100 x 2,160   = $151.2
General liability insurance 4.5% of base wages = 4.5/100 x 2,160   = $97.2

Fringe benefits are $5.30 per hour. Carpenter works 48 hours a week. Hence fringe benefits are 48 x 5.30 = $254.4
a)   Total cost per week to hire a carpenter = 2,340 + 70.2 + 140.4 + 151.2 + 97.2 + 254.4 = $ 3,053.4
b)   New hourly rate of the carpenter = $ 3,053.4/48 = $ 63.6
(Note: Carpenter was paid for 52 hours but he worked only 48 hours.)

Alternative Solution: Above problem can be solved using base rate and actual rate. Following is the solution using rates.

STEP 1:   Base rate = $45/hour

STEP 2:   Find the actual rate:
Carpenter works 48 hours but gets paid for 52 hours, since he get paid time and half for Saturday.

Actual rate = Base rate x hours paid/hours worked
Actual rate = 45 x 52/48 = $48.75

STEP 3:   Add insurance and benefits to the actual rate;

Hourly rate (Actual)	= $ 48.75
Unemployment insurance 3% of actual rate = 3/100 x 48.75	= $ 1.4625
Social security tax 6% of actual rate = 6/100 x 48.75	= $ 2.925
Worker's compensation insurance 7% of base rate = 7/100 x 45	= $3.15
General liability insurance 4.5% of base rate = 4.5/100 x 45	= $2.025
Fringe benefits	= $5.30 per hour

Total hourly rate = 48.75 + 1.4625 + 2.925 + 3.15 + 2.025 + 5.30 = 63.61
Total weekly wages = 48 x 63.61 = 3,053

1.3 Project Schedule
1.3.1 Activity Identification and Sequencing
It is important to sequence construction activities. Walls cannot be built without constructing the footings. Footings cannot be built before the earthwork is finished. The contractor needs to identify activities that need to be completed and sequence them.
Site clearing → Excavation for footings → Formwork for footings → Footing construction

1.3.2 Activity on Node Networks and CPM Network Analysis
There are two types of networks. They are;
- Activity on node networks
- Activity on arrow networks

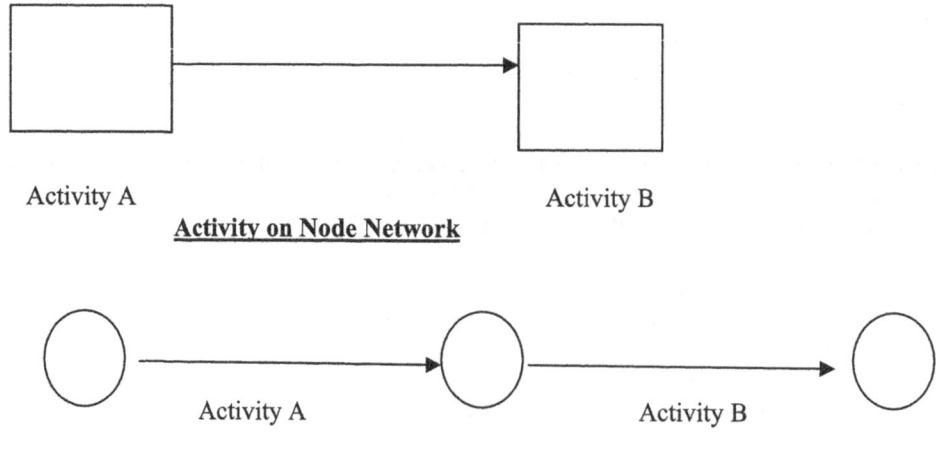

Activity A   Activity B
Activity on Node Network

Activity A   Activity B

Activity on Arrow Network

Activity on Node Networks: Let us first look at activity on node networks. It is fair to say that activity on node networks are widely used than activity on arrow networks.
Critical path method is the most popular technique adopted for scheduling using both networks.

In critical path method, all activities have following attributes.
Start time
Finish time
Duration

Let us look at the following example.

Practice Problem: A site contractor has forecasted that he would be able to complete site clearing in 10 days and construct footings in 9 days.

Site clearing (10 days)   Construction of footings (9 days)

Site clearing is the first activity. It can be started on day 1. Some authors start the activity in day 0. However, NCEES examples book uses day 1 as the first day.
Site Clearing: Early Start (ES) = 1
Site Clearing: Early Finish (EF) = 1 + 10 = 11

Construction of footings: Early Start (ES) = 11
Construction of footings: Early Finish (EF) = 11 + 9 = 20

	Activity name	
	Duration	
ES		LS
EF		LF

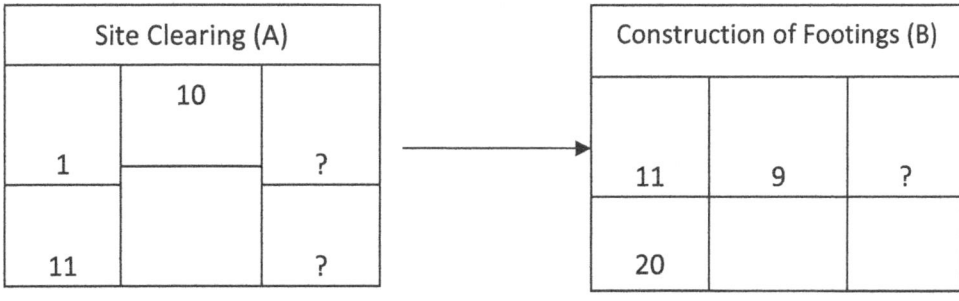

Late start (LS) and late finish (LF) has to be completed.
What is the late finish of activity B?
Early finish time of activity B is 20. Since there is no any other information available, late finish also would be 20.

Late finish of footing construction (LF) = 20
Late start of activity B = LF - duration = 20 - 9 = 11

Now late finish of activity A can be found.
Late finish of activity A = 11
Late start of activity A = LF - duration = 11 - 10 = 1

These values can be included in the schedule.

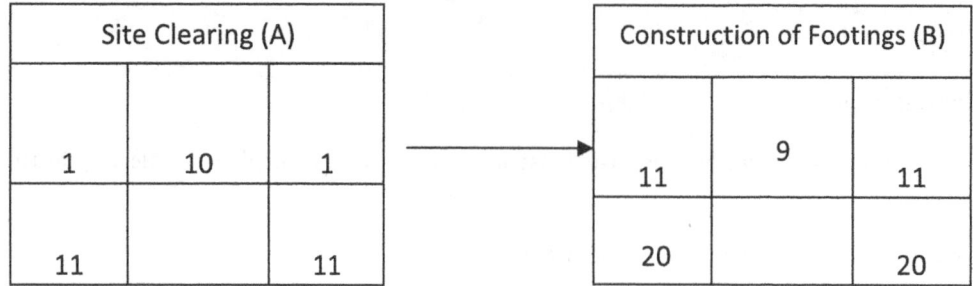

Site Clearing (A)		
1	10	1
11		11

Construction of Footings (B)		
11	9	11
20		20

Practice Problem: A building contractor can complete construction of building walls in 20 days. He is planning to start construction of the roof immediately after construction of building walls. This contractor has estimated that he needs 12 days to construct the roof.
Activity (A) = Construction of walls
Activity (B) = Construction of roof

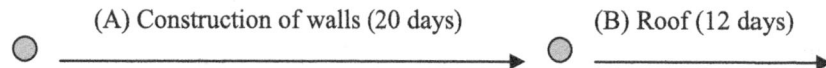

(A) Construction of walls (20 days)   (B) Roof (12 days)

Forward Pass: Forward pass is going forward starting from first activity. During forward pass, early start time (ES) and early finish time (EF) of activities will be completed.
Late start time (LS) and late finish time (LF) of activities will NOT be completed during forward pass.
LS and LF are completed during backward pass.
Construction of walls (A):   ES (Early start) = 1,   EF (Early finish) = $1 + 20 = 21$

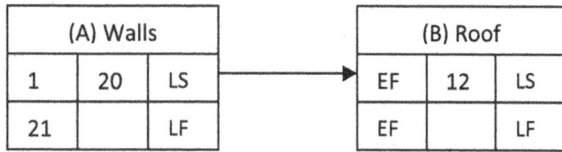

(A) Walls		
1	20	LS
21		LF

(B) Roof		
EF	12	LS
EF		LF

The contractor can start the roof construction immediately. Hence, early start (ES) of roof construction is 21.
Construction of the roof "B" :   ES = 21,   EF (Early Finish) = $21 + 12 = 33$
Early finish time (EF) of roof construction is $21 + 12 = 33$
Now these numbers can be inputted into the CPM diagram.

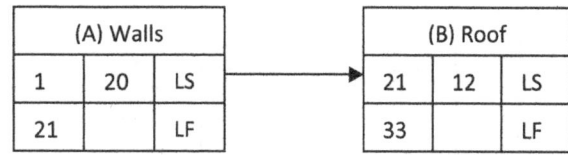

(A) Walls		
1	20	LS
21		LF

(B) Roof		
21	12	LS
33		LF

Now the forward pass is completed.

Backward Pass:
Early finish time of activity B = 33. Hence LF = 33
Now you will be able to compute the late finish time of activity "A".
Late finish time of activity A = Late start time of activity B = 21
Late start time of activity A = Late finish time of activity A – duration of A = $21 – 20 = 1$

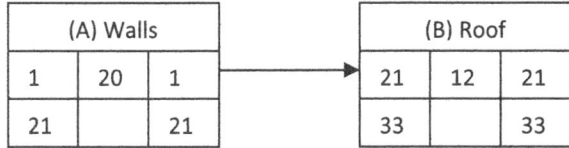

Predecessor and Successor:

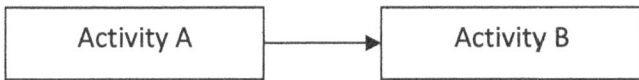

Activity A is the predecessor and activity B is the successor.

Dependence:

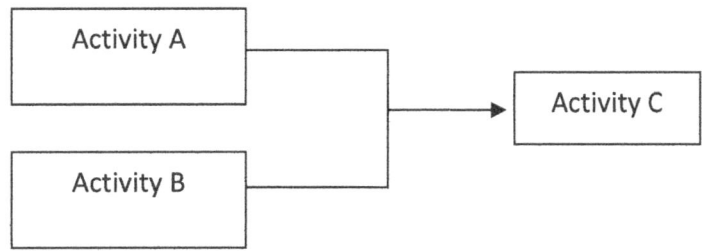

Activity C is dependent upon activity A and activity B.
Activity B does NOT depend on activity A.

Practice Problem: A contractor has to complete construction of walls and fabrication of the roof truss before constructing the roof. The roof truss is constructed outside, brought in, and installed.
Activity A = Wall construction (Duration = 20 days)
Activity B = Fabrication of the roof truss (Duration = 25 days)
Activity C = Installation of the roof (Duration = 15 days)

Activity A – No predecessors
Activity B – No predecessors
Activity C - Predecessors (A and B)
Conduct the forward and backward passes.

Solution:
STEP 1: Draw the activity diagram. (Input the durations)

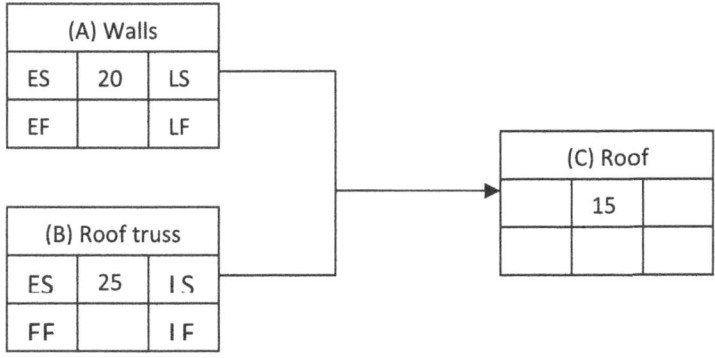

STEP 2:   Forward pass: (Find early start time and early finish time of each activity)

Activity A:
Early start time (ES) = 1
Early finish time (EF) = 21

Activity B:
Early start time (ES) = 1
Early finish time (EF) = 26

Activity C:
Activity C cannot be started until both activities A and B are completed.   Activity A is completed on 21 and B is completed on 26.
Early start time of activity C (ES) = 26
Early finish time of activity C (EF) = 26 + 15 = 41
Forward pass is completed. Input the above values in CPM diagram.

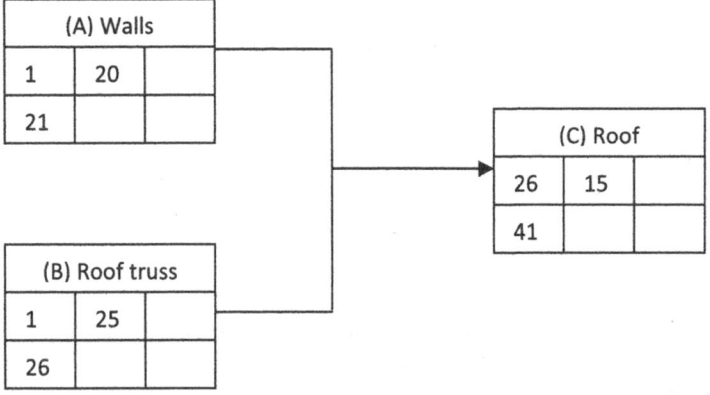

STEP 3:   Backward pass: (Late start time and late finish time of each activity)
Start to compute the late start time and late finish time of each activity starting from activity C.

Activity C: Since there is no any other information available, late finish time of activity C is 41.
Late start time of activity C = 41 – 15 = 26.

Activity B: Activity C cannot be started until B is finished. Late start time of activity C is 26. Hence, activity B has to be finished by 26.

Late finish time of activity B (LF) = 26
Late start time of activity B (LS) = 26 – 25 = 1
Activity A: Activity C cannot be started until A is finished as well. Late start time of activity C is 26. Hence, activity A has to be finished by 26.
Late finish time of activity A (LF) = 26
Late start time of activity A = 26 – 20 = 6
Look at the late start time of activity A.
Activity A **can** be started on day 6, without affecting the schedule.

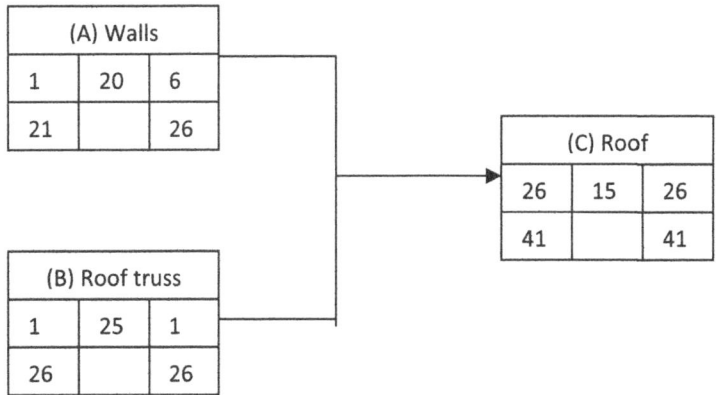

Practice Problem: Following information is given. Draw the CPM diagram.

Activity A = No Predecessors (Duration = 12 days)
Activity B = No Predecessors (Duration = 15 days)
Activity C = Predecessors (A and B), (Duration = 16 days)
Activity D = Predecessors (A and B), (Duration = 18 days)
a)   Draw the CPM diagram.
b)   Conduct the forward and backward passes.

Solution: STEP 1: Draw the activity diagram. (Input the durations)

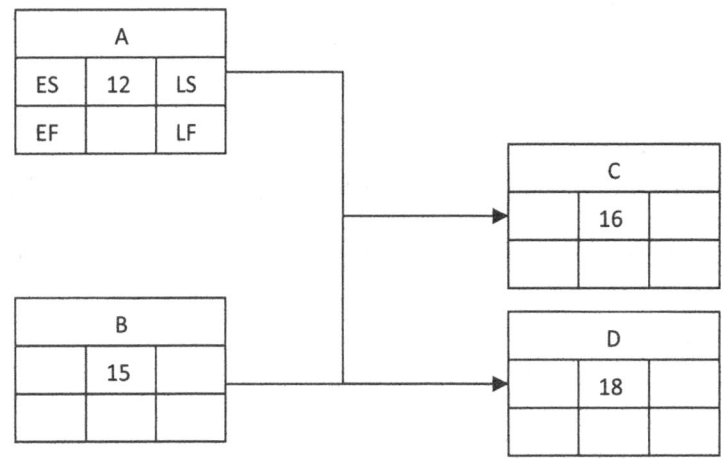

STEP 2: Forward pass: (Early start time and early finish time of each activity)
Activity A:
Early start time (ES) = 1
Early finish time (EF) = 1+ 12 = 13

Activity B:
Early start time (ES) = 1
Early finish time (EF) = 1 + 15 = 16

Activity C:
Activity C cannot be started until both activities A and B are completed. Activity A is completed on 13 and B is completed on 16.
Early start time of activity C (ES) = 16
Early finish time of activity C (EF) = 16 + 16 = 32

Activity D:
Activity D cannot be started until A and B are completed. Activity B is completed on 16 and activity A is completed on 13.
Early start time of activity D (ES) = 16
Early finish time of activity D (EF) = 16 + 18 = 34
Forward pass is completed. Input the above values in CPM diagram.

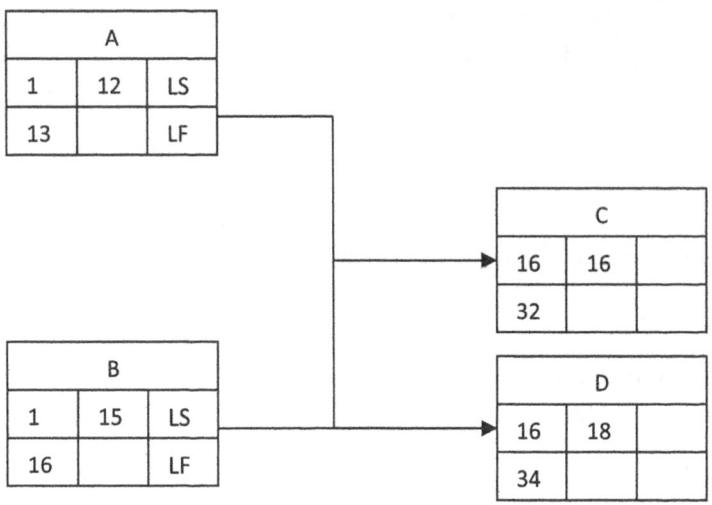

STEP 3:   Backward pass: (Late start time and late finish time of each activity)
Start to compute the late start time and late finish time of each activity from activity D.

Activity D: Late finish time of activity D is 34.
Late start time of activity D = 34 – 18 = 16

Activity C: Activity C can be completed by 32. However, it can be delayed until day 34 without affecting the project schedule. Why?
Activity D will be completed by day 34 for the project to be completed. Hence, if required, activity C can be delayed until day 34.
Late finish time of activity C (LF) = 34
Late start time of activity C (LS) = 34 – 16 = 18
Let us input the numbers we acquired so far in the CPM diagram.

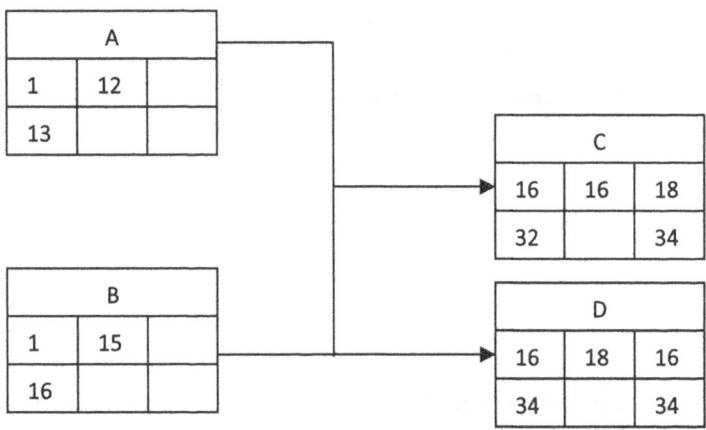

Activity B: Activity C and D cannot be started until B is finished. Late start time of activity C is 18 and late start time of activity D is 16. Hence, activity B has to be finished by 16.

If activity B is finished after 16, activity D cannot be started by 16.

Late finish time of activity B (LF) = 16

Late start time of activity B = 16 – 15 = 1

Activity A: Activity A has to be completed to start activities C and D.

Late start time of activity C is 18 and late start time of activity D is 16.   Hence, activity A has to be finished by day 16.

Late finish time of activity A (LF) = 16

Late start time of activity A (LS) = 16 – 12 = 4

Now you can input the numbers in the CPM diagram.

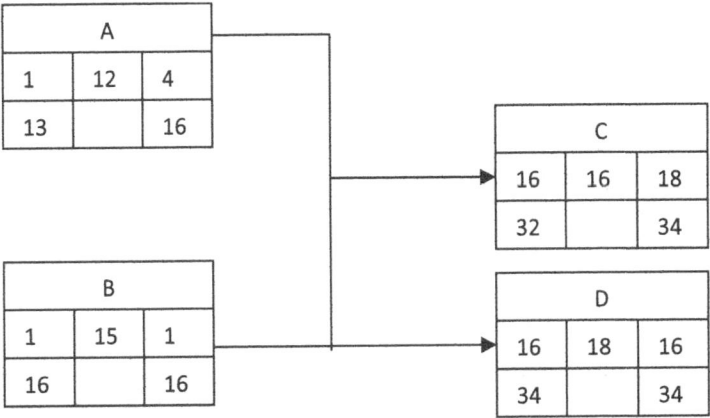

What is the critical path of the above project?

Activities A and C can be delayed without delaying the project. Activities B and D cannot be delayed.

Hence the critical path of the project is

B ⟶ D

Practice Problem: Complete the network shown. Durations are as shown.

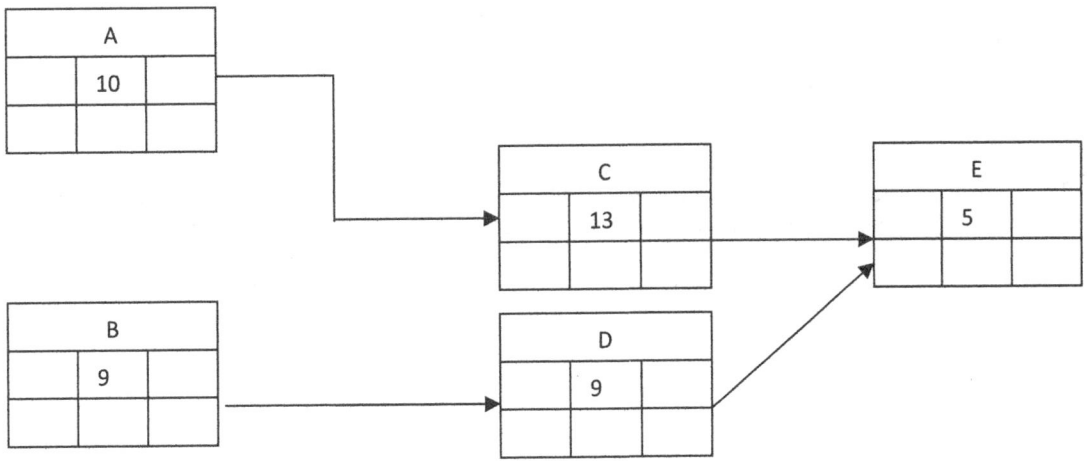

Solution: Complete the forward pass: During the forward pass early start times and early finish times are found.

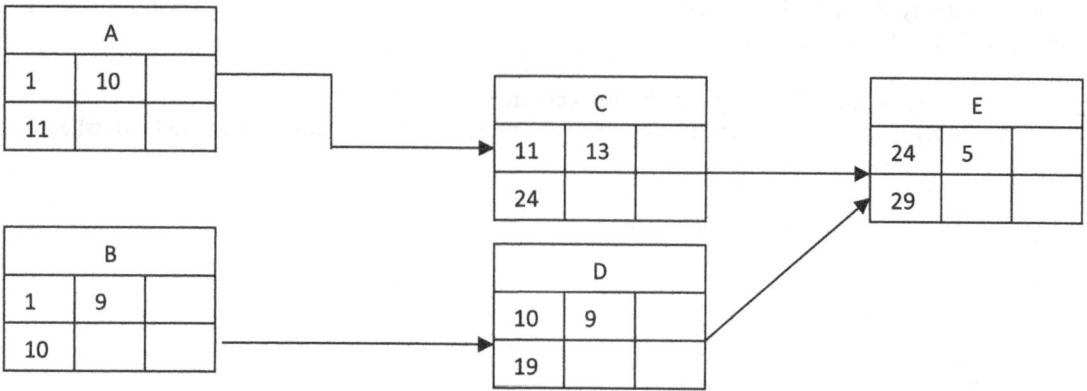

Complete the backward pass:

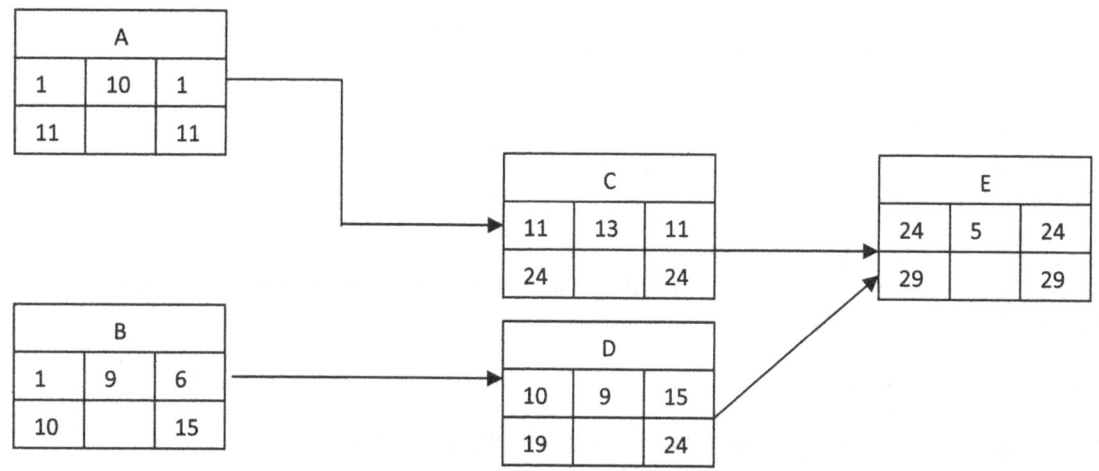

Critical path is the path with zero float.

Critical Path: A ----> C ------> E

Finish to Start Lag Time: So far, we considered that next activity would start just after the previous activity is finished. There are some situations where this is not possible. Concrete footing has to be constructed and a steel column has to be placed on the footing. After construction of the footing, there is a lag of 10 days for the concrete to cure.

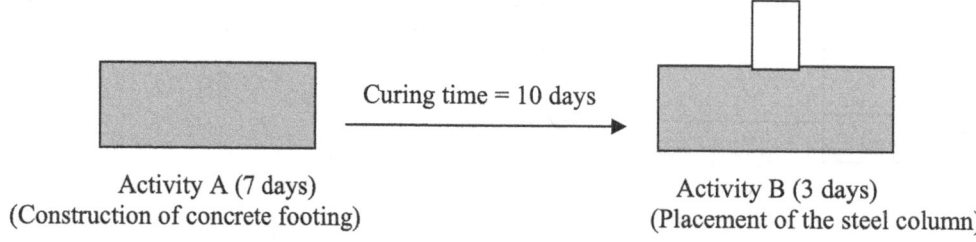

Activity A (7 days)   Activity B (3 days)
(Construction of concrete footing)   (Placement of the steel column)

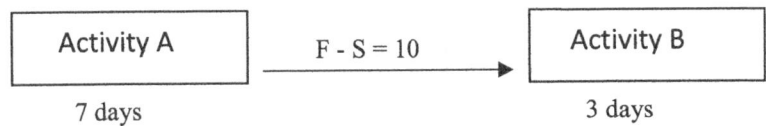

In the above example, steel column cannot be placed on the concrete footing until 10 days after the construction of the footing. This can be shown symbolically as below.

Activity A	F - S = 10 →	Activity B
7 days		3 days

F - S = 10 indicates that finish to start there is a delay of 10 days. In this case, it is for the curing of concrete.
It is important to note that this equation is valid for both early and late times.

F – S = 10
ES of activity B – EF of activity A = 10
(Early finish of activity A to Early start of activity B, there is a delay of 10 days).
And also
LS (activity B) – LF (activity A) = 10
(Late finish of activity A to Late start of activity B, there is a delay of 10 days).

Construction of footing (A)			
1 (ES)	7		LS
8 (EF)			LF

F - S = 10 →

Place the steel column (B)			
18 (ES)	3		LS
21 (EF)			LF

Early start of activity A = 1
Early finish of activity A = 1 + 7 = 8

Early start of activity B = 8 + 10 = 18   (10 is the finish to start delay)
Early finish of activity B = 18 + 3 = 21

Now let us find LS and LF of two activities. Note that finish to start lag exists for both forward pass and backward pass.
LF of activity B is 21.
LS of activity B = 21 - 3 = 18

LF of activity A = 18 - 10 = 8   (remember there is a 10 day delay between activity A and B).
LS of activity A = 8 - 7 = 1

Construction of footing (A)		
1	7	1
8		8

F - S = 10

Place steel column (B)		
18	3	18
21		21

Start-to-Start Relationships:

A steel contractor is erecting steel columns. He is supposed to erect 100 steel columns. He could deliver all 100 steel columns and start erecting steel columns.

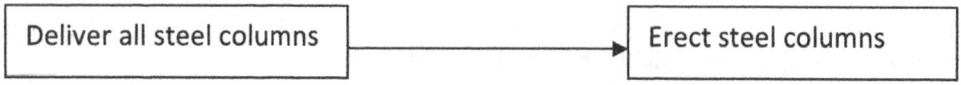

However, the steel contractor is planning to deliver portion of steel columns and start erecting prior to completion of delivery of all steel. He is planning to deliver portion of steel columns in first two days and start erecting right away. While erecting, he will keep delivering steel to the site. Hence, two activities, delivering steel and erecting steel would happen simultaneously.

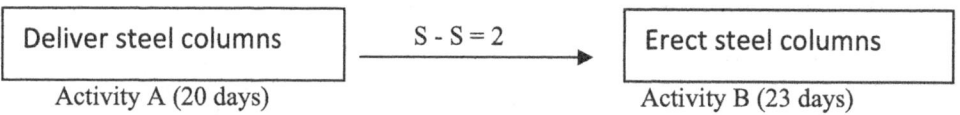

S - S = 2 means two days after start of activity A, activity B would start. After that, activity A and activity B would happen simultaneously.

S - S = 2 can be represented as follows.
Early start of activity B = Early start of activity A + 2
It will be true for late start as well.
Late start of activity B = Late start of activity A + 2

Now let us fill the boxes.
Early start of activity A = 1
Early finish of activity A = 1 + 20 = 21

Early start of activity B = 1 + 2 = 3
(Contractor is planning to start activity B, just two days after starting activity A).
Early finish of activity B = 3 + 23 = 26

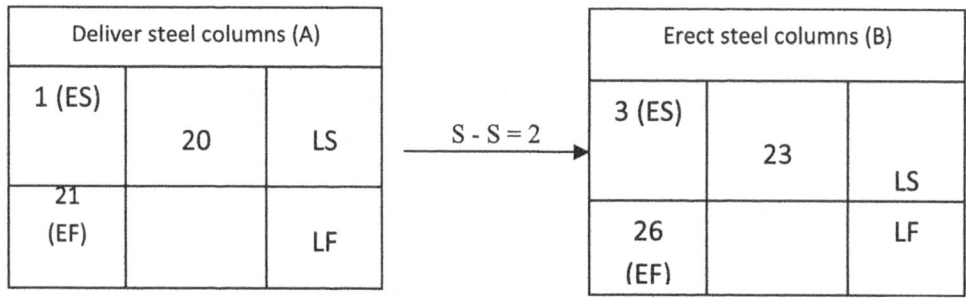

Now let us find LS and LF of two activities.
LF of activity B is 26.

LS of activity B = 26 - 23 = 3
LS of activity B = LS of activity A + 2
(Remember two activities are connected by S - S relationship.)
Hence;   LS of activity A = LS of activity B - 2 = 3 - 2 = 1
LF of activity A = 1 + 20 = 21

Deliver steel columns (A)		
1 (ES)	20	
		1
21		
(EF)		21

S - S = 2 →

Erect steel columns (B)		
	23	
3 (ES)		3
26		26
(EF)		

Finish-to-Finish Relationships: Let us consider the same example above. A steel contractor is planning to start delivering steel columns and start erecting steel columns while steel been delivered. The contractor is planning to complete erecting all steel columns 12 days after completion of delivery of all steel. This can be represented as follows.

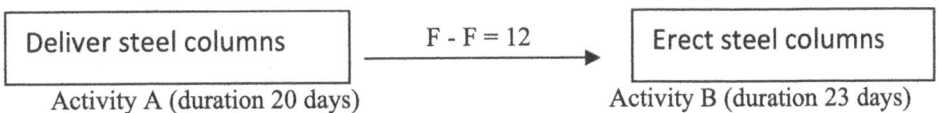

| Deliver steel columns | F - F = 12 → | Erect steel columns |

Activity A (duration 20 days)   Activity B (duration 23 days)

F - F = 12 means 12 days after activity A is **finished**, activity B would be **finished**.
Early finish (EF) of activity B = EF of activity A + 12
Similarly,
Late finish (LF) of activity B = LF of activity A + 12
Let us fill the boxes.
Early start of activity A = 1
Early finish of activity A = 1 + 20 = 21

Early finish (EF) of activity B = EF of activity A + 12
Early finish (EF) of activity B = 21 + 12 = 33
ES of activity B = 33 - duration = 33 - 23 = 10

Deliver steel columns (A)		
1 (ES)	20	LS
21		
(EF)		LF

F - F = 12 →

Erect steel columns (B)		
10 (ES)		
	23	LS
33 (EF)		LF

Backward pass:
LF of activity B = 33
Now we can find the LF of activity A.
Since,
Late finish (LF) of activity B = LF of activity A + 12

LF of activity A = LF of activity B - 12 = 33 - 12 = 21
LS of activity A = 21 - 20 = 1
LS of activity B = LF of activity B - duration = 33 - 23 = 10
Now all boxes can be filled.

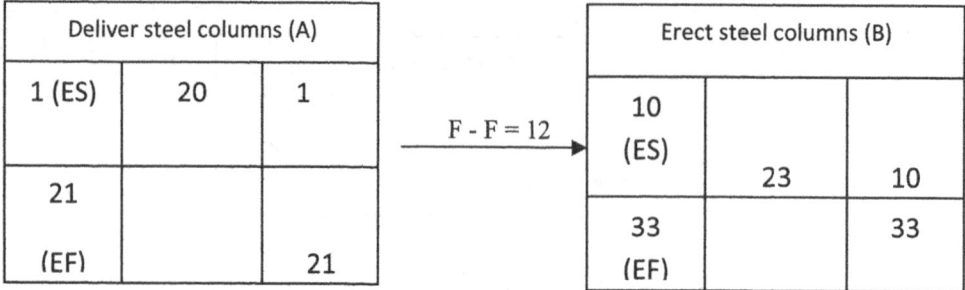

S - S and F - F Relationships Together: S - S and F - F relationships can be used together. Let us look at the same steel column example.

Contractor is planning to start delivering steel columns and start erecting steel columns while steel been delivered. Contractor is planning to start erecting steel columns 5 days after start delivering steel columns. At the same time contractor is planning to finish erection of steel columns 15 days after completion of delivery of all steel. This can be represented as follows.

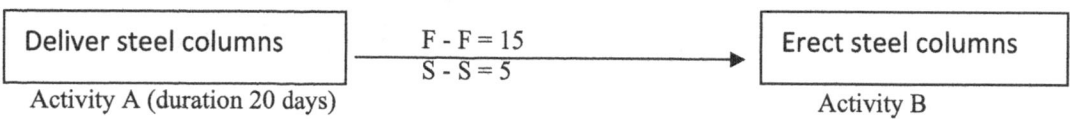

Note that duration of activity B is not given. If duration of activity B is given, there would be a conflict. Duration of activity B is not needed if two relationships are given.
We can write the following;
ES of activity B = ES of activity A + 5
LS of activity B = LS of activity A + 5
EF of activity B = EF of activity A + 15
LF of activity B = LF of activity A + 15

Forward pass:
ES of activity A = 1
EF of activity A = 1 + 20 = 21
ES of activity B = ES of activity A + 5 = 1 + 5 = 6
EF of activity B = EF of activity A + 15
EF of activity B = 21 + 15 = 36
Now these numbers can be represented in the boxes.

Deliver steel columns (A)				Erect steel columns (B)		
1 (ES)	20	LS	F - F = 15	6 (ES)		
21			S - S = 5			LS
(EF)		LF		36 (EF)		LF

Backward Pass:
LF of activity B = 36
Now we can use the F - F relationship.
LF of activity B = LF of activity A + 15
LF of activity A = LF of activity B - 15
LF of activity A = 36 - 15 = 21
LS of activity A = 21 - 20 = 1

Now we can use the S - S relationship.
LS of activity B = LS of activity A + 5
LS of activity B = 1 + 5 = 6
Now all boxes can be filled.

Deliver steel columns (A)		
1 (ES)	20	1
21		21
(EF)		(LF)

F - F = 15
S - S = 5

Erect steel columns (B)		
6 (ES)		6
		(LS)
36		36
(EF)		(LF)

S - F Relationships: Start to finish relationships rarely occur in the construction industry. Hence many textbooks ignore start to finish relationships. Many software programs do not allow start to finish relationships. Start to finish relationship can be represented as follows.
ES of activity A + 5 = EF of activity B
Can you think of a practical application of such a situation?

Negative Lags: Above paragraph, we discussed positive lag between activities. Positive lag occurs due to curing of concrete or lead time of an item tc. Next, we consider **negative lag**.

Finish to Start (Negative Lag):   It is possible to have a negative lag between two activities. Let us assume that a contractor has to complete two activities, erection of steel and painting. Contractor can start painting prior to finish of steel erection. Contractor can erect 50% of steel and then start painting while steel erection is going on.

F - S = -5
Above equation means contractor is planning to erect steel. At the same time, he is planning to start painting 5 days prior to completion of steel erection.
Finish to start difference is -5 days. It does not matter whether it is early or late start and finish times. Nevertheless, one has to be consistent. If it is early start time, then finish should be early finish time. Similarly, if F is late finish time then S should be late start time. Hence, two equations can be developed for early and late times.

EF - ES = -5
And
LF - LS = -5
Above equation means, that contractor is planning to start painting 5 days before the finish of steel erection.

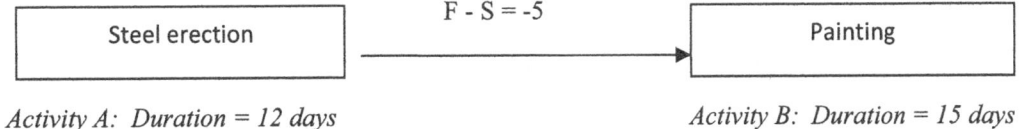

Steel erection	F - S = -5	Painting

Activity A: Duration = 12 days   *Activity B: Duration = 15 days*

ES of Activity B = EF of activity A – 5
(Activity B would start 5 days before activity A is finished).
And
LS of Activity B = LF of activity A – 5
Lets' do the forward pass:

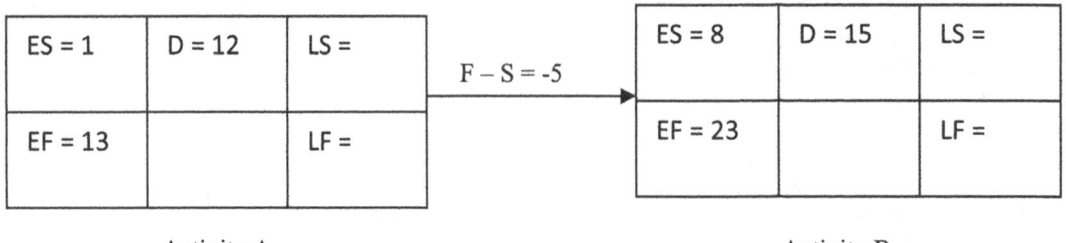

ES = 1	D = 12	LS =
EF = 13		LF =

Activity A

ES = 8	D = 15	LS =
EF = 23		LF =

Activity B

F – S = –5

F – S = –5
Early finish of activity A is 13.
ES of Activity B = EF of activity A – 5
ES of Activity B = 13 – 5 = 8

Next, we can do the backward pass:

ES = 1	D = 12	LS = 1
EF = 13		LF = 13

ES = 8	D = 15	LS = 8
EF = 23		LF = 23

F – S = –5

LS of Activity B = LF of activity A – 5
LS of activity B = 8
Hence
8 = LF of activity A – 5
LF of activity A = 13

Finish-to-Finish (Negative Lag): It is possible to have a negative lag between two activities with finish-to-finish relationship. Let us assume that a contractor has to build a retaining wall and paint it. The client does not want to paint the whole wall. The client wants only 75% of the wall to be painted. He believes that 25% of the wall is out of public view and need not be painted.

In this scenario, wall painting cannot be started until wall is built.
Wall painting **can be** completed prior to completion of the wall. (This is possible since only 75% of the wall need to be painted).

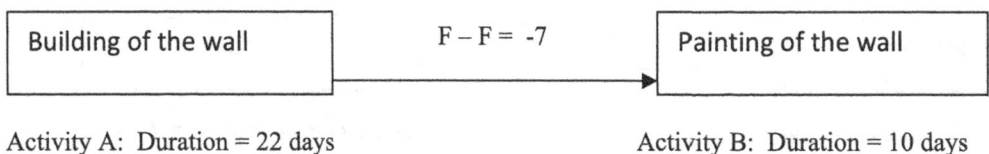

| Building of the wall | F – F = –7 | Painting of the wall |

Activity A: Duration = 22 days   Activity B: Duration = 10 days

EF of Activity B = EF of activity A – 7
(Activity B finishes 7 days before activity A).
And
LF of Activity B = LF of activity A – 7

Let us do the forward pass:
Activity A: ES = 1 and EF = 23
EF of Activity B = EF of activity A – 7 = 23 – 7 = 16
ES of activity B = EF of activity B – duration = 16 – 10 = 6

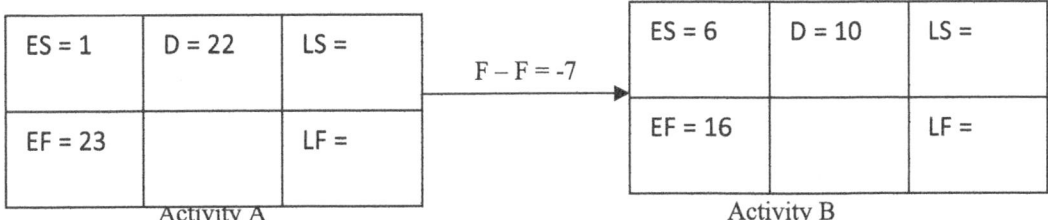

Backward Pass:
LF of Activity B = 16
LF of Activity B = LF of activity A – 7
LF of activity A = LF of Activity B + 7 = 16 + 7 = 23
LS of activity A = 23 – duration = 23 – 22 = 1

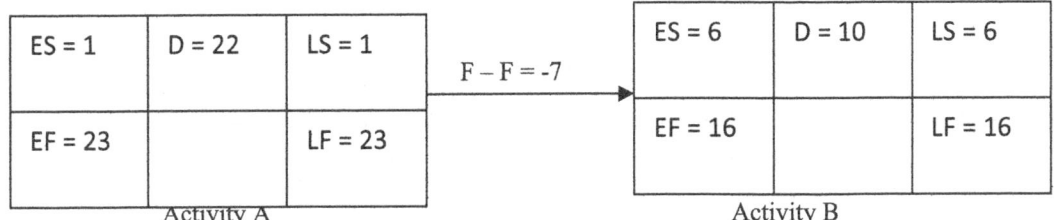

Practice Problem: Complete the network below. Relationship between activity C to E is F – F = -4. Durations are as shown.

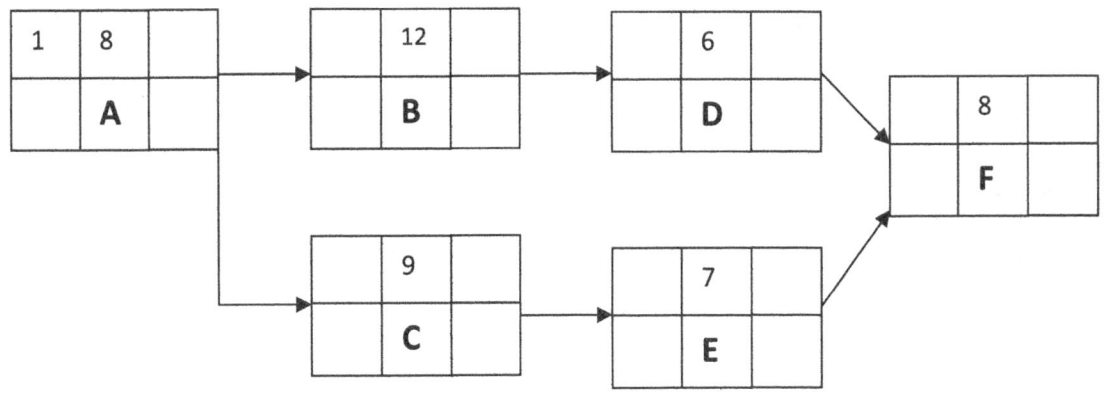

Solution: STEP 1:   Complete the forward pass.

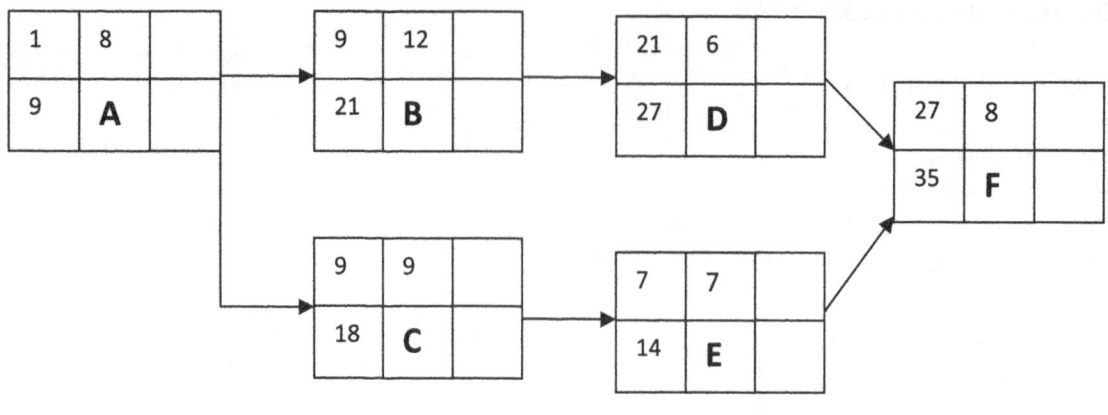

F – F = -4

Forward pass of activities A, B, C and D are straightforward.

Activity C to E: Activity C to E, one has to worry about the relationship given.
EF of activity C is 18.
Hence, EF of activity E should be 18 – 4 = 14. Note that the relationship is finish to finish and it is negative.
If EF of activity E is 14, then ES of activity E = 14 – duration = 14 – 7 = 7.

Activity F cannot be started until activity D and E are completed. Activity E is completed by 14 and activity D is completed by 27. Hence, ES of activity F is 27.

STEP 2:   Backward Pass:

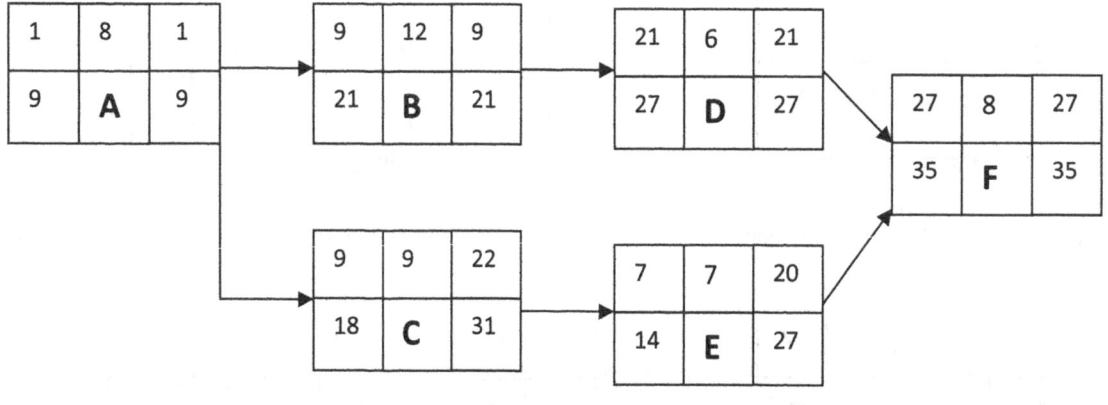

F – F = -4

Backward Passes from F to D and F to E are standard.

Backward Pass from E to C: LF of activity E is 27. Hence, LF of activity C is 31. This may be little confusing.
Look at EF of activities C and E.
F – F = -4
LF of activity E = LF of activity C – 4
27 = LF of activity C – 4
LF of activity C = 27 + 4 = 31

1.3.3 FLOATS: (Total Float, Free Float and Independent Float)

ES = Early Start , LS = Late Start   EF = Early Finish,   LF = Late Finish,

D = Duration

Activity No:		
ES	D	LS
EF		LF

Floats: Three types of floats are identified.

Total Float:

$$\text{Total Float = LF – EF}$$

LF – EF is same as LS – ES.

This can be shown as follows.   LF = LS + D   (D = Duration)

EF = ES + D

LF – EF = (LS + D) – (ES + D) = LS – ES

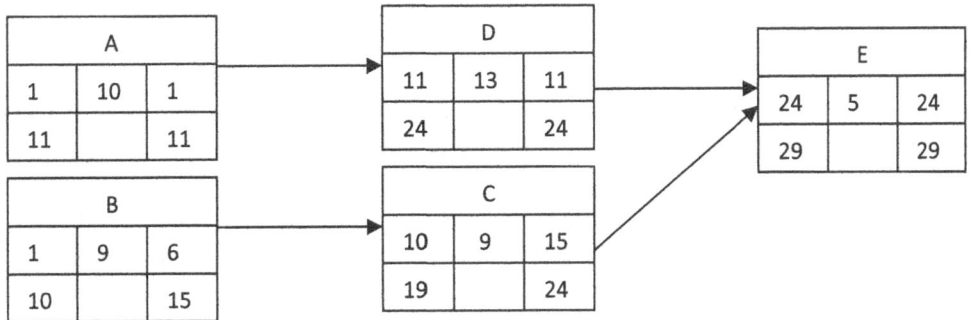

Total float of activity A = LF – EF = 11 – 11 = 0
Total float of activity B = LF – EF = 15 – 10 = 5
Total float of activity C = LF – EF = 24 – 19 = 5
Total float of activity D = LF – EF = 24 - 24 = 0
Total float of activity E = LF – EF = 29 -29 = 0
Note: All floats are zero for critical path activities.

Free Float:

$$\text{Free Float of Activity A = ES}_{\text{successor}} \text{ – EF of activity A}$$

Free Float of A = ES $_{successor}$ – EF of Activity A
Successor of A is D. Hence ES $_{successor}$ = 11
EF of Activity A = 11
Free Float = 0

Free Float of B = ES $_{successor}$ – EF of Activity B
Successor of B is C. Hence ES $_{successor}$ = 10
EF of Activity B = 10
Free Float = 0

Free Float of C = ES $_{successor}$ – EF of Activity C
Successor of C is E. Hence ES $_{successor}$ = 24
EF of Activity C = 19
Free Float = **24 – 19 = 5**

Free Float of D = ES $_{successor}$ – EF of Activity D
Successor of D is E. Hence ES $_{successor}$ = 24
EF of Activity D = 24
Free Float = 24 – 24 = 0

Free float and total float are zero along critical path.

Note that total float of B is 5 but free float is zero.
Total float and free float of C is 5.

Discussion of Total Float and Free Float: Let us assume that a certain project is of interest to the senator of the state and the governor of the state. Governor tells the project manager that he does not want to move any of the early start times of major activities. On the other hand, senator tells the project manager that he don't give a damn about early start times of activities but cares only of the final completion date.

If the project manager was to work without changing early start times of activities then he has to work with free floats.
If the project manger is interested only of the completion date then he can work with total floats.

For an example, if the project manager were to delay activity B in previous example by 5 days, he will delay the early start time of activity C, but will not delay the completion date. (Free float of activity B is zero while total float is 5). The governor of the state would be a very angry man since activity C cannot be started at the early start time as scheduled. On the other hand, senator may not have a problem with delaying activity B by 5 days. Senator would ask the governor why you care about early start time of activity C. The governor would say what if there is an unseen situation in activity C and has to be delayed? Then the whole project would be delayed.

Why some do not want to change early start times of activities?
There is a fear that if all the available slack (Total float) is utilized at the start of a project, there would not be any slack at the end. If any contingency were to occur at the end, the project would be delayed. Hence, some executives do not want to change early start times of activities.

Independent Float:

> **Independent Float of activity A = ES $_{successor}$ – LF $_{Predecessor}$ – Duration of A**

If you need to find the independent float of activity A, obtain the early start time of successor. Then obtain the LF time of predecessor. Use the above given equation to find the independent float.

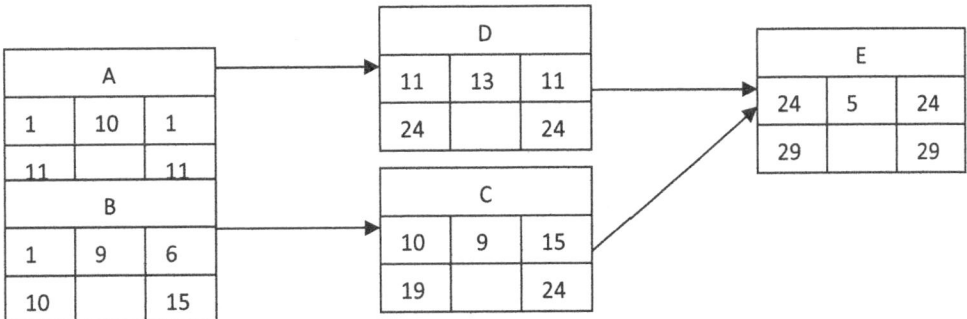

Independent float of activity A = There is no predecessor. Hence, no independent float
Independent float of activity B = There is no predecessor. Hence, no independent float.
Independent float of activity C = ES of activity E – LF of activity B – Duration of C = 24 – 15 – 9 = 0
Independent float of activity D = ES of activity E – LF of activity A –Duration of D = 24 – 11–13 = 0
Independent float of activity E = There is no successor. Hence, no independent float.

Discussion: Independent float indicates the slack that each activity has so that it will have absolutely no impact on preceding and successive activities. In other words, preceding activity can be finished at late finish time and succeeding activity can start at early start time. In most cases, independent float is zero.

1.3.4 Activity on Arrow Networks;

In activity on arrow diagrams, activities are represented using arrows. Nodes are considered events. The very first event is the "Start" event. Very last event is the "End" event. A given activity cannot start until the event prior to that activity is accomplished. An event to be completed, all activities coming to that event (node) should be completed.

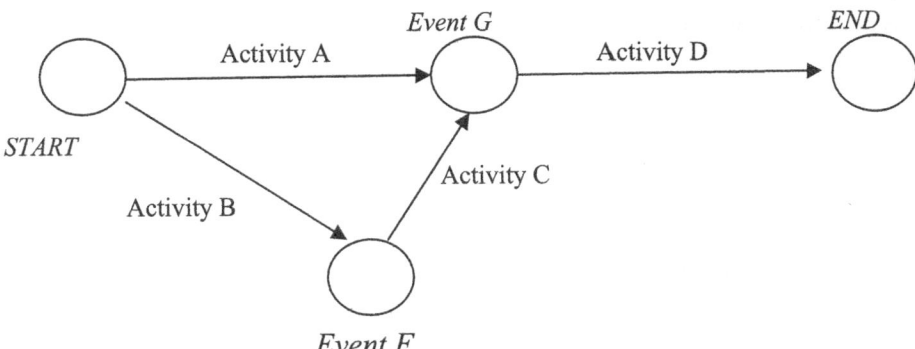

Above is an activity on arrow network. It has three activities. (activities A, B and C). It also has three events. (START event, Event F, Event G and END event).

Activity D cannot start until event G is accomplished. Event G is accomplished when both activities A and C have been completed.

Practice Problem)   Activity on arrow diagram is shown below. Find the duration of the project.

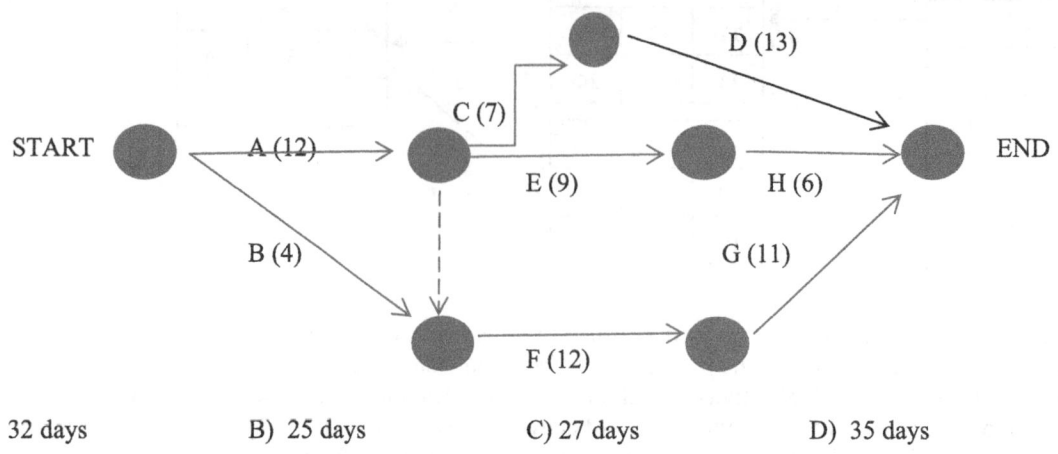

A) 32 days   B) 25 days   C) 27 days   D) 35 days

Solution):   Activities are shown on arrows in activity on arrow diagrams

In activity on arrow diagrams, arrows are used to show activities.
In activity on arrow diagrams, nodes are called events.
First node is always the "START" event and last node is the "END" event.
On the other hand, in activity on node diagrams, activities are shown on the node.

Dummy Activities: Broken arrows are used to indicate dummy activities. Activity F cannot be started until activities A and B are completed. If the dummy activity is not shown, activity F can be started as soon as activity B is completed.

There are number of paths exist from start to end.

Duration of path A, C, D = 12 + 7 + 13 = 32
Duration of path A, E, H = 12 + 9 + 6 = 27
Duration of path A, F, G = 12 + 12 + 11 = 35
Duration of path B, F, G = 4 + 12 + 11 = 27

To complete the project, longest path needs to be completed.

Duration of the project = 35   (Ans D)

Practice Problem: Find the early start and late start of activity D for the project given in the previous problem.

A) ES = 12, LS = 15   B) ES = 13, LS = 22   C) ES = 19, LS = 22
D) ES = 22, LS = 24

Solution:

Activity D, cannot be started until activity C is completed.

Early start of activity D = 12 + 7 = 19   (ES = 19)

Find the late start (LS) of activity D

Finding late start of an activity can be tricky.

We found that project duration to be 35 days.

Now we need to find what is the latest day that activity D can be started without delaying the project.

Latest day that activity D can be started without delaying the project = 35 – 13 = 22

35 is the duration of the project and 13 is the time need to complete activity D.

If activity D is started on day 22, there would not be any delay to the project. LS = 22

(Ans C)

Practice Problem: Find the total float of activity D.

A) 1   B) 3   C) 9   D) 2

Solution: Total float of an activity is given by the following equation;

$$\text{Total Float of an Activity} = LS - ES$$

LS = Late start;   ES = Early start

LS of activity D = 22

ES of activity D = 19

Float = LS – ES = 3

Activity Time Analysis: Activity times can be changed in order to change the critical path. Activity time of an activity can be either increased or decreased. Activity time can be decreased by increasing manpower. Similarly, activity time can be increased by reducing resources to that particular activity.

1.3.5 Resource Leveling

We may develop a fast schedule that completes the project on time. But what about the resources? Does the contractor have resources (Equipment and work force) to do two or three activities at the same time? If the contractor does not have enough equipment and labor to do two or three simultaneous activities, then resources should be increased. This can be done by renting new machines and hiring new personnel. Renting more machines will be an expensive thing to do. Hiring new people can also be a problem. It is not easy to find people who have suitable expertise.

Hence, the next option is to manipulate the activities

Construction resources are labor, material and equipment. In construction scheduling, conflicts can arise when activities compete for common resources that are available in limited quantities. After development of the critical path schedule, resource utilization chart is developed.

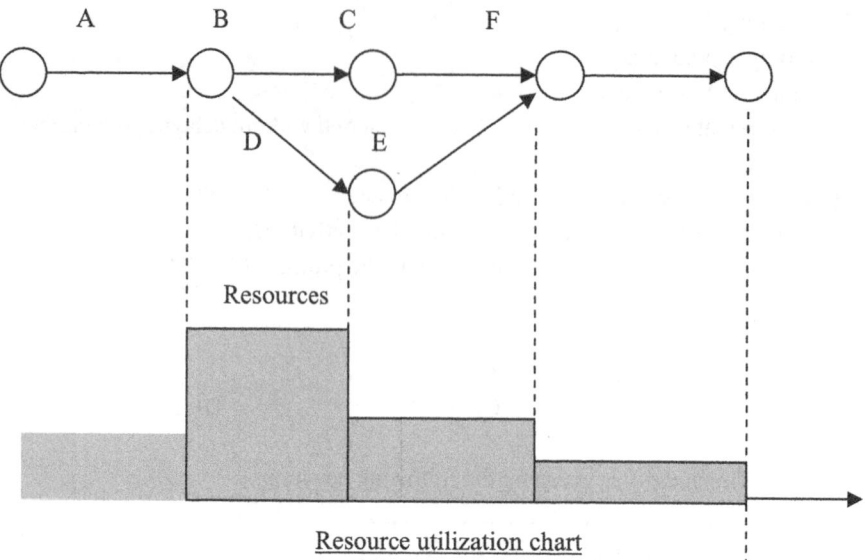

Resource utilization chart

Resources need to be allocated evenly, during the lifetime of the project. In the above figure, when activities B and D are conducted, demand for resources increases. It is important to level the resources during the project duration. In many cases, it is not an easy task. In the above figure, it is possible to stretch activities B and D and shorten the duration of activities A and F.

Increase the duration of activities B and D. This can be done by reducing the daily quantity of resources.

Decrease the duration of activities A and F. This can be done by increasing the daily quantity of resources.

It is possible to level resources by manipulating duration of activities. In some cases, it is possible to move activities around for the purpose of resource leveling.

Various computer algorithms are developed to level resources without affecting the schedule.

Competition for resources among activities:

Resource Leveling Procedure: Following procedure is normally adopted.

Construct the critical path schedule

Develop the resource schedule

Move around the activities to level the resources

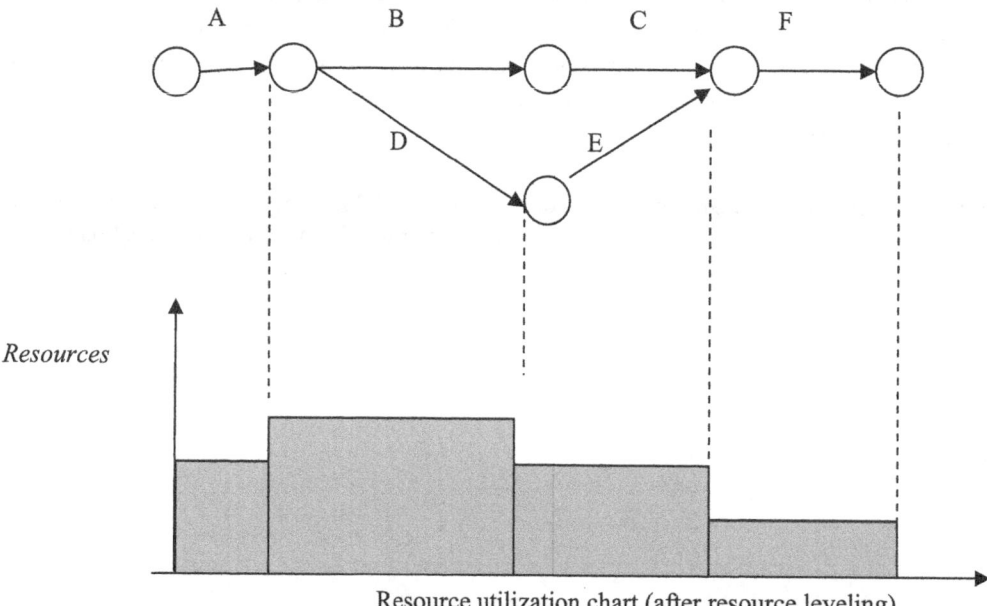

Resource utilization chart (after resource leveling)

Practice Problem: A contractor has developed durations and resources required for construction of two slabs.

Resources required for each activity is given below;

Activity 1:	Formwork slab in building A:	1 foreman, 8 carpenters, 6 laborers
Activity 2:	Rebar installation of slab in building A:	1 foreman, 5 iron workers, 3 laborers
Activity 3:	Concreting of slab in building A:	1 foreman, 8 concrete masons, 6 laborers
Activity 4:	Formwork of slab in building B:	1 foreman, 8 carpenters, 6 laborers
Activity 5:	Rebar installation of slab in building B:	1 foreman, 5 iron workers, 3 laborers
Activity 6:	Concreting of slab in building B:	1 foreman, 8 concrete masons, 6 laborers
Activity 7:	Opening ceremony	

Activity 1:	Duration	7 days
Activity 2:	Duration	10 days
Activity 3:	Duration	7 days
Activity 4:	Duration	12 days
Activity 5:	Duration	11 days
Activity 6:	Duration	7 days
Activity 7:	Duration	1 day

Logic of activities:

Activities 1 and 4 can start at any time.
Predecessor of activity 2 is activity 1.
Predecessor of activity 3 is activity 2.

Predecessor of activity 5 is activity 4.
Predecessor of activity 6 is activity 5
Predecessors of activity 7 are 3 and 6.

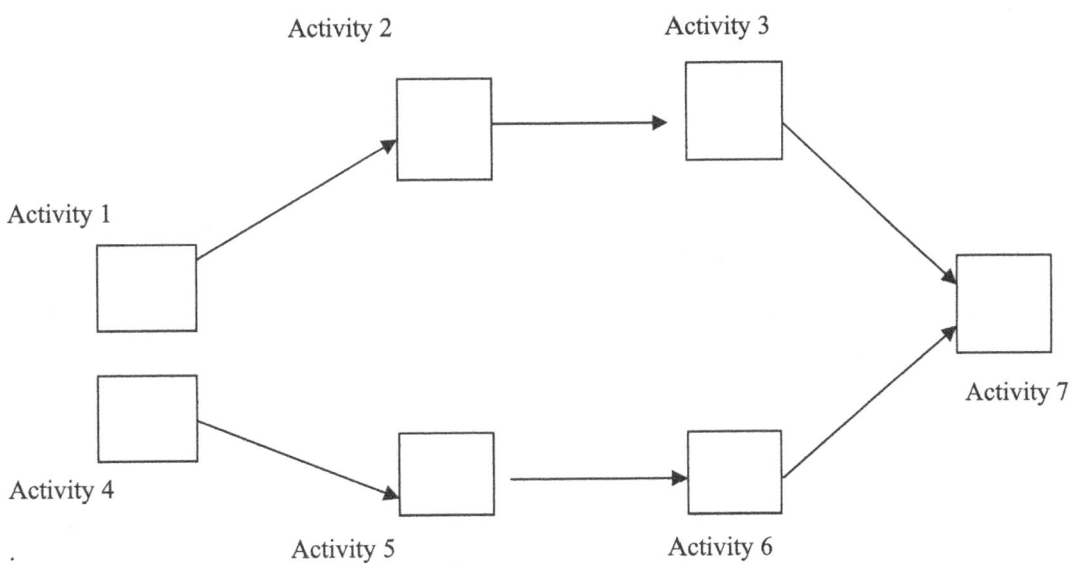

Maximum resources available at a given time;
2 Foremans
8 carpenters
10 iron workers

16 concrete masons
12 laborers
What is the project duration?

Solution: There are only 8 carpenters available. Hence, activities 1 and 4 cannot go parallel.
There are 10 ironworkers available. Hence, there are enough ironworkers to conduct activities 2 and 5 in parallel.
There are 16 concrete masons available. Hence, activities 3 and 6 can be done parallel.
Hence, new network can be drawn as shown.

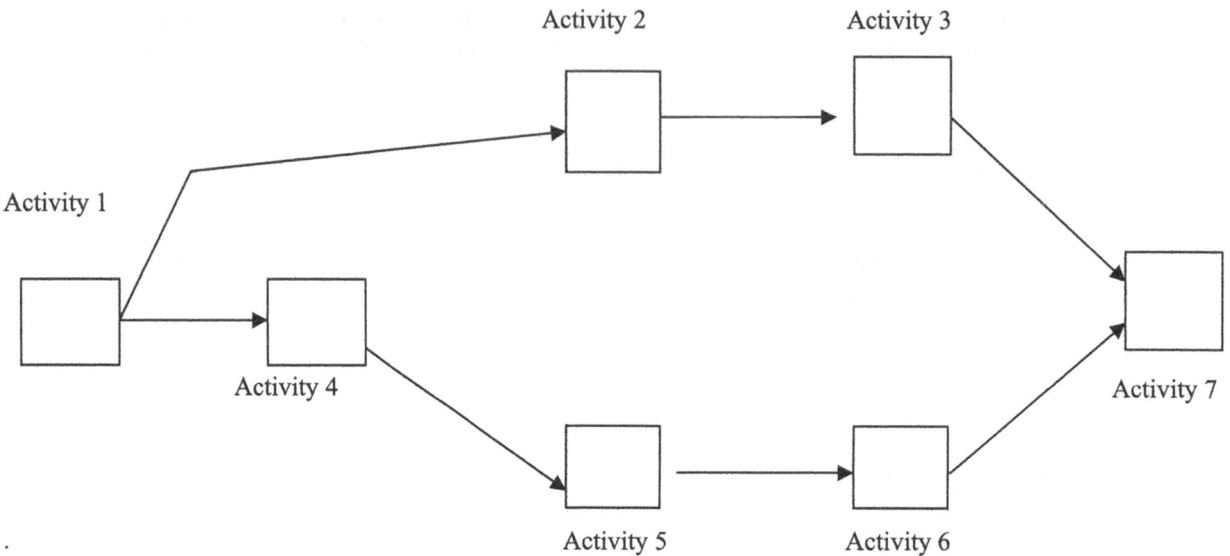

There are two paths.
Path 1, 2,3, 7
Path 1, 4, 5, 6, 7

Duration of path 1, 2, 3, 7 = 7 + 10 + 7 + 1 = 25 days
Duration of path 1, 4, 5, 6, 7 = 7 + 12 + 11 + 7 + 1 = 38 days
Project duration = 38 days

1.3.6 Time - Cost Tradeoff:

Acceleration of a project can be done by increasing the labor. Unfortunately, in some situations increasing manpower may bring diminishing returns. This could be due to number of reasons.
More management staff is needed to manage a bigger crew. The crew may not have enough space to work.
There may not be enough machinery to support a bigger crew. Delays would have a bigger cost impact due to the larger crew size.

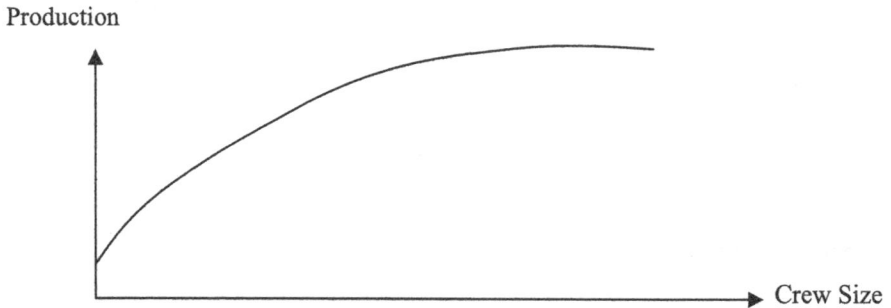

2.0 <u>Means and Methods</u> - 3 Questions

<u>History of Construction:</u>

<u>Early Period:</u>   Probably around 5,000 years ago, cities started to appear in Mesopotamia, India, Egypt and China. Egyptians took construction to a very high level by building large-scale pyramids. During later periods, construction activities became more and more diverse. In addition, functions of structures became diverse as well. Chinese great wall served as a protection barrier while Roman aqua ducts were built to transport water. Pyramids were built for spiritual reasons.

<u>Development of the Number System:</u>   Early engineers used very little mathematics for construction work. Any mathematics that was used mainly limited to geometry. Simple computations such as addition, subtraction, multiplication and division were extremely cumbersome using Egyptian, Greek or Chinese numbers. They did not have the numbers we use today. New era of construction industry started due to the discovery of Hindu Arabic number system by Aryabhatta around 500 AD. One hundred years later Brahmagupta (600 AD) gave arbitrary rules for the four fundamental operations. (Addition, subtraction, multiplication and division). Brahmagupta's system was transported to Europe through Arabia and to rest of the world. Today arithmetic has become an integral part of construction. Interestingly even today, most construction work does not need any mathematics beyond the four fundamental operations.

<u>Machinery:</u>   Next major development to benefit construction industry was the use of machinery. Earlier construction workers used wedges, pulleys and other hand operated tools. Today construction is mostly an affair of labor and machinery. Backhoes, trucks, loaders, forklifts, cranes, derricks, conveyor belts, jackhammers, compactors, rollers, compressors are some of the few machines used in construction sites.

<u>Internal Combustion Engine:</u>   Development of the internal combustion engine provided a new dimension to the construction industry. Nicklaus Otto, Gottlieb Daimler, Karl Benz and James Atkinson invented internal combustion engine around 1890. Internal combustion engine or its variations are needed for today's machinery.

<u>Electricity:</u>   Discovery of electricity by Michael Faraday, James Joule, Thomas Edison and Nicolai Tesla is another major development that affected the construction industry. It is unthinkable to do any construction activities without the use of electricity. Power drills, jackhammers, cutting machines, hoists, and conveyor belts are operated using electricity.

<u>Computer:</u>   Computers have become an integral part of the construction industry. Scheduling, cost estimating, design and construction management highly dependent on computers. Development of the computer by US engineers such as Attanasof, Allen Turing and Von Neumann brought a new dimension to the construction industry.

Building Construction: All construction engineers will encounter building construction work eventually. Construction of a building starts with a need. A company may need more office space or new stores. Management of that company would meet an architect and explain their needs. Architect would come up with a set of architectural drawings. Owner would look at it and make comments. Owner may want the meeting hall to be larger or more bathrooms be added or change appearance of the building. Based on owner needs, architect would redraw the plans. After the architectural drawings were finalized, structural drawings and civil drawings will be prepared by structural engineers and civil engineers. Mechanical drawings and electrical drawings also would be prepared. Today network drawings are prepared for communication and computer networks.

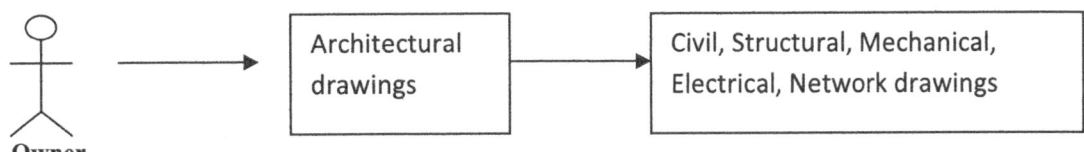

After all necessary drawings and specifications are prepared, bids would be called. Based on bids, a contractor would be selected.

Construction – Early Phase: Prior to any construction work, site has to be prepared. Site preparation work includes, cutting trees, constructing temporary roads for delivery trucks and concrete trucks, construct temporary parking lots for workers, dewatering to remove water from excavations, setup office trailers, security fences, erosion control mechanisms such as silt fences, setup temporary phone lines and power.

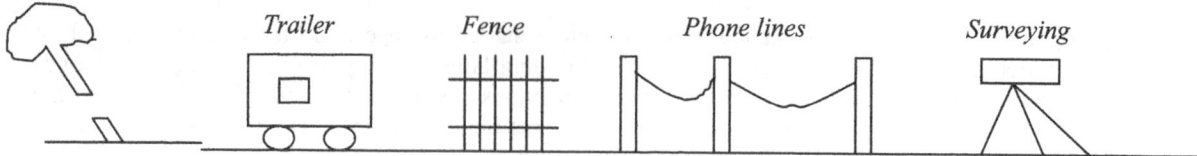

Cutting trees, setting up trailers, security fences, temporary power lines, surveying
Site Preparation Work

Soil Grading and Fill: After site preparation work, site has to be graded. High ground has to be cut and low ground has to be filled.

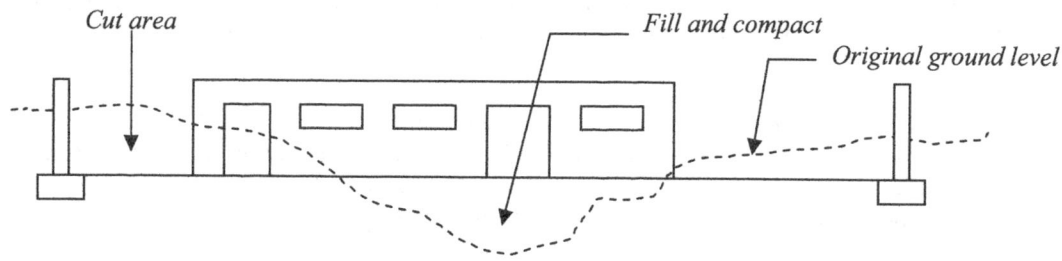

Proposed construction and original ground level

Soil grading is the process of cutting the original ground to the proposed level.

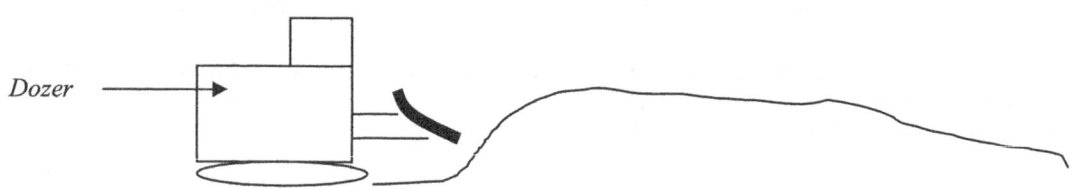

Fill and Compaction: As you could see from the above figure, some areas have to be filled and compacted. Prior to any fill activity, soils that will be used have to be approved by a geotechnical engineer. Usually well graded sandy soil or stones are used for fill work. (Well graded means particles are of various sizes. Poorly graded soil has same size particles). Clay soils and silty soils are considered unsuitable for fill and compaction.

Removal of Rock: Rock removal is done using blasting and rock breakers. Generally, igneous rocks are hard and difficult to rip. Sedimentary rocks are relatively easier. Rippability of rock depends on rock type and degree of weathering,

Pile Driving: Some buildings require piles when existing ground conditions are not suited for shallow foundations. Piles transfer loads to a lower level.

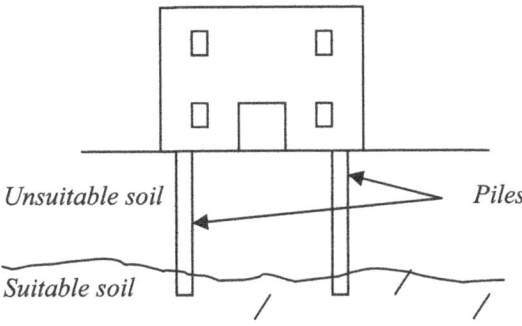

Pile diving is done using pile drivers.

Excavation: Excavation is needed for basements, shallow foundations, sewer pipes, drainage pipes, manholes and swimming pools.

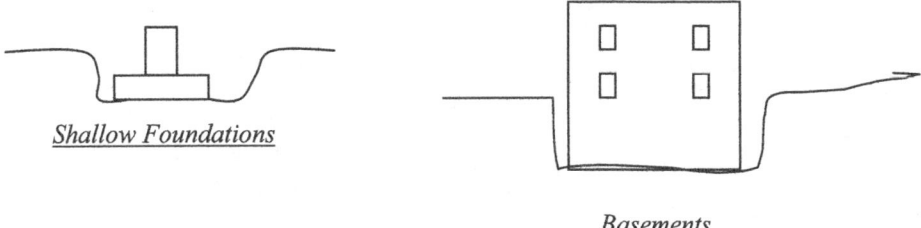

Excavation work is usually done using backhoes.

Concreting: Concreting work will start after completion of site preparation work, excavations and pile driving is completed.

Concreting work can be divided into

- Formwork preparation
- Installation of reinforcement bars
- Concreting and curing

Formwork is needed to hold the concrete in place. Concreting can be done by bringing pre mixed concrete from a concrete yard. Also concrete can be prepared on site using a mixer. Concrete can be pumped or lifted to high elevations.

Steel Erection: Some buildings are designed using steel beams and columns. Steel members are connected using bolts or welding. Rivets were used in the past and not used anymore.

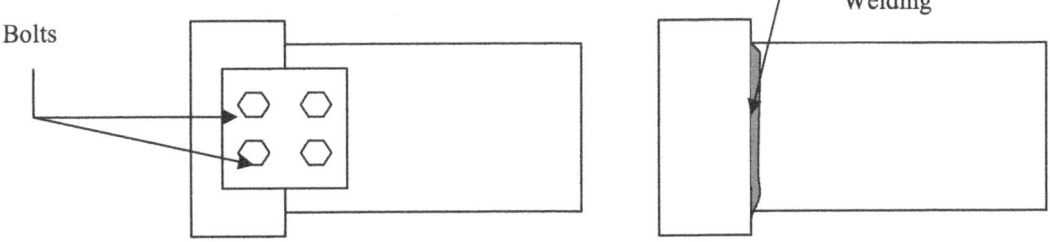

Moving and Lifting of Material (Rigging): Material such as concrete, timber, steel brought to the site using trucks. Then cranes are used to move them inside the site. There are many different types of cranes available.

- Tower cranes – Large tower crane can cover a large area. Material on one corner of a site can be moved to the other end in a second.
- Crawler Mounted Cranes – These are good for small sites.
- Gantry cranes – These cranes move on rail

2.1   Site Work:
2.1.1 Permanent and Temporary Site Work:

In construction work, many things need to be done outside the main building or structure that is planned. Look at the photograph below. It shows an undeveloped plot of land.

Let us assume a developer is planning to build a shopping mall in the land shown in the above photograph..

Can you prepare a list of site work items that need to be done to convert the undeveloped land shown in the previous photo into this building?

First, the developer needs to cut the trees and clean the top soil. This is not a complicated activity. However, there are many state and federal permits have to be obtained prior to cutting trees. Original land contains a small water logged area. This may be considered as a wetland. The developer needs to find out from relevant authorities whether he could fill that area. In addition, the land has to be leveled by filling soil or cutting soil.

Site work can be divided into permanent site work and temporary site work. A sheetpile wall may be erected to hold back the soil during construction. This could be considered as temporary retaining wall. In addition, some retaining walls can be a permanent structure that is part of the design. A contractor may decide to have a quick and dirty drainage lines during construction. Later he may build the permanent drainage lines to the site. Some areas may be temporarily paved for delivery trucks to come and go.

Some temporary site work;

- Site clearing or grubbing
- Demolition of abandoned structures
- Fill depressed areas
- Excavations
- Cut and fill
- Breaking rock
- Mass grading
- Fine grading
- Compaction of soil
- Removal of existing utilities
- Install temporary lighting, water or gas supply
- Provide temporary drainage to the site
- Provide temporary paving

- Temporary retaining walls, coffer dams, sheet pile walls
- Temporary sediment and erosion control structures (rip rap, silt fences etc)
- Soil stabilization (Vibroflotation, dynamic compaction, soil surcharging)

Some permanent site work;
- Retaining walls
- Roads
- Parking lots
- Construct permanent utilities (water supply, electricity, gas, cable, communication etc)
- Planting trees
- Ponds and canals
- Landscaping

2.1.2 Site Clearing

Site clearing is also known as "grubbing". Site clearing involves cutting trees, removal of bushes, removal of top soil and roots etc. Typically, backhoes, dozers and tree cutting machines are used for site clearing.

Some specialized equipment used for site clearing;

- Stump splitters
- Stump pullers
- Clearing rakes
- Grapples

Site Clearing Equipment:

Left: Backhoe clearing the site
Right: Stump splitter removing tree stumps

Left: Equipment removing grass and grading the soil
Right: Stump puller (This machine can remove deep rooted stumps)

Left: (Grapple):   Grapples are effective in removing boulders, roots and vegetation
Right: *Clearing Rake*:   Clearing rakes are used to remove small trees and vegetation. Some rough grading also   can be done with these machines

2.1.3 Demolition of Existing Structures and Utilities

In many situations, old abandoned structures need to be demolished. Typically, demolition involves demolition of concrete, steel structures, fences, masonry structures, roofs etc.

Left: Jack hammer: (Jack hammers are widely used for breaking concrete)
Right: Jack hammer mounted in a loader

Left: Wrecking balls are widely used for demolition of buildings
Right: Shear - These machines can cut thru pipes and metals

2.1.4 Temporary Drainage

Once the trees and grass is removed, rain water tends to make the site muddy. Working becomes highly inefficient in a muddy site. Hence, temporary drainage should be provided. Temporary drainage is provided thru backfilling, gravel beds, perforated pipes and trenches. Typically, one has to locate low areas where ponding could occur. These areas can be backfilled and bring it up. In addition, gravel could be placed and a perforated pipe can be installed.

Temporary drainage in a construction site

2.1.6 Sheet piles:

I have never seen a construction site without sheetpiles. Sheetpiles are extremely versatile. They can be used as retaining walls, excavation support, trenching, cofferdams and temporary bearing platforms.

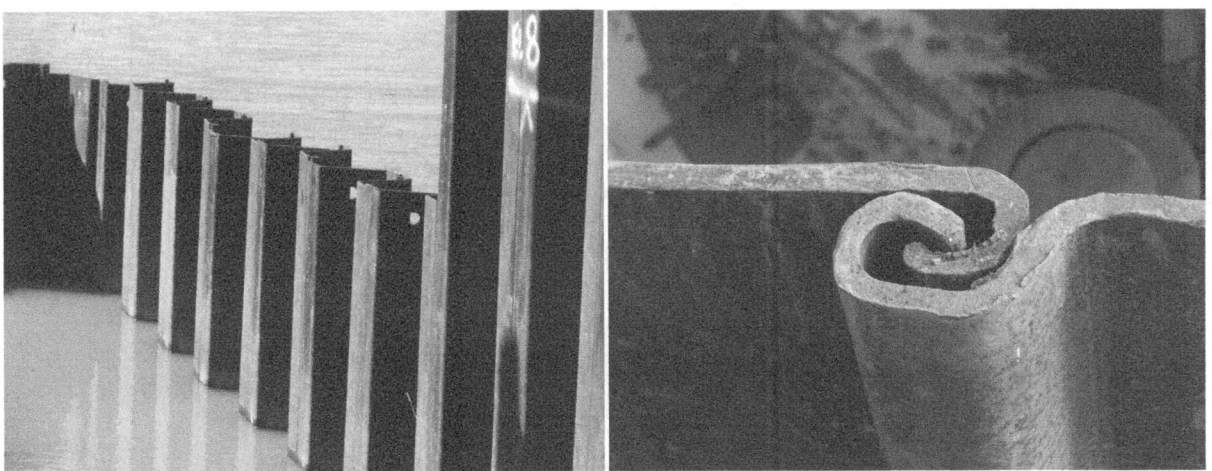

Left: Cofferdam built using sheetpiles
Right: Sheet pile connection detail

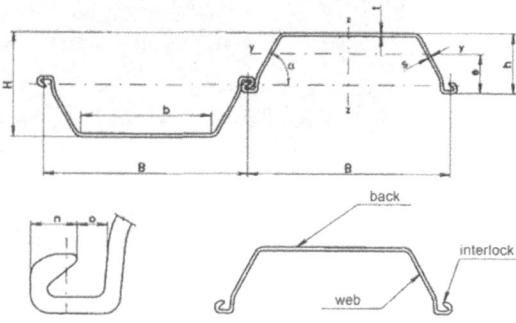

Left: Vibratory hammers are used to drive sheet piles
Right: Sheet pile configuration

2.1.7 Soil Stabilization:

Some sites are not suitable to have shallow foundations. In such cases, piles need to be driven. Piles generally come with a heavy price tag. Instead, one can try to improve the soil bearing strength. This is known as soil stabilization. Many methods are used for soil stabilization. Vibroflotation, dynamic compaction, surcharging, Wick Drains and pressure grouting are some processes.

Note: Please see my geotechnical module for complete description of soil stabilization.

2.1.8 Site Work - Permanent Construction;

Any building needs roads and parking lots. Roads and parking lots are constructed by bringing the soil to the required grade and then providing a gravel base. This is known as sub base. After the gravel layer (sometimes crushed stone also used), asphalt base course is provided. On top of the asphalt base course, asphalt surface layer is provided. Surface course is designed to have better friction between tires and asphalt. Base course is designed to provide rigidity to the road.

Left: Layer of gravel or crushed stones is spread out prior to installation of asphalt
Right: Compaction of asphalt surface course. Gravel sub base and Asphalt base course is seen in the picture.

2.1.9 Permanent Drainage:

Drainage is provided thru installation of storm water pipes and manholes. Trenches need to be dug to install storm water pipes. Excavation support is provided with trench boxes or shoring.

2.1.10   Construction of Utilities (Water Pipes, Sewer Pipes, and Electrical Conduits)

Any facility requires water, electricity, sanitary, cable and communication lines. These utilities are part of site work. Other site work items include retaining walls, storm water detention systems, playgrounds and landscaping.

Storm water detention ponds -   During storms, all the storm water ends up in storm pipes. This could create overflow of manholes and flooding. Hence, large sites are required to maintain storm water detention ponds. During a storm, water is drained to the storm water detention pond. Later when the storm is over, detention pond would discharge to the storm water pipes.

2.1.11   Landscaping:

Landscaping is the process of creating an aesthetic and natural environment around the facility. Generally, trees, flowers, water ponds and plants are used to create a pleasant environment.

Well Designed Landscaping

2.1.12   Earthwork Construction and Layout:

Excavation, transporting and compacting of soil is known as earthwork. Let us look at a construction of a new road. Naturally, existing ground may not be level. High ground needs to be cut to level the road. In addition, there could be some depressions. These depressions need to be filled. Technical term for filling depressions is "backfill". When a depressed area is backfilled, it is necessary to compact with rollers.

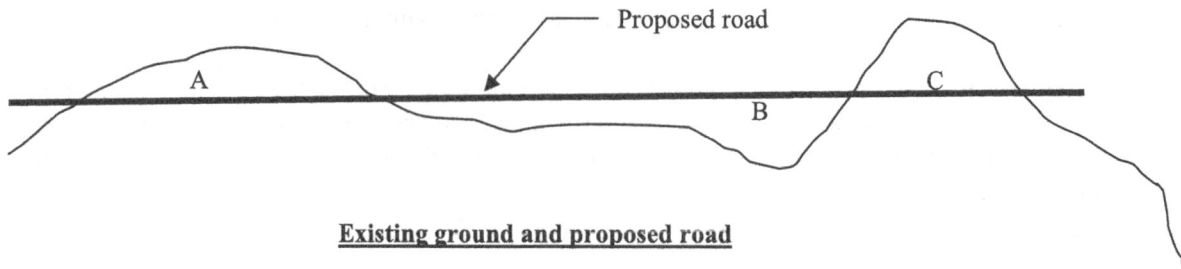

Existing ground and proposed road

In the above figure, areas marked "A" and "C" need to be cut. Area marked "B" needs to be backfilled and compacted.

Different machines are used for cutting, transporting, backfilling and compaction. Cutting of ground is done with dozers and backhoes. Transportation of soil can be done using dozers, loaders, trucks and scrapers. Compaction of soil is done with rollers. After compaction, the soil needs to be graded. This is done with graders.

Backhoe and a dozer combo (Above Figure).

Dozers are very versatile. One would rarely see a construction site without a dozer. Dozers can cut ground and also transport soil for short distances.

Loader- Loaders can be used to cut soil and also transport soil. When it comes to cutting ground, loaders are not efficient as dozers.

Scraper – Scrapers scrape soil and store in the underbelly of the machine. Scrapers can transport soil economically for 2 miles.

Left: <u>Typical Dump Truck</u>
Right: <u>Large Dump Truck</u>

Backhoes: Backhoes are used to cut ground. Backhoes are not good at transporting soil.

Dozers: Dozers are also used to cut and level ground. Dozers can push soil for few yards. Dozers are not economical to transport soil for long distances.

Loaders: Loaders can be used to cut ground. Loaders are not efficient as dozers in cutting ground. Nevertheless, loaders have a storage bucket that can be used to transport soil.

Scrapers: Scrapers can scrape soil and store in the storage area. Scrapers can transport large quantity of soil.

Graders: Grading is the process of leveling the ground. Graders are used to level the ground. Some graders are equipped with laser technology to fine grade the soil.

Grader - *Grader has a blade that can level the soil. Graders are not used for cutting.*

2.3 Concrete Construction:

Mixing, Transportation and Placement of Concrete: (ACI 304R)

It is important to make sure that the design mix has not undergone major changes when the concrete truck arrives at the job site. One major problem that occurs during transportation is segregation. It is common sense that heavy particles such as aggregates tend to settle at the bottom if given the chance. Hence, it is important that drums of concrete trucks rotate during transportation.

Another major parameter that affects the strength is the water content. Higher water content would give rise to low strength and low durability. It is not possible to reduce the water content indefinitely, since workability will be reduced. It would be very difficult to get a smooth finish if the water content is too low.

Concrete Plants: Concrete is a mixture of cement, sand, coarse aggregates, admixtures and water. In some cases as discussed before fly ash and various other pozzalans are also added. General functioning of a concrete plant is shown below.

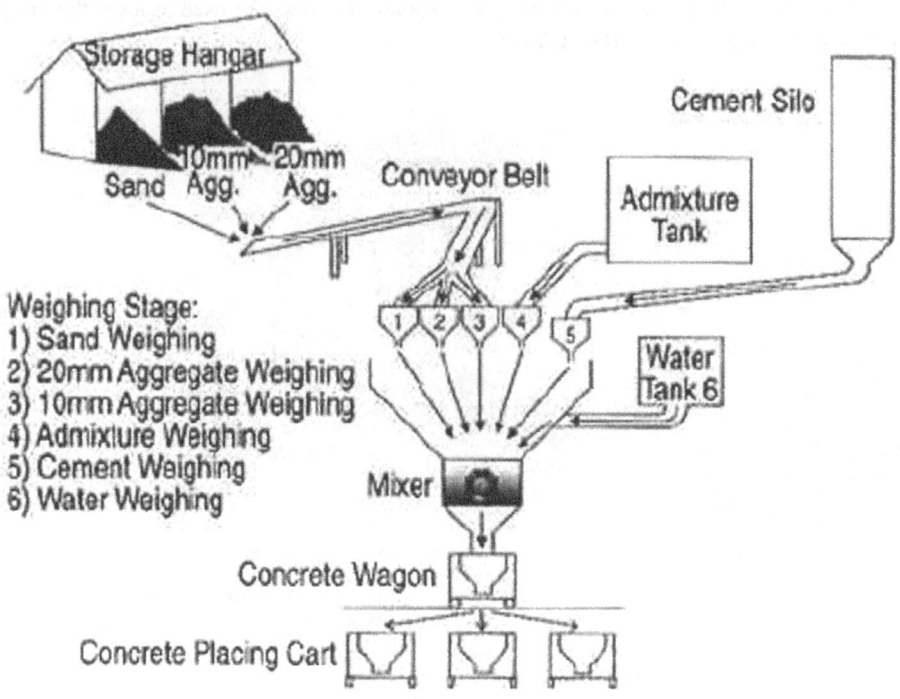

General outline of concrete batch plant

Concrete plant needs to have storage facilities for cement, fine aggregates, coarse aggregates, chemical admixtures and water. These ingredients need to be transported to the mixer. Typically, conveyor belts are used to transport materials in a concrete plant. Modern mixers are computer controlled. The operator input the proportions of material that need to be mixed. This data is used by a control system and correct amount of material is transported to the mixer. Concrete trucks are placed just under the mixer. In some plants, mixing is done inside the truck while travelling.

Concrete mixer is just above the concrete truck. Cone like figure on right is the cement silo. There is a conveyor belt from the cement silo to the mixer. Trapezoidal shaped container on left will be filled with sand and stones. Water usually comes from the street.

Concrete truck mixer

ACI 304R prefers all the water to be added at the plant so that water content can be properly controlled. Some water can be added at the job site to obtain the correct slump.

Concrete Mixing at the Job Site: For small projects, mixing of concrete can be done at the job site. Portable mixers of various sizes are available.

Mixing concrete at the job site

Concrete Placement:
Once the concrete arrives at jobsite, concrete placement can be done with many different equipment.

<u>Concrete Buckets</u>:   Concrete trucks would dump concrete into buckets. Concrete buckets are lifted by a crane to the proper elevation for concreting.

<u>Concrete bucket lifted by a crane</u>

It is important to clean the concrete bucket end of each day so that the openings are not obstructed by hardened concrete. ACI 304R recommends side slopes of the bucket to be at least 60 digress from the horizontal.

<u>Concrete Buggies</u>:   Concrete buggies are used for horizontal transportation of concrete. Concrete tends to segregate during transportation of concrete using buggies. Hence, it is not a very good way of transporting concrete. ACI limit the concrete transportation distance to 200 ft for manually operated buggies and 1,000 ft for power buggies. To minimize segregation, rails should be provided for the buggies. Hence, transportation can be made smooth. If rails cannot be provided, the surface should be made smooth as possible.

<u>Left</u>:   Concrete manual buggy (maximum travel distance 200 ft)
<u>Right</u>:   Concrete power buggy   (maximum travel distance 1,000 ft)

<u>Concrete Chutes</u>:   Concrete chutes are typically used to transport concrete from a higher elevation to a lower elevation. ACI does not give a maximum allowed length that can be used to transport concrete using chutes.

Concrete chute

Concrete chutes should have rounded corners. Slope should be steep enough for the concrete to travel freely.

Concrete Pumping: Concrete pumping has become very popular in large projects. Concrete pumping requires less labor and easy to control.   Height of pumping is dependent on the size of the pump. Concrete pump design mixes typically have a higher slump compared to regular concrete mix. To obtain a high slump, one needs to increase the water content. Increasing the water content would affect the strength. Hence, water-reducing admixtures are used to obtain a higher slump without increasing the water content.

Concrete pump

Concrete pumping pipes can be rigid pipes or flexible pipes. Rigid pipes will have fewer problems during pumping. The major disadvantage of rigid piping is difficulty of handling. On the other hand, workers can take the flexible pipes to the location of placement without much difficulty. Concrete pumps have a maximum rate of flow and a maximum pressure. Both cannot be achieved at the same time. One ft of additional vertical height is equal to 3 to 4 ft of additional horizontal distance.

Concrete Vibration (Concrete Consolidation) ACI 309:

ACI document 309 deals with concrete consolidation or vibration.

Following is a list of benefits one may obtain due to proper vibration of concrete.

- Higher compressive strength
- Higher bond between rebars and concrete
- Increase bond at cold joints
- Reduction of honeycombing and air pockets inside concrete
- Avoid segregation of cement paste and aggregates

Adverse Conditions due to Inadequate Vibrating;

Inadequate vibration would cause

- Honeycombing
- Voids due to high air entrapment - Proper vibration would allow air to escape.
- Sandstreaking - Sandstreaking is loss of cement water paste due to excessive bleeding. When cement water paste bleeds out between the form and mass of concrete sand lines would be exposed. Proper

vibration can be useful in avoiding sandstreaking. Not enough fines in the concrete mix also could lead to sandstreaking.

- Placement lines - Concrete is not placed with one truck. There could be a time lag up to half hour in some cases between trucks. One may see placement lines if concrete is not properly vibrated.

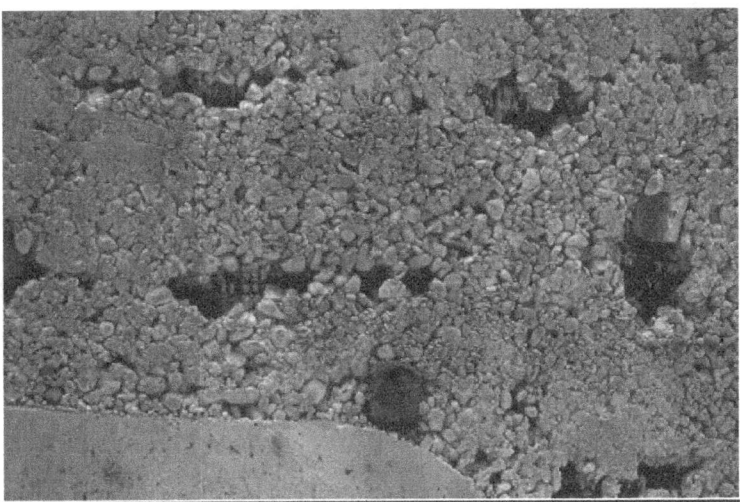

Honeycombing and large voids due to air entrapment in concrete

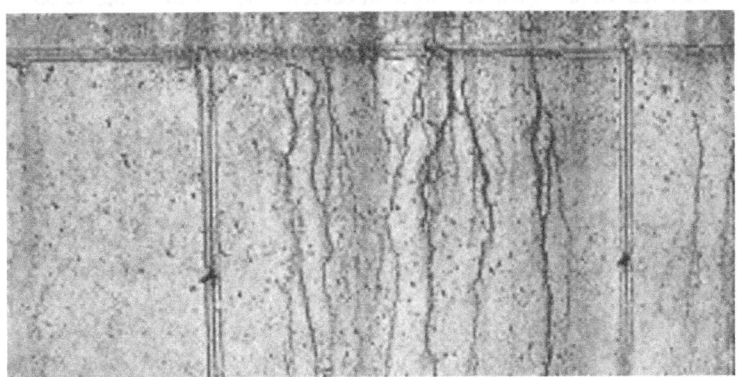

Sandstreaking

Concrete Finishing:

Concrete finishing is also known as concrete screeding is the process of achieving a smooth concrete surface. Most common screeding apparatus is the bull float. Bull float can be used for average size concrete slabs.

Bull Float

Concrete hand trowels are used for areas where bull float cannot reach. Curbs, inside edges of a slab, round surfaces etc. are finished with concrete hand trowels.

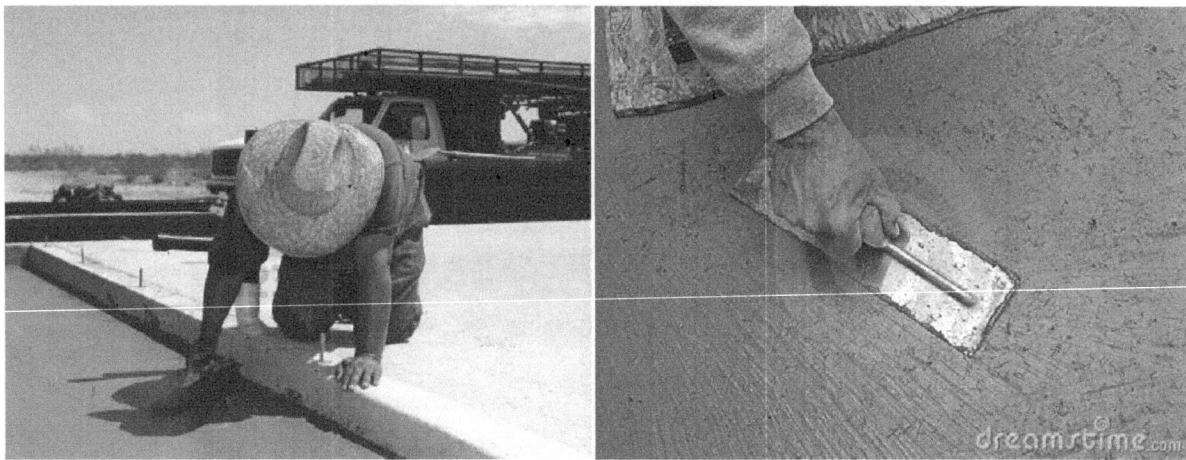

Concrete Hand Trowels

Bull float and hand trowels may be time consuming for larger slabs. In such situations, concrete trowel machines are used. Trowel machines come in various sizes. Small scale trowel machine is shown below.

Small trowel machine

Concrete Broom Finish:

Figure: Concrete Broom Finish

Some instances, concrete is finished with a broom to get a rough surface. This type of finish is required when epoxy or other layers were to be placed on concrete. The rough surface obtained will be useful to provide friction so that whatever the layer going on top could properly adhere to concrete.

Ride on Trowel Machines:

Large projects typically uses ride on trowel machines. A person can ride these machines and trowelling can be done faster.

Ride on Trowel Machine

Power screeds are also famous among concrete contractors.

Concrete power screed

Precautions should be taken to protect the concrete from high temperatures during the curing period. Concrete can be kept moist by continuously spraying water. One can also use wet blankets to keep the concrete at moderate temperatures.

<u>Concrete Elements</u>:   Typically, there are many different types of concrete elements in a structure. Some of the concrete elements are beams, columns, slabs on grade, structural slabs, slabs on metal decks, concrete walls, retaining walls, concrete piles, pile caps, topping slabs, piers, footings, curbs, stairs, equipment pads, conduit encasements and concrete filled metal pan stairs.

Concrete Accessories:

Reinforcement Supports: Rebars need to be supported in slabs and beams to obtain proper concrete cover requirements. Concrete cover is needed to protect rebars from water and outside chemicals.

Different types of rebar supports available in the market. In most cases, bricks are used as rebar supports.

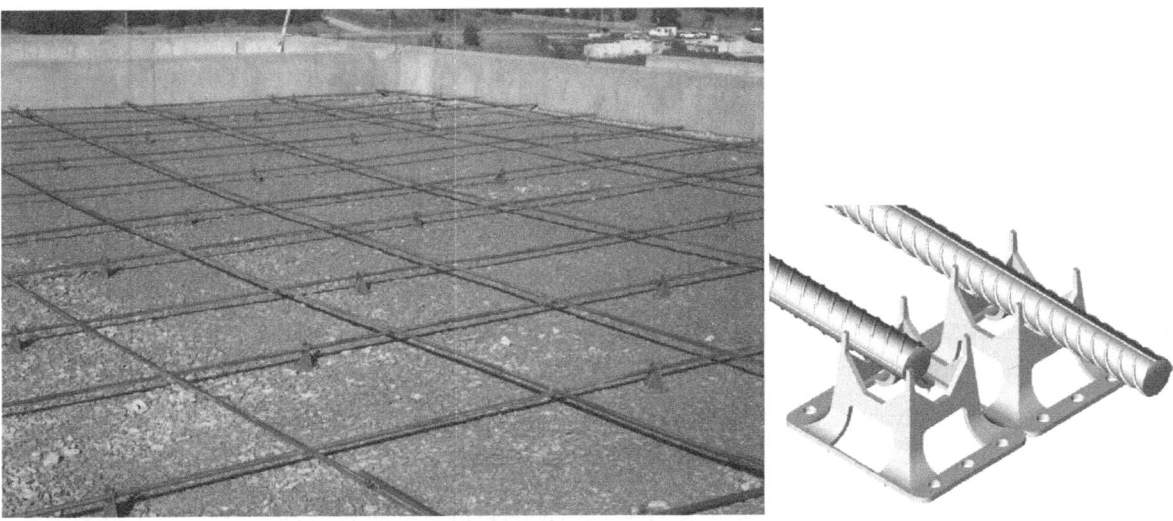

Rebar chairs

<u>Curing Compounds</u>:   Curing is the process of concrete hardening or hydration. The hardening process needs water. The concrete shall be kept in moist condition during curing. This can be achieved by flooding the concrete with water or sprinkling water during curing. Other methods include wet blankets and curing compounds.

Layer of water   Wet blankets

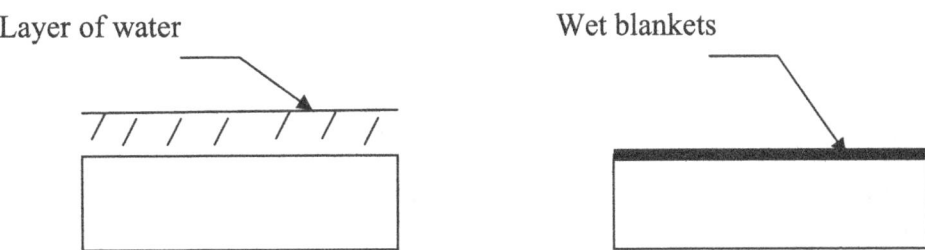

Left: Flooding the concrete with water
Right: Place a wet blanket and sprinkle water

Most contractors would prefer to apply a curing compound to the concrete. Curing compounds are a liquid that would seal the concrete surface and keep the water inside. After concrete is hardened, the curing compound would break down and sealing effect will disappear.

Application of a curing compound is shown in the photograph. Curing compounds come in 55-gallon barrels.

Curing compounds essentially act as a sealant during curing. It will not allow water to evaporate from the concrete.

<u>Bonding Admixture (Bonding Agent)</u>:   Fresh concrete do not bond well to old concrete.   Bonding agents are used to bond new concrete to old concrete, metal to concrete, concrete to topping slabs etc.

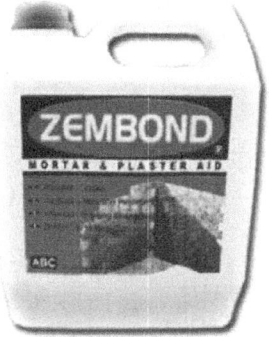

Left: Application of a bonding agent to old concrete
Right: Bonding agent container

Waterstops:

Waterstop is an expanding material that is installed in footings and in walls to stop water from entering the concrete. Waterstops are made of PVC, stainless steel and swellable clays.

Swellable waterstop is shown above. When water is encountered, this waterstop will expand and stop water from getting inside. It is possible in some occasions to damage the concrete due to swelling of waterstop.

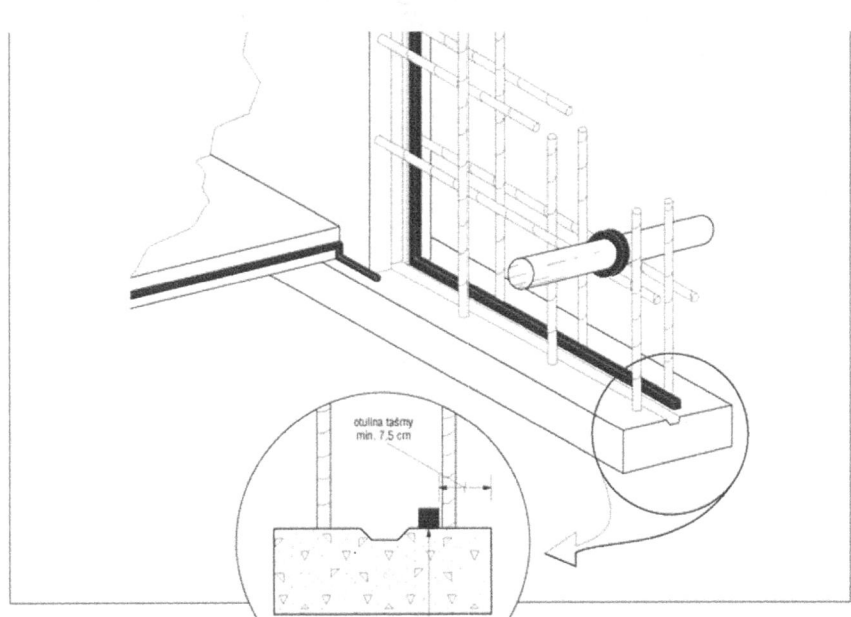

Waterstop installed in a shallow foundation is shown above. It is not advisable to install the waterstop inside the keyway since that would reduce the shear capacity of the keyway.

Most waterstop materials would expand when it encounter wet concrete. Due to the expansion, it would create a barrier for water.

Concrete Formwork:

In many projects, formwork cost exceeds the cost of concreting. Hence, one may need to pay proper attention to formwork. Formwork in the past was mostly constructed using timber. Still timber is widely used. Timber formwork though cheaper than metal or plastic formwork, cannot be reused several times due to damage. Hence, other products can be more economical than timber formwork.

Wall Formwork:

Wall formwork using timber is shown in the above figure.
Main items in a timber wall formwork system are;
  1) Sheathing
  2) Wales
  3) Posts and Buttresses

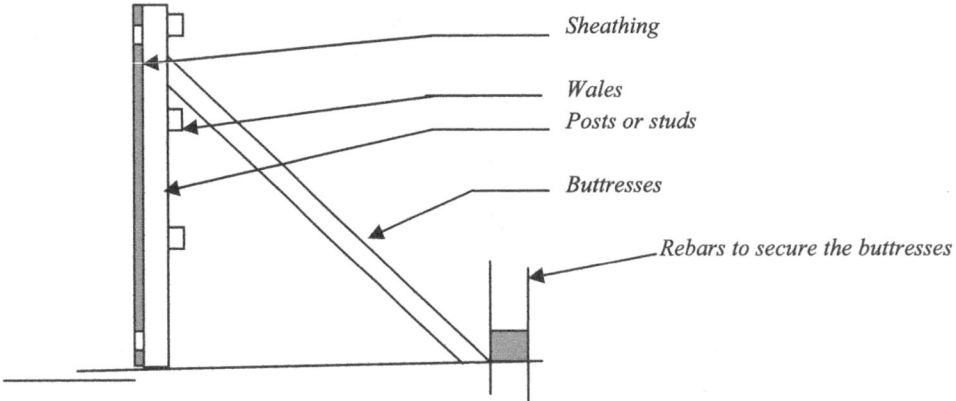

Timber formwork may be economical for small-scale projects.

Prefabricated Formwork: For larger projects, it is more economical to rent or buy prefabricated formwork. In addition, it may be faster to install them than constructing timber formwork.

Prefabricated wall formwork

Formwork Material:

Plywood Forms: Plywood is the cheapest. They can be reused many times.
Steel Forms: Steel forms can be reused up to 300 times. Cost of steel forms are much higher than plywood forms.
Steel forms come in different sizes and shapes hence work in the field can be reduced.
Aluminum Forms: Aluminum is lighter than steel. Hence, Aluminum forms can be erected with few men.

2.4 Steel Construction:

Steel is still widely used in construction. Steel has many advantages over concrete. After 9/11, many proponents of concrete stated that if the world trade center towers were built using concrete, they would not have collapsed. Concrete is an extremely good fireproof material. Under high temperatures, steel connections tend to fail. Concrete beams and columns do not need additional fireproofing. On the other hand, steel beams and columns need to be fireproofed.

Left: *Steel beams and columns prior to fireproofing*
Right: *Steel columns after fireproofing*

On the other hand, performance of steel structures during earthquakes is much better. This is because steel structures can be designed with more flexibility than concrete structures. Nevertheless, some argue with modern design techniques even concrete structures can be made very safe under earthquake loadings.

It is generally believed that cost of steel framed structures to be much higher than concrete structures. However, some experts believe that with the advent of high strength steel combined with state of the art design techniques, one may able to design cheap steel structures. One of the main advantages of concrete is the availability. Concrete is available thru out the year in most parts of the country. On the other hand, availability of steel can be an issue if not ordered ahead of time. Concrete structures can be built faster than steel frames. However, if one can get design, detailing and fabrication done in time, steel erection can go faster. On the other hand, any errors during detailing or fabrication can delay the project. Another advantage of concrete is that any shape an architect imagines can be easily achieved. Same cannot be said with steel. Though complex shapes are possible with steel, fabrication issues can delay the progress. However, one can argue that most aesthetic buildings are steel framed structures.

<u>Left</u>:   Steel building
<u>Right</u>:   Concrete building

Steel Construction Process:

Design Drawings and Shop Drawings:

Once the architect has completed the architectural drawings, structural engineers would design columns and beams. Structural engineers would size up the columns and beams and provide the type of sections needed at each location. In addition, they would provide the size of anchor bolts and various other structural information. Design drawings and specifications would be part of the contract. Contractor who won the contract would provide the design drawings to a steel fabricator. Steel fabricator would use the design drawings and develop shop drawings. During the shop drawings stage many issues that were not considered during the design phase would be considered. What is the best method to attach a gusset plate to a beam? Should the plate be welded in the field or in the shop? Welding a steel member in a shop is always cheaper than welding in the field. On the other hand, if welded in the field, workers can make minor adjustments to the piece to fit into the structure. In many instances, design engineer would delegate design of connections to steel fabricating shop. This is known as design delegation. However, as per law, this would not release the design engineer from the responsibility. Since two parties are involved, any issues arising due to bad connection design would be the responsibility of both the fabricator and the design engineer.

<u>Steel Design Drawings</u>:   Design drawings would tell the erector what beams and columns need to be used. Beam elevations and column elevations are also given.

Building grid lines and beams and column schedule is shown below.

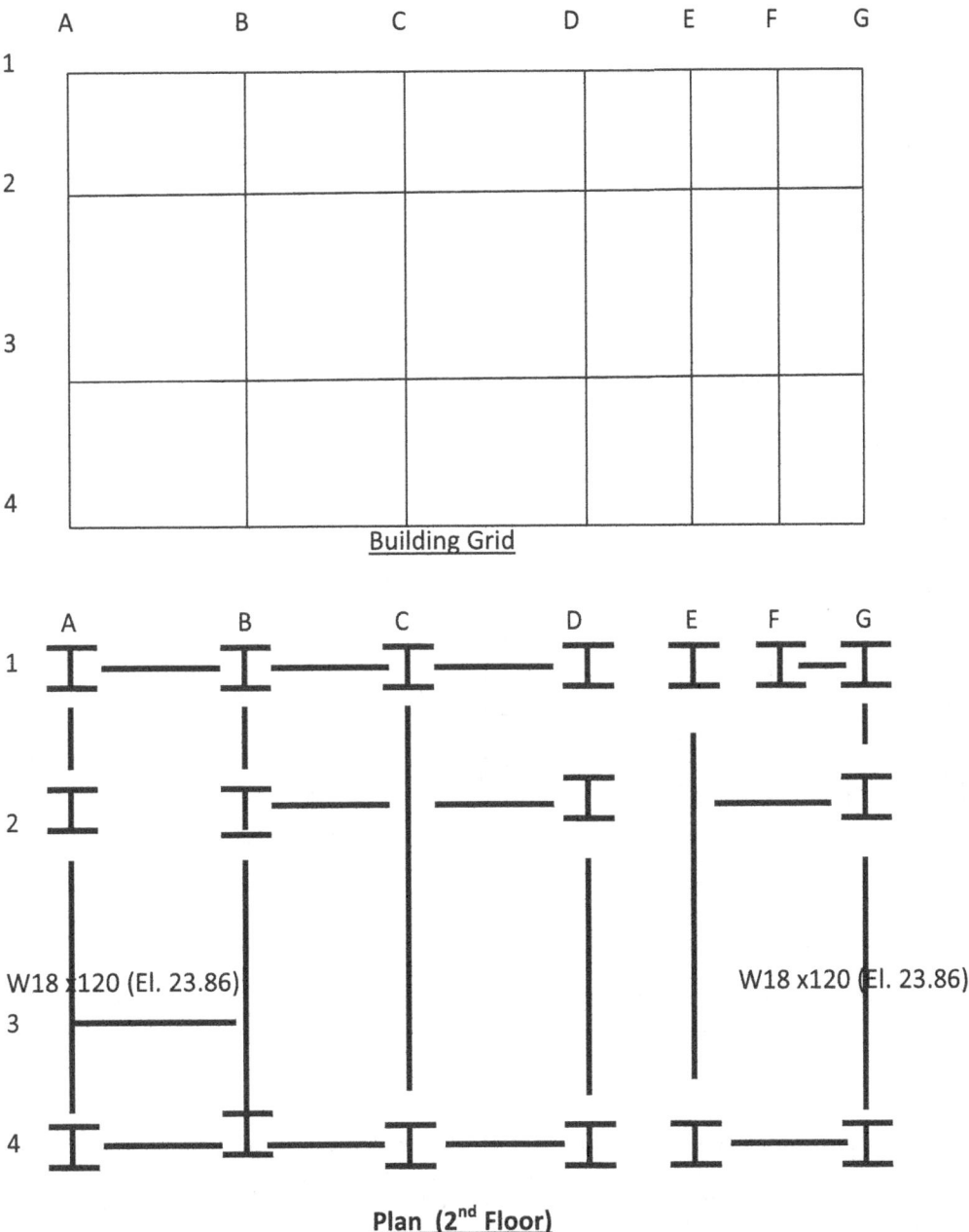

Building Grid

Plan (2nd Floor)

Above figure shows the plan view of second floor. Plan view of each floor should be provided. Size of each beam and top of the beam elevation should be given in the drawing. In the above figure, size of two beams shown. In addition, connection details also should be provided. Bolt pattern or weld information need to be provided.

Column schedule;

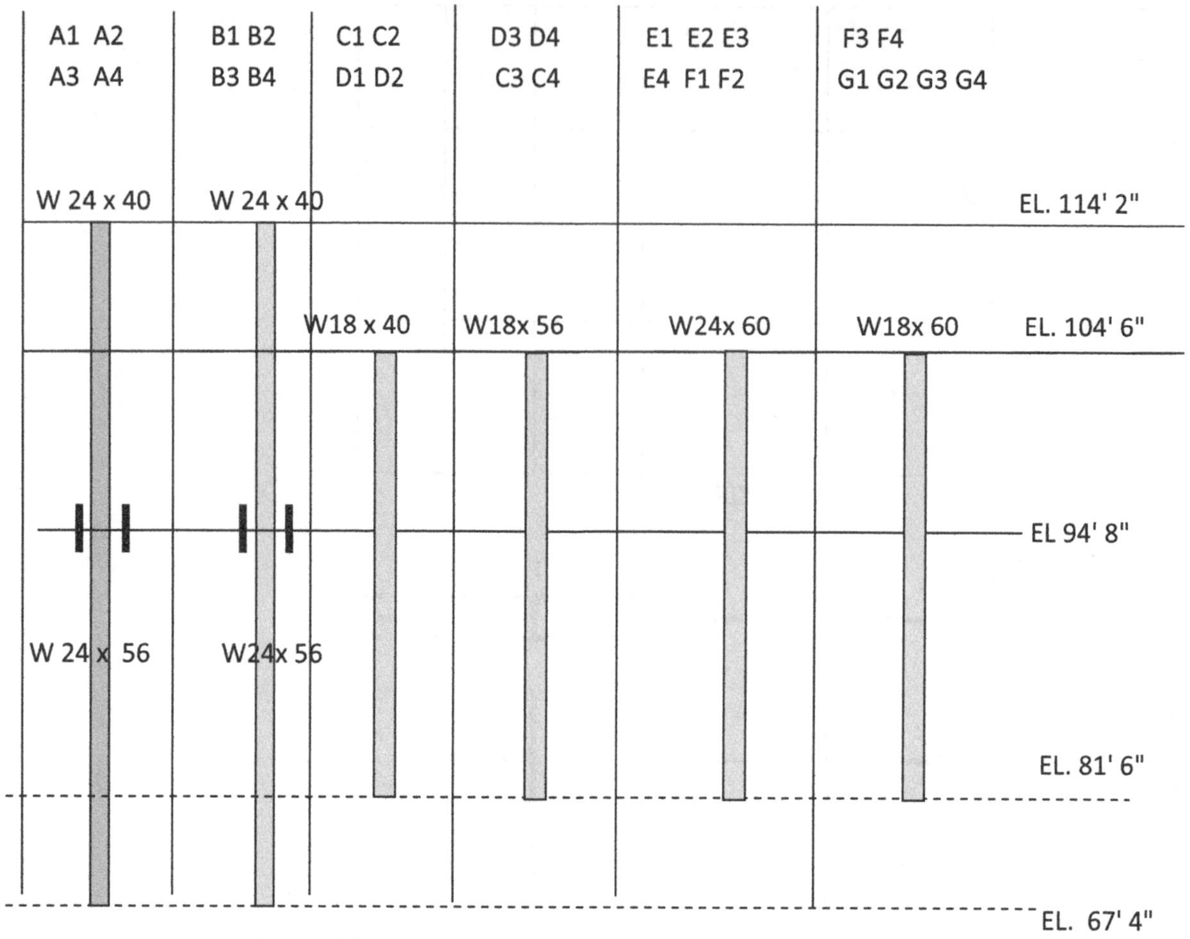

A1 A2 A3 A4	B1 B2 B3 B4	C1 C2 D1 D2	D3 D4 C3 C4	E1 E2 E3 E4 F1 F2	F3 F4 G1 G2 G3 G4	
W 24 x 40	W 24 x 40					EL. 114' 2"
		W18 x 40	W18x 56	W24x 60	W18x 60	EL. 104' 6"
						EL 94' 8"
W 24 x 56	W24x 56					EL. 81' 6"
						EL. 67' 4"

Column schedule should provide column location, column size, column start and end elevations. (i.e. In the above figure at C1, install a column W18 x 40 at an elevation of 81' 6". Top of column elevation is 104' 6".

Erection Drawings:

Once the shop drawings are approved by the design engineer, the fabricator would develop erection drawings. Each steel member is given a piece number. Erection drawings would very specifically give where each steel member goes. The workers would pick the marked steel member and find out where it would be erected. Then they would erect the pieces as shown on erection drawings. When steel members come to the site, they would be sorted out as per location that they are supposed to be erected. This is known as shaking. Steel members would be transported to the location of erection using a crane.

Steel Erection Process:

Steel erection has different phases and different crews. Some of the steel working crews are;
- Connecting crew
- Bolting crew
- Detailing crew
- Welding crew
- Decking crew
- Rigging crew

Very first crew to go up is the connecting crew. These workers connect beams and columns with few bolts. Connecting is considered to be the hardest job in steel work. Connecting crew connect steel beams and columns while they are loose and dangling in air.

Following is a write up by a steel worker;
"*As the steel pieces goes up, every time a beam is set onto a column, two pieces of steel meet in thin air. It is windy up there, and frames tend to sway without walls to stiffen them. A "connector" has to be at the top of that column, ready to pin the beam to it — and he may be 30 floors above the street. The work is simple to understand, but that does not make it easy. It is dirty, difficult and dangerous, and it takes a very determined man to do it. There are no gray areas. The reality of the work hits a man like a baseball bat each day. He can either do it or he can't*".

Steel Connecting Crews

Bolting Crew: Once the connecting crew had connected steel beams and columns with few bolts, bolting crew comes in. Bolting crew would bolt rest of the bolts and tighten them. Since the steel is, already connected, steel pieces are not moving in air. Bolting crew need to know which bolt goes where. In addition, there is a way of fastening them. In addition, bolting men need to carry hefty amount of bolts with them in addition to various wrenches.

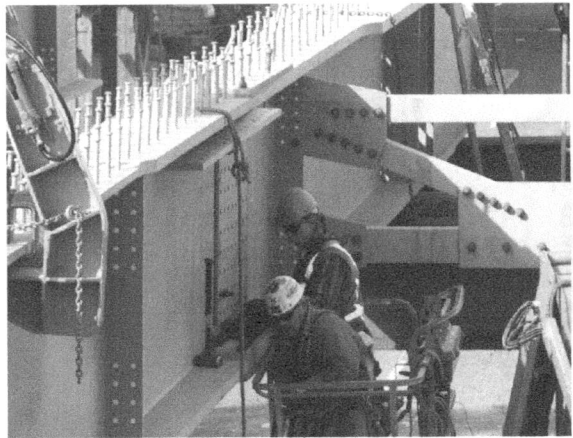 

Left: Steel workers bolting a beam
Right: Bucket of bolts handed down to a worker

Detailing Crew: After the bolting crew had completed bolting, detailing crew would follow up. As the name implies they would go through all the details and make sure the connection is completed as per design specifications. Any changes that the design engineer had come up with will be done by the detailing crew.

<u>Welding Crew:</u>   Some connections also need welding. Welding crew will conduct any welding required.

<u>Decking Crew:</u>   After all the connections are completed, decking crew will install the deck.

<u>Metal deck installation</u>

<u>Rigging Crew:</u>   Riggers would tie up beams and columns and signal the crane operator to take steel pieces to steel workers. Riggers should know various knots and chokers.

<u>Iron worker putting a choker to steel pieces to be lifted</u>

<u>Steel Wire Ropes:</u>   Steel wire ropes are made of strands. Strands are made of wires. There is a fiber or steel core at the center. 6 x 24 - FC means there are 6 strands. Each strand has 24 wires and FC means fiber core.

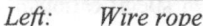
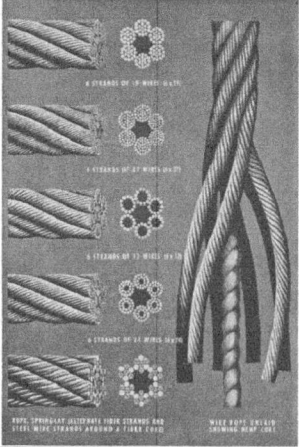

<u>Left:</u>   *Wire rope*   <u>Right:</u> *Different combinations of strands, wires and cores*

2.5 Temporary Structures

2.5.1 Falsework and Scaffolding:
Falsework is built to support formwork.

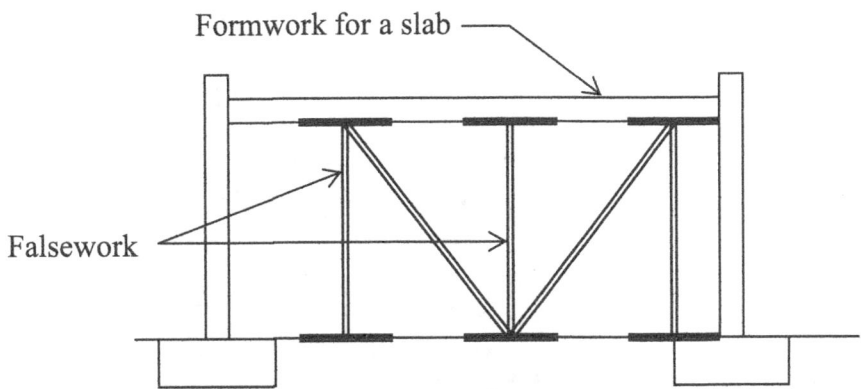

Falsework supporting formwork

Falsework needs to be designed by professional engineers since failure of falsework may cost lives of workers who work underneath or above.

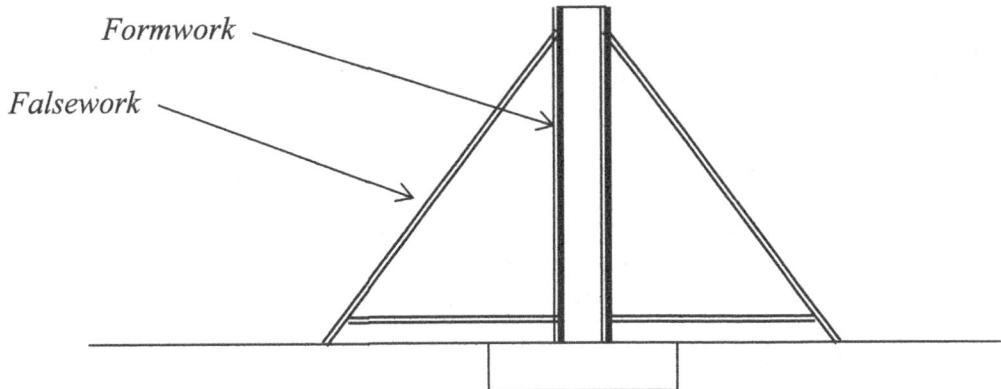

Formwork and falsework for a wall

Scaffolding: Scaffolding is constructed for men to work and store construction material such as windows, timber, concrete and steel.

Scaffolding for a painting project

As per Means guidelines, tubular scaffolding is cost effective up to 60' high or five stories. Above this, it is usually better to use hung scaffolding if construction permits. Swing scaffolding operations may interfere with tenants. In this case, the tubular is more practical at all heights.

In repairing or cleaning the front of an existing building, the cost of tubular scaffolding per SF of building front increases as the height increases above the first tier. The first tier cost is relatively high due to leveling and alignment.

Scaffolding Types: Tube and coupler scaffolds are the most popular scaffold type today. Wooden scaffolding was popular in the past. In Asia, still bamboo is the most widely used material to build scaffolding. Tube and coupler scaffolding may be expensive but could save time and probably will last during the life of the project.

- Tube and coupler scaffolds
- Carpenter's bracket scaffolds
- Outrigger scaffolds
- Bricklayer's square scaffolds
- Putlog scaffold
- Hanging scaffold or suspended scaffold

Figure: Tube and coupler scaffold

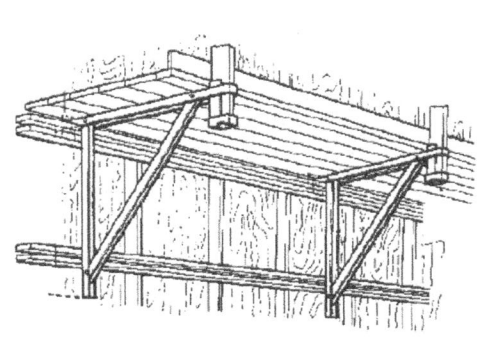

Figure: Carpenter's bracket scaffold

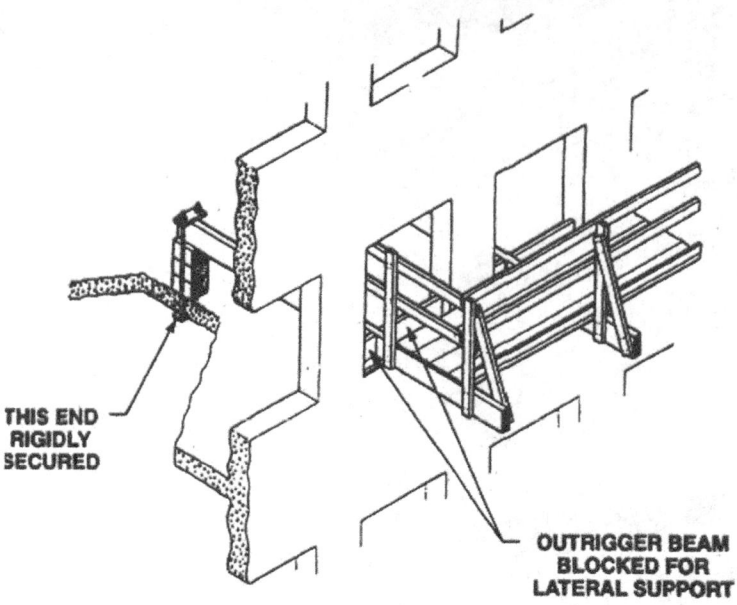

Figure: Outrigger scaffold

Outrigger scaffolds are built using outriggers that juts out from buildings. To build this type of scaffolds, one need openings in the building to place outriggers.

Hanging scaffold:

Figure: Hanging scaffold

Hanging scaffolds are hung from the roof. Roof assembly of a hanging scaffold is shown below.

Figure: Hanging scaffold roof assembly. The cantilevered beam supports the hanging scaffold below.

(Metal Modular Scaffolding):

Pre-made modules are becoming common in many construction projects.

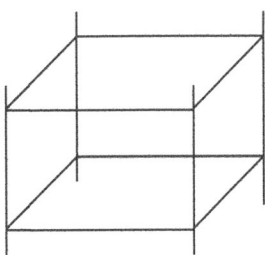

Scaffolding modules

Wood or metal Board

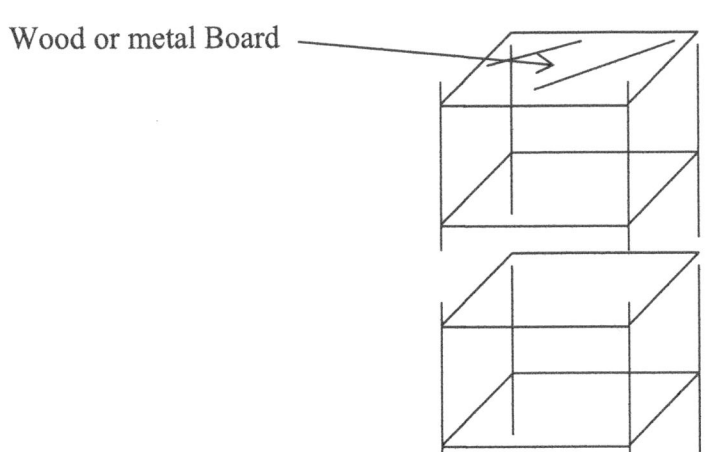

Scaffolding modules are fitted together to reach high elevations

Boards: Boards are made of metal or wooden planks attached to the scaffolding for people to stand and work.   Uprights also known as standards and poles are used to carry the load to base. False uprights are mainly used near entrances to the work platform. False uprights do not transfer any vertical loads to the ground. Though false uprights may provide lateral support to handrails, they do not provide any lateral supports to the scaffold system as a whole.

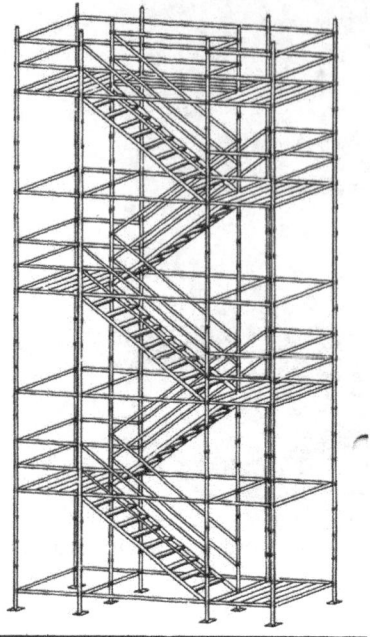

3.0 <u>Soil Mechanics</u>   -   6 Questions

Geotechnical engineers have to deal with soil and rock. Let us first look at strength of soil. Strength of soils comes from two parameters. They are friction and cohesion.

<u>Friction</u>:   Friction is a physical process. Friction of soil occurs due to friction between soil particles. Let us look at the figure below. On left there is a pyramid of balls made of glass. On right there is a pyramid of balls made of oranges.

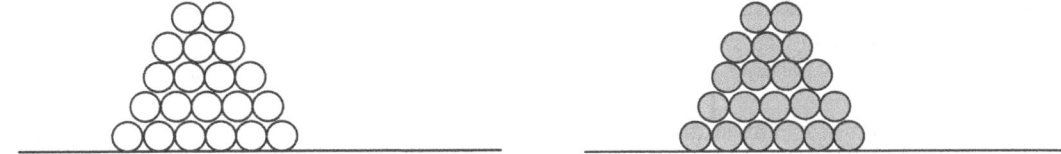

It is obvious that the pyramid of balls built with oranges would be much more stable than the pyramid built with glass. The friction between two orange surfaces is much higher than two glass surfaces. Same could be said of soils. When the friction between individual soil particles are high, the load that can be placed on soil will be greater. The issue is how to measure the friction in soil. Methods have been developed to measure the friction in soils. We would discuss these methods later in the book.

Above figure shows a failed building. The building load was too much for the soil. The friction in soil was not enough to accommodate the building load.

<u>Cohesion</u>:   As mentioned earlier, friction is a physical process. In addition to friction, there is another process. This is the cohesion between particles. Cohesion is an electro-chemical process. Cohesion occurs in clay soils. Cohesion between particles occur due to different chemical properties in particles. Different concentrations of ions will give rise to negative and positive charges.

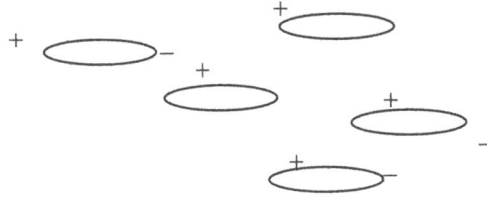

Positive and negative charges in soil particles give rise to cohesive attraction between particles.

Clay particles are shown in the above figure. Unlike sand particles, clay particles are not round but platy. Also, compared to sand particles, clay particles are extremely small and cannot be seen with the naked eye.

Gravel, Sand, Silt and Clay: Gravel particles are larger than sand particles. Sand particles are larger than silt particles. Generally, silt particles are larger than clay particles. Silt and clay particles cannot be seen with the naked eye.

Above figure shows silt particles. Silt particle are not visible to naked eye.

Pure Silt has no Cohesion: One important fact is pure silt has no cohesion. Silt is considered to be frictional soil. But in real life, in many cases silt particles are mixed with clay particles. Silty clays will have both cohesive properties and frictional properties.

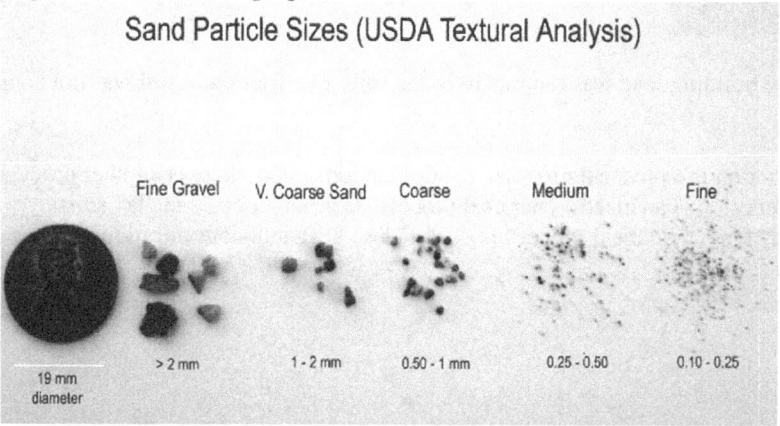

Above figure shows sand particle sizes relative to a penny. Silt and clay particles are not shown in the figure.

Origin of Rocks and Sand:
Which came first? Rocks or sand? The answer is rocks. There were rocks on earth before sands. Let us see how this happened. The universe was full of dust clouds. Dust particles due to gravity became small size objects. They were known as planetesimals. These planetesimals bashed on to each other and larger planets such as earth was formed.

Dust cloud ⟶ *Small planets (planetesimals)* ⟶ *Planets*

 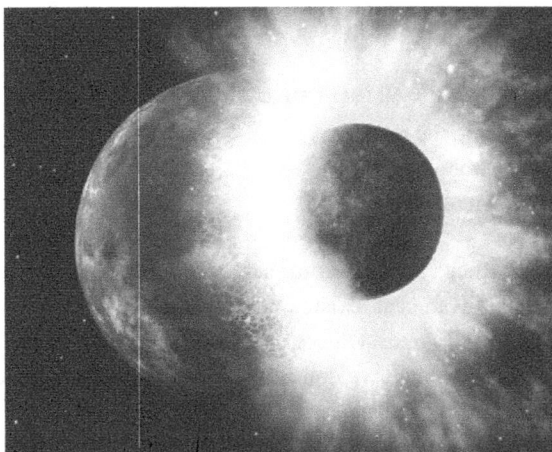

Figure on Left: *Most famous dust cloud (Horse head nebula)*:
Figure on Right: *Planetisimals colliding with each other*

Many dust clouds can be seen in the space. Horse head nebula is the most famous dust cloud. Dust clouds gave rise to millions of smaller planets that bashed on to each other. The early earth was a very hot place due to myriad of collisions. The earth was too hot to have water on the surface. All water evaporated and existed in the atmosphere.

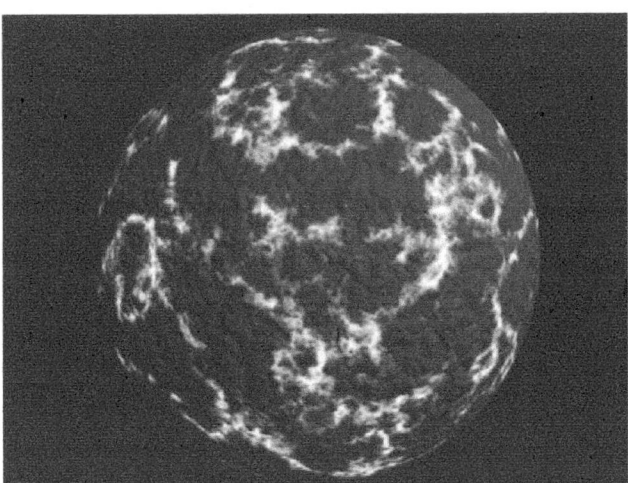

Artistic representation of early earth. The whole earth was covered with an ocean of lava:

Above figure shows a representation of the early earth. The whole earth was covered with a lava ocean. There was no water on surface of earth due to high temperatures on earth. But there was water in the atmosphere as water vapor.

Earth Cools Down: Many millions years later, earth cooled down. Water vapor that was in the atmosphere started to fall on to the earth. Rain started to fall on earth for the first time. This rain could have lasted for millions of years. Oceans were formed. Molten lava ocean became a huge rock. Unfortunately we have not found a single piece of rock that formed from the very first molten lava ocean.

Rock Weathering: Due to flow of water, change in temperatures, volcanic actions, chemical actions, earthquakes and falling meteors broke the rock into much smaller pieces. Many millions of years later rocks have broken down to pieces so small we differentiate them from rocks by calling them sand. Hence for our first question, which came first? rock or sand? Now you know the answer. The winner is rock.

Dust cloud ⟶   *Small planets (planetesimals)* ⟶ *Planets* ⟶ *Lava ocean* ⟶   *Lava ocean cools down to*
form bedrock ⟶ *First rains* ⟶ *Oceans* ⟶ *Weathering of rocks* ⟶ *Formation of sand and clay particles*

Rock Types:

<u>Brief Overview of Rocks:</u>   All rocks are basically divided into three categories.
- Igneous rocks
- Sedimentary rocks
- Metamorphic rocks

<u>Igneous Rocks</u>: Rocks that left from the very first lava ocean is lost completely. This is due to the fact that the earth is very old. Original lava ocean existed nearly 4.5 billion years ago. Since then many volcanoes have erupted and many oceans have come and gone. Oldest rocks identified are dated to 4.3 billion years. That is 200 million years after the original lava ocean. Though we don't have any rocks that formed from the original lava ocean, we have plenty of rocks that were formed due to volcanoes. Rocks formed due to cooling of lava is known as igneous rocks. Earth diameter is measured to be approximately 8,000 miles and the bedrock is estimated to be only 10 miles. The earth has a solid core with a diameter of 3,000 miles and the rest is all lava or known as the mantle. Occasionally, lava comes out of the earth during volcanic eruptions and cools down and becomes rock.

Some of the common igneous rocks are:
- Granite
- Diabase
- Basalt
- Diorite

Igneous rocks have two main divisions. They are intrusive and extrusive igneous rocks.

<u>Extrusive Igneous Rocks</u>:   We all have seen on TV lava flowing on surface of the earth. When lava is cooled on the surface of the earth extrusive igneous rocks are formed. Typical examples for this type of rocks are andesite, basalt, obsidian, pumice and rhyolite.

<u>Intrusive Igneous Rocks</u>: This type of rocks are formed when lava flow occurs inside the earth. Typical examples are diorite, gabbro, granite.

<u>Basalt:</u> Extrusive igneous rock type

Granite: _Intrusive igneous rock type_

Sedimentary Rocks: Volcanic eruptions are not the only process of rock origin. Soil particles constantly deposit in lakebeds and ocean floor. Many million years later these depositions solidified and converted into rock. Such rocks are known as sedimentary rocks.

Some of the common sedimentary rocks are:
- Sandstones
- Shale
- Mudstone
- Limestone
- Chert

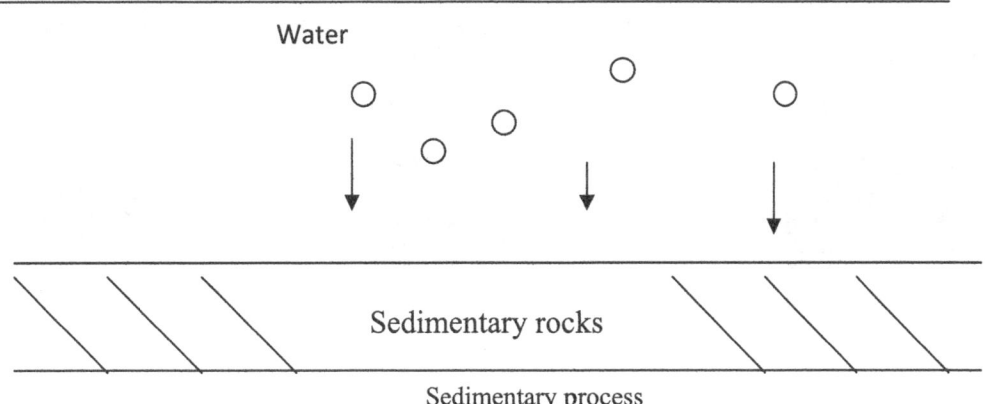

Sedimentary process

Imagine a bowl of sugar left all alone for few months. Sugar particles would bind together due to chemical forces acting between them and solidify. Similarly sandstone is formed when sand is left under pressure for thousands of years. Limestone, siltstone, mudstone and conglomerate forms in a similar fashion.

Lime -----------> Limestone
Mud -----------> Mudstone (Also known as shale)
Silt   -----------> Siltstone
Mixture of sand, stone and mud -----------> Conglomerate

Photo on Left: Sandstone   _Photo on Right:_ Mudstone

Columns made of conglomerate in Palace of Charles V, Spain. Congloemerate is a mixture of sand, gravel and clay.

Metamorphic Rocks: Other than these two rock types, geologists have discovered another rock type. The third rock type is known as **metamorphic rocks**.   Imagine a butterfly metamorphosing from a caterpillar.   Caterpillar will transform into a completely different creature.

Caterpillar metamorphoses to a butterfly:

Early geologists had no problem in identifying sedimentary rocks and igneous rocks. They were not sure what to do with this third type of rocks. They did not look like solidified lava. Also they did not look like formed due to sedimentation. Metamorphic rocks were formed from previously existed rocks due to heavy pressure and temperature. Volcanoes, meters, earthquakes and plate tectonic movements could generate huge pressures and very high temperatures.

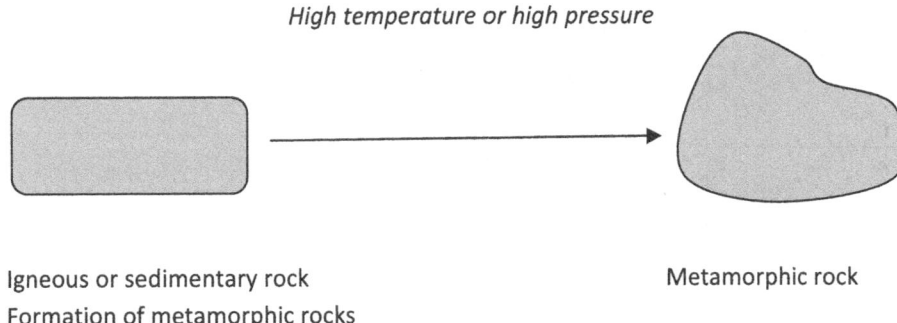

High temperature or high pressure

Igneous or sedimentary rock   Metamorphic rock

Formation of metamorphic rocks

Both sedimentary rocks and igneous rocks could transform into metamorphic rocks. For an example when shale or mudstone is subjected to very high temperatures, it would become schist. When limestone is subjected to high temperatures it would transform into marble. Some metamorphic rocks and their parent rocks are shown below;

Metamorphic rock	Parent rock
Gneiss	Shale or Granite
Schist	Shale
Marble	Limestone

Shale produces the greatest diversity of metamorphic rocks. Metamorphic rocks slate, phyllite, schist and gneiss all come from shale.

1.4   Soil Strata Types:

So far we learnt that early earth was extremely hot and was covered with an ocean of lava. Next, rocks were formed due to cooling of lava. Rocks were broken into pieces due to weathering. Rock pieces are today known as soil.

Soil strata are formed due to forces in nature. Following main forces can be identified for formation of soil strata.

- Water   (Soil deposits in river beds, ocean floor, lake beds and river deltas)
- Wind   (Deposits of soil due to wind)
- Glacial (Deposits of soil due to glacier movement)
- Gravitational forces (Deposits due to landslides)
- Organic   (Deposits due to organic matter such as trees, animals and plants)
- Weathered in-situ (Rocks or soil can weather at the same location).

1.4.1   Water:

Alluvial Deposits (River Beds):   Soils deposited in river beds are known as alluvial deposits. Some text books use the word "Fluvial deposits" for the same thing. Size of particles deposited in river beds depends on the speed of flow. If the river flow is strong, only large cobble type material can deposit.

Alluvial deposits:

Above two photos are from alluvial or deposits in river beds. One on left has large size gravel type particles. The photo on right has small particles in the range of fine sand. We can clearly say that the photo on above left shows an alluvial deposit in a river that had a strong flow.

Marine Deposits: Soil deposits in ocean beds are known as marine deposits or marine soils. Though oceans can be very violent, sea beds are very calm for the most part. Hence very small particles would deposit in sea beds. Texture and composition depends on proximity to land and biological matter.

Lacustrine Deposits (Lake beds): Typically very small particles deposit in lake beds due to tranquility of water. Lakes deposits are mostly clays and silts.

1.4.2   **Wind Deposits (Eolian Deposits):**
Wind can carry particles and create deposits. These wind borne deposits are known as eolian deposits.
- Loess: Loess soil is formed due to wind effects. Main characteristic of loess is that it does not have stratifications. Instead the whole deposit is one clump of soil.
- Eolian sand (Sand dunes):   Sand dunes are known as eolian sands. Sand dunes are formed in areas that has desert climate.
- Volcanic ash: Volcanic ash is carried away by wind and deposited.

1.4.3   **Glacial Deposits:**
Ice ages come and go. Last ice age came 10,000 years ago. During ice ages, glaciers would be marching down to southern part of the world from the north. During last ice age, New York city was under 1,000 ft of ice. Glaciers bring in material. When the glaciers are melted away, material that was brought in by glaciers would be left.

In the above photograph, glacier has retreated. Nevertheless it had left large rocks, sand and clay. Glacial deposits are known as glacial till or glacial moraine.

Characteristic of Glacial Till (Moraine): Glacial till contain particles from all sizes. Boulders to clay particles will be seen in glacial till. When the glacier is moving it would scoop up any material on the way.

Glacio-Fluvial Deposits: During summer months, glaciers would melt. Melt water would run away with material and deposit along the way. Glacio-fluvial deposits are different than glacial till.

Glacio-Lacustrine Deposits: Deposits made by glacial melt water in lakes. Due to low energy in lakes, glacio-lacustrine deposists are mostly silts and clays.

Glacio-Marine Deposits: As the name indicates, glacial deposits in ocean is known as glacio-marine deposits.

1.4.4   Colluvial Deposits:
Colluvial deposits are formed due to landslides.

Above photo shows a landslide. The deposits due to landslides contain particles from all sizes. Large and heavy particles would be at the bottom of the mountain.

Foundations: So far we have studied soil and rock types and now let us pay our attention to foundations. Any Civil Engineering structure will exert pressure on the ground. The pressure will cause the ground to settle or in some cases to cause complete failure. Complete sudden failure is known as bearing failure.

Foundations:

Sudden bearing failure of a building

"However we can save 700 lira by not doing a geotechnical investigation"

Pisa tower in Italy has been settling for centuries. That is not a bearing failure. Settling is due to consolidation of soil under the footing. Different soils have different properties. Hence, it is important to know what type of soil resides underground.

3.1   Drilling and Sampling Procedures:

Subsurface conditions in a site are extremely important for geotechnical engineers. The knowledge of subsurface soil conditions, depth to various layers of soil, depth to groundwater and strength parameters of soil need to be investigated. Field visit is recommended prior to start of a drilling program.

3.1.1   Field Visit:

During the field visit, one may be able to observe surface soil conditions. If any slopes are exposed, the engineer may be able to know the subsurface soil conditions.

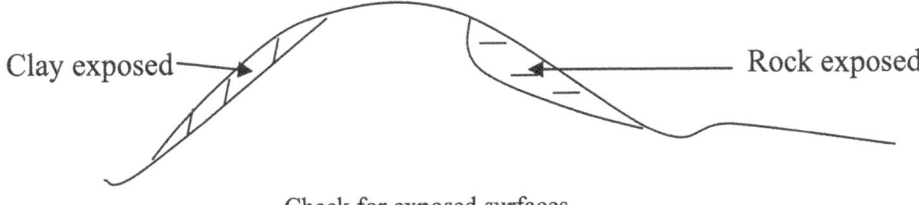

Clay exposed   Rock exposed

<u>Check for exposed surfaces</u>

Engineers during the field visit will be able to see any man made fill material on the surface. Man made fill material indicates disturbance to the site by previous human activities. Surface water flow is a good indication of groundwater flow. In many situations, surface water flow and groundwater flow are in the same direction. There could be hillocks and depressions. Depressions in the ground indicate soft soil conditions or weak bed rock.

Depression indicates soft soil below

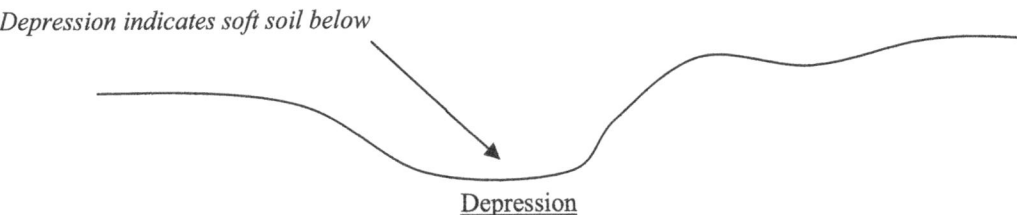

<u>Depression</u>

Slopes, hills, surface topography, streams, soft grounds, fill areas, potential contaminated locations, existing utilities and possible obstructions for future site investigation activities needs to be investigated.

<u>Groundwater Information:</u>   Groundwater information is extremely important in geotechnical engineering. Depth to groundwater can be found by installing groundwater monitoring wells. There are indirect ways to figure out the depth to groundwater. If a stream is nearby, one can assume groundwater to be shallow.

<u>Hand Auguring:</u>   Hand augers can be used during the field visit to obtain underground information. Hand augers can be used to depths up to 5 feet.

<u>Slopes:</u>   Construction in slopes is not economical. During the field visit, it is important to locate slopes and note them down.

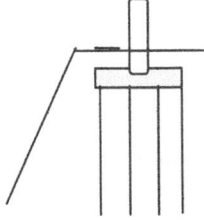

<u>Piles may be necessary near slopes</u>

<u>Nearby Structures:</u>   Construction work can be subjected to various limitations due to nearby existing facilities. Noise levels near hospitals, courthouses and schools need to be controlled. Pile driving, truck traffic and jack hammering could create a major noise disturbance. If the proposed building has a basement, underpinning of nearby buildings may be necessary.

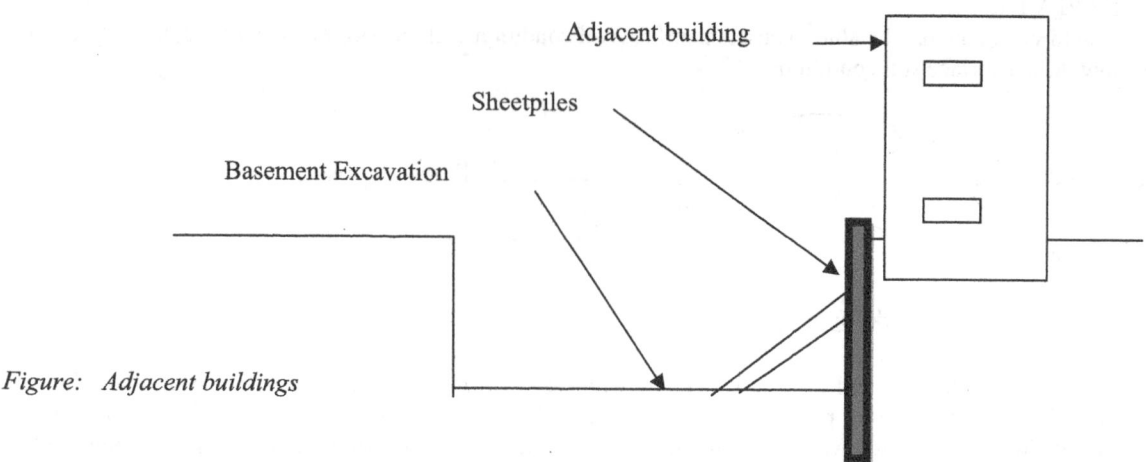

Figure: Adjacent buildings

<u>Contaminated Soils</u>:   Contaminated soils can increase the construction cost. Identifying possible contamination sources is important during the field visit.

<u>Man Made Fill Areas</u>:   Most urban sites are subjected to past human activities. Humans leave garbage and various other fills. It is not possible to compact some manmade fills such as rubber tires, clothes and plastics. Fill areas need to be investigated thoroughly during subsurface investigation phase of the project.

<u>Field Visit Checklist</u>:
          a)   Identify man made fill areas, depth and type of fill material
          b)   Investigate for possible utilities, overhead power lines and gas lines.
          c)   Note down topographical features such as depressions, hills, slopes, marsh areas and streams.
          d)   Nearby facilities (Hospitals, schools, courthouses etc).
          e)   Sloping ground – Construction in slopes are costly.

3.1.2   <u>Drilling Phase</u>:   Drilling program is conducted to obtain subsurface information. Following information is useful for the geotechnical engineer.

- Identification of subsurface soil and rock conditions. (Sand, clay, silt etc)
- Depth and thickness of each soil strata
- Cohesion (C) and friction angle (ϕ) of soil. (These two parameters are responsible for soil strength. Cohesion is obtained by conducting an unconfined un-drained strength tests. Friction angle is obtained by finding the SPT (N) value of the soil).

3.1.3   <u>Drilling Methods</u>:
- Augering
- Mudrotary drilling

<u>Auguring</u>:   Augering is a very common method of conducting subsurface investigations. Auger is a screw like rod that can be pushed into the ground. Hand augers were around for centuries. After the advent of machines, large augers are fitted with machines.

Hand auger

Augering

Mud Rotary Drilling: Mud rotary drilling process is different than augering. Instead of augers, roller bits are used to drill. Water is circulated through the rods and roller bit to keep the roller bit cool so that it will not over heat and stop functioning. Water is mixed with slurry to keep the soil walls from falling. Usually drillers mix Bentonite slurry (also known as drilling mud) to the water to thicken the water. Drilling mud goes through the rod and the roller bit and comes out from the bottom removing cuttings. The mud is captured in a basin and re-circulated.

Left: Mud rotary drill rig;   **Right**:   Roller bit

3.2   Drilling and Boring Program: Geotechnical engineer should identify locations of borings. Usually borings are conducted near columns and other footing locations. Number of borings that need to be constructed may sometimes can be regulated by local codes. For an example, New York City building code requires one boring per every 2,500 square feet of building footprint. Typically, borings are constructed to 10 ft below the bottom level of the foundation.

Hand digging prior to drilling: In many cases utilities need to be protected. In such situations, hand digging the first 6 feet prior to drilling boreholes is conducted.

3.3   Test Pits: In some situations, test pits would be more advantageous than borings. Test pits can provide information down to 15 ft below the surface. Unlike borings, soil can be visually observed from the sides of the test pit.

Test pits: Unlike in a boring, soil is clearly visible in a test pit.

3.4   Soil Sampling Using Split Spoons: Split spoon samples are obtained during boring construction. Split spoon samples are typically 2 inch diameter and has a length of 2 ft. Soils obtained from split spoon samples are adequate to conduct sieve analysis, soil identification and Atterberg limit tests. Consolidation tests, tri-axial tests and unconfined compressive strength tests need large quantity of soil and in such situations, Shelby tubes are used. Shelby tubes have a larger diameter than split spoon samples.

SPT Test: (Standard Penetration Test):   SPT (N) value is an important parameter that is been used to find the friction angle of sandy soils. SPT value is obtained by dropping a 140 lb (63.5 kg) hammer from a height of 30 inches (760 mm). Split spoon samples are taken during the process.

Un-opened split spoon samples on the left and opened split spoon sample on right

3.5   Soil Mechanics Analysis:
3.5.1 Pressure Distribution:
Pressure distribution in soils is a complicated process. For most practical purposes, many engineers assume the pressure distribution to be approximately 2:1.   (2 Verticals to 1 horizontal).

Practice Problem: Below figure shows a (4ft x 4 ft) column footing loaded with 80 kips. Bottom of the footing is at 3ft below the ground surface.

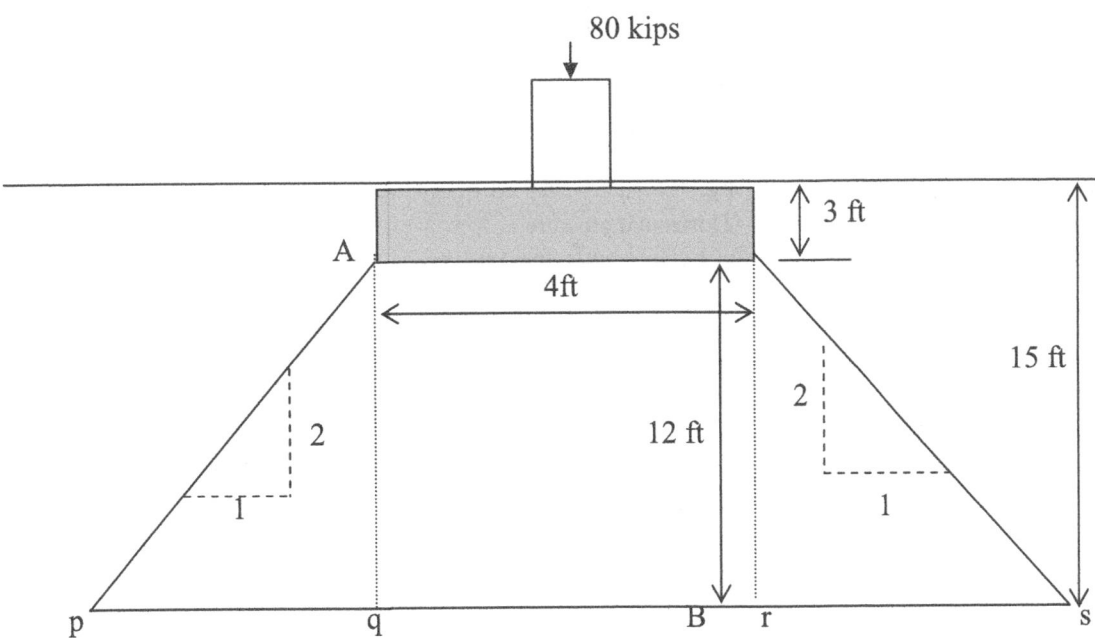

1) Find the pressure due to footing load at bottom of footing. (level A)

A)   1,140 lbs/ft^2.   B) 2,150 lbs/ft^2.   C) 5,000 lbs/ft^2.   D) 2,167 lbs/ft^2.

2) Find the pressure due to footing load at 15 ft below the ground surface (level B).
A)   2,340 lbs/ft^2.   B) 312.5 lbs/ft^2.   C) 978 lbs/ft^2.   D) 2,987 lbs/ft^2.

Solution:
a)   Pressure at bottom of footing (Point A):

1) Pressure at bottom of footing due to column load = 80/(4 x 4) kips/ft^2 = 5 kips/ft^2.
Ans C
 2) The pressure due to footing load 15 ft below the ground surface (level B):

Find the area of the square at level B.
Point B is 15 ft below the ground surface. (Given). Point B is 12 ft below the bottom of footing.

Length pq = 12/2 = 6 ft (assuming a 2:1 distribution).
Length qr = 4 ft
Length rs = 12/2 = 6 ft
          Length ps = 6 + 4 + 6 = 16 ft
          Area of the square at level B = 16 x 16 = 256 ft^2.
          Pressure at level B = 80/256 = 0.3125 kips/ft^2.
          Pressure at level B = 312.5 lbs/ft^2.   (Ans B)

3.6 Lateral Earth Pressure:
Once the vertical effective stress is found, it is a simple matter to calculate the lateral earth pressure.
Consider a basin full of water. Let us look at water pressure at point A.

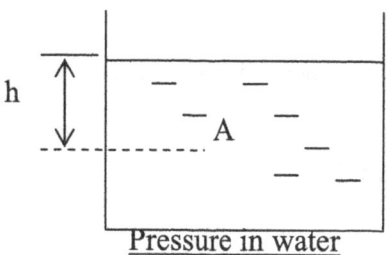

Pressure in water

Vertical pressure at point "A "   = γ_w x h
Horizontal pressure at point "A "   = γ_w x h
In the case of water, Vertical pressure = Horizontal pressure
In the case of soil, vertical pressure is not equal to horizontal pressure. In soil, horizontal pressure (or stress) is different from the vertical stress.

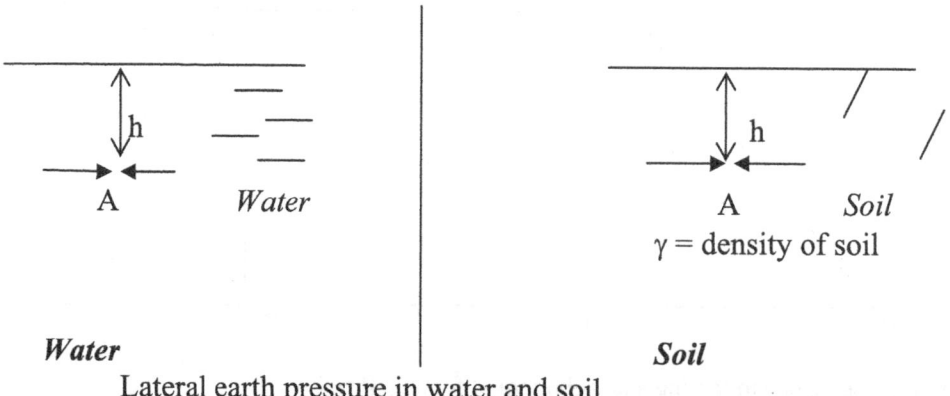

Water   Soil
Lateral earth pressure in water and soil

In the case of soil, experimentally it has been found that the horizontal pressure is given by the following equation.
Horizontal pressure at point "A" in soil $= K_0 \times$ (vertical effective stress)
K_0 = Lateral earth pressure coefficient at rest.
Following equation is used to compute K_0.
$$K_0 = 1 - \sin \varphi \qquad (\varphi = \text{Friction angle of soil})$$

Horizontal pressure at point "A" in soil $= K_0 \times \gamma \times h$
     γ = Density of soil
     h = Height of soil
When the soil can move, K_a and K_p values should be used instead of K_0.

     K_0 = At rest condition, when soil does not move = $1 - \sin \varphi$
     K_a = Active condition, when soil moves to relax the stress = $\tan^2 (45 - \varphi/2)$
     K_p = Passive condition, when soil moves to increase the stress = $\tan^2 (45 + \varphi/2)$

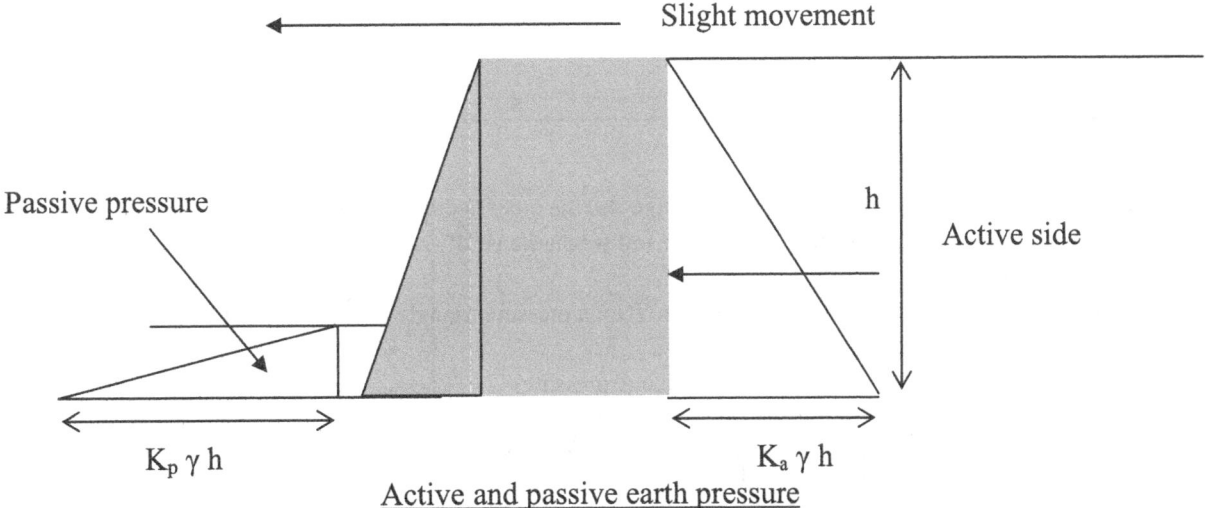

Active and passive earth pressure

The retaining wall will slightly move to the left due to the earth pressure. Due to this slight movement, pressure on one side will be relieved and the other side will be amplified. K_a is known as the active earth pressure coefficient and K_p is known as the passive earth pressure coefficient. Passive earth pressure coefficient is larger than the active earth pressure coefficient. K_0, K_a and K_p are given by the following equation.

$$K_0 = 1 - \sin \varphi, \qquad K_a = \tan^2 (45 - \varphi/2), \qquad K_p = \tan^2 (45 + \varphi/2)$$
$$K_a < K_0 < K_p$$

Practice Problem: Sandy soil has a friction angle of 35^0. Find K_0, K_a and K_p for this soil.

Solution:

     $K_0 = 1 - \sin \varphi$
     $K_0 = 1 - \sin (35) = 0.426$
     $K_a = \tan^2 (45 - \varphi/2) = \tan^2 (45 - 35/2) = 0.271$
     $K_p = \tan^2 (45 + \varphi/2) = \tan^2 (45 + 35/2) = 3.69$

It is clear that K_p is much larger than K_0. At the same time, K_0 is larger than K_a.

3.6.1 Soil Surcharge Pressure:

When a soil is placed behind a retaining wall as shown in below figure, the pressure diagram is considered a rectangle. Assume the surcharge pressure to be q psf. Then the horizontal earth pressure diagram will be as shown in the next figure.

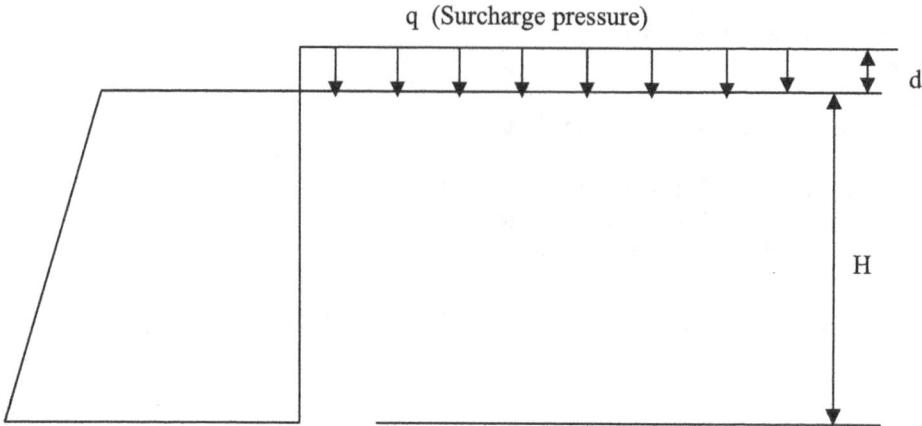

There are two horizontal forces. One is due to surcharge and the other is due to regular earth pressure.
Soil surcharge is shown above. If the thickness of the soil surcharge is "d", surcharge pressure (q) is $\gamma \cdot d$
q = Surcharge pressure = $\gamma \cdot d$
Surcharge pressure will generate a pressure **rectangle**. (Not a pressure triangle as in the case of backfill pressure).

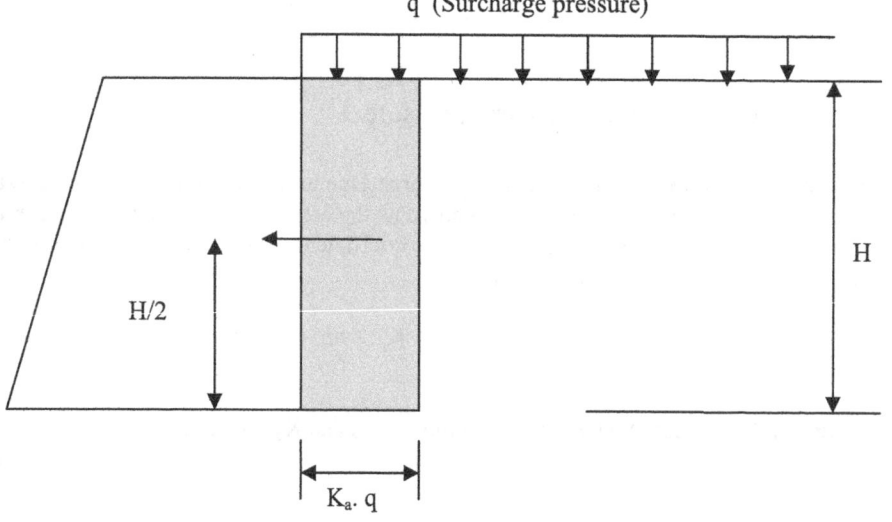

Pressure exerted on the wall due to surcharge = $K_a \cdot q$

Note that for surcharge pressure calculations, height of the retaining wall (H) is not a factor. H comes into picture when computing the force exerted on the retaining wall. Note that force and pressure are two different parameters.

Force on the retaining wall only due to surcharge   = Area of the pressure rectangle
                                                = $K_a \cdot q \cdot H$
The resultant of the pressure rectangle acts at the center of the rectangle.
Overturning moment due to the surcharge load = $(K_a \cdot q \cdot H) \times H/2 = K_a \cdot q \cdot H^2/2$

In reality there is pressure due to backfill as well. If there is water, water also would exert a pressure on the retaining wall.

Hence;
Total pressure on a retaining wall = Pressure due to surcharge + Pressure due to backfill + Pressure due to water

Practice Problem: Vehicle traffic weight near a retaining wall is considered equal to 2 ft of fill material. Assume density of fill to be 105 pcf. Lateral active earth pressure coefficient of the soil is 0.42. Density of backfill soil behind the retaining wall = 110 pcf.
   A) Find the total sliding force of the retaining wall
   B) Find the overturning moment of the retaining wall.

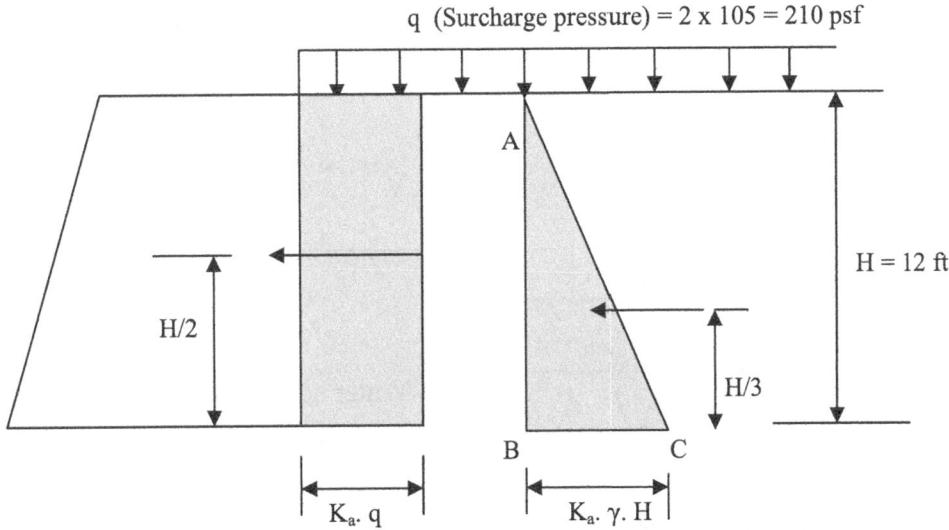

Solution:

STEP 1: Find the lateral earth pressure due to surcharge;
q = Surcharge = 2 x 105 = 210 psf
Lateral earth pressure due to surcharge = K_a. q = 0.42 x 210 = 88.2 psf
STEP 2: Horizontal force due to surcharge = Area of the rectangle = 88.2 x 12 = 1,058.4 lbs/ft
Horizontal force due to surcharge is a rectangle.

STEP 3: Overturning moment due to surcharge = Force x length to center of gravity = 1,058.4 x 12/2 = 6,350.4 lbs. ft/ft
STEP 4: Find the earth pressure due to backfill soil;
Earth pressure at bottom of the wall = K_a. γ. H = 0.42 x 110 x 12 = 554.4 psf
STEP 5: Find the horizontal force due to backfill soil;
Force due to backfill soil = Area of the triangle ABC = 554.4 x 12/2 = 3,326.4 lbs/ft
STEP 6: Find the overturning moment due to backfill soil;
Overturning moment due to backfill soil = Force x Length to center of gravity of the triangle
Overturning moment due to backfill soil = 3,326.4 x 12/3 = 13,305.6 lbs.ft/ft
STEP 7: Find the total horizontal force on the retaining wall;
Total horizontal force on the retaining wall = Horizontal force due to surcharge + Horizontal force due to backfill soil
Horizontal force due to surcharge = 1,058.4 lbs/ft (See Step 2).
Horizontal force due to backfill soil = 3,326.4 lbs/ft (See step 5).

Total horizontal force on the retaining wall = Force due to surcharge + Force due to backfill = 1,058.4 + 3,326.4 = 4,384.8 lbs/ft

STEP 8: Find the total overturning moment on the retaining wall;

Total overturning moment on the retaining wall = Overturning moment due to surcharge + Overturning moment due to backfill soil

Overturning moment due to surcharge = 6,350.4 lbs.ft/ft (See Step 3).

Overturning moment due to backfill soil = 13,305.6 lbs.ft/ft   (See step 6).

Total Overturning moment = 6,350.4 + 13,305.6 = 19,656 lbs.ft/ft

3.7   Soil Consolidation and Foundation Settlement:

What would happen when you place a foundation on top of a clay layer? It is obvious that the clay layer would compress and as a result, the foundation would settle. Why would the foundation settle due to an applied load?

Clay is a mixture of soil particles, water and air. When clay soil is loaded, water particles inside the soil mass would be pressurized and tend to **squeeze** out. Imagine what would happen if you keep a weight on top of a wet sponge? Due to the weight, water would squeeze out from the wet sponge.

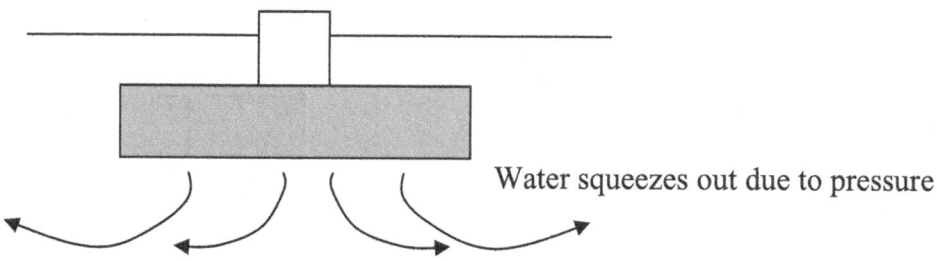

Water squeezes out due to pressure

The settlement due to water squeezing out is known as primary consolidation.   When a clay is loaded, excess pore pressure will be developed. Squeezing out of water is caused due to dissipation of excess pore pressure. When all the water in the soil is squeezed out, one may declare that 100% primary consolidation has achieved. In reality, 100% primary consolidation usually can never be achieved. For all practical purposes 90% primary consolidation is taken as the end of the primary consolidation process.

Tower of Pisa (Clay consolidation)

Secondary Compression: Now let us look at the secondary compression.
After all the water is squeezed out, one would expect the settlement to stop. Interestingly that is not the case. The settlement would continue after all the excess pore water is squeezed out because soil particles start to re-arrange their orientation. Let us see what this means.

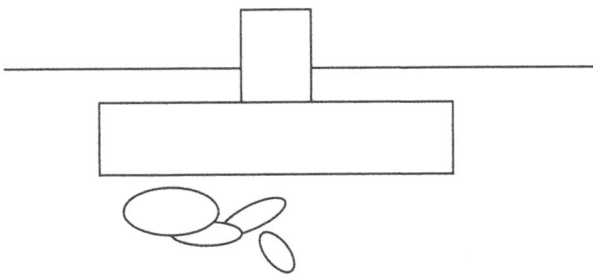

Soil particles below a footing

These soil particles try to re-arrange themselves to a more stable configuration.

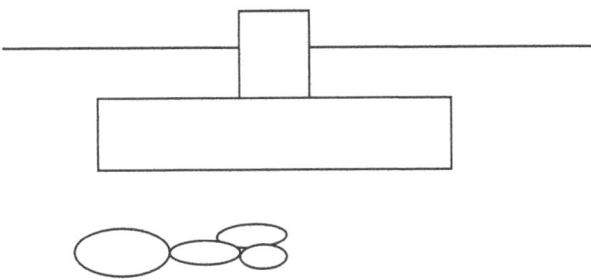

Stable re-arrangement of soil particles

Re-arrangement of soil particles would create further settlement of the foundation. In reality secondary compression starts as soon as the foundation is constructed. However, for simplicity, usual practice is to compute the secondary compression after the primary consolidation is completed.

Summary of Concepts Learned:
- Primary consolidation occurs due to dissipation of excess pore pressure.
- In reality, the primary consolidation never ends. For all practical purposes 90% consolidation is taken as the end of the process.
- Secondary compression occurs due to re-arrangement of soil particles due to the load. Secondary compression has nothing to do with the pore water pressure. Secondary compression occurs due to re-arrangement of soil particles

3.7.1 Normally consolidated clays and overconsolidated clays:

Settlement due to consolidation occurs in clayey soils. Clay soils usually originate in lake beds or in ocean floors. Sedimentation of clay particles occurs in calm waters.

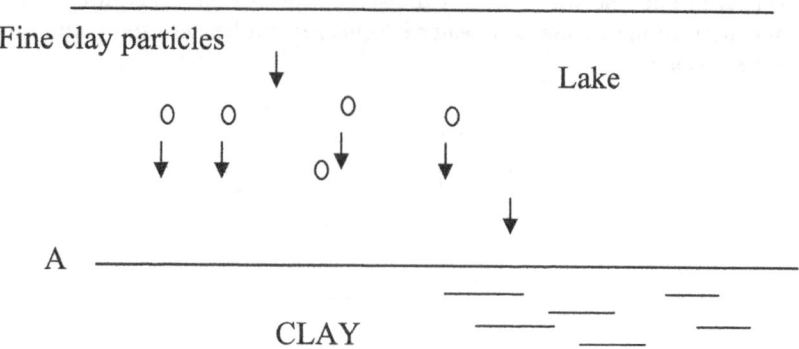

Fine clay particles

Lake

A

CLAY

Formation of a clay layer due to sedimentation

Years down the road, lake would dry up and the clay layer would be exposed to the atmosphere.

Water body disappeared.

A (Original surface)

CLAY

Normally consolidated clay

The above clay layer was never been subjected to any loads other than the load due to water body. Such clays are known as normally consolidated clays. The earth is a dynamic planet. Changes occur on earth surface every day. Hurricanes, tsunamis, landslides, coming and going of glaciers and volcanoes are some of the events that would occur on earth surface. During the last ice age, large area of northern hemisphere was under few hundred feet of ice. When glaciers are formed during an ice age, the clays in that region would be subjected to a tremendous load. Due to the ice load, the clay would undergo settlement.

Glacier

A (Original surface)

B (Settled location due to glacier)

CLAY

Clay layer would settle from point "A" to point "B" due to the glacier.

Point "A" is where the surface of clay layer immediately after the clay layer was formed.

Glaciers normally melt away and disappear at the end of ice ages. Today glaciers can be seen only in north and south poles. After the melting of glaciers, load on clay will be released. Hence the clay layer would re-bound. Clays that had been subjected to high stresses in the past are known as pre consolidated clays or overconsolidated clays.

Glacier had melted

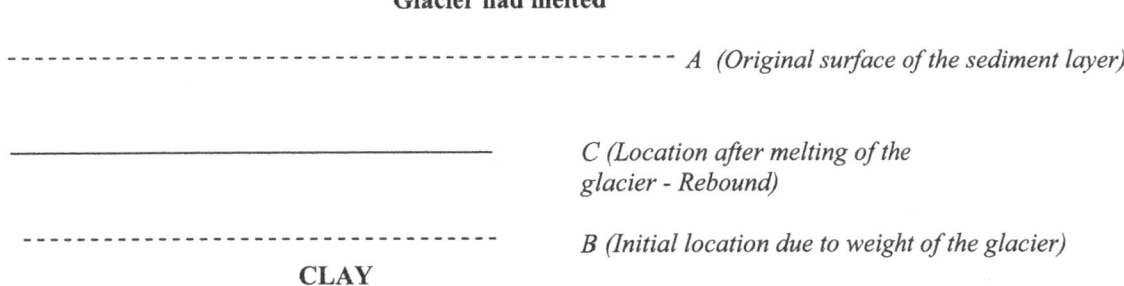

When the glacier is melted, the clay layer will re-bound to point "C".

Now let us assume that a footing was placed on the clay layer. The clay would settle due to the footing load.

The clay layer would undergo settlement due to the footing load. New location of the top surface of the clay layer is shown below. The clay layer would settle from C to D due to footing load.

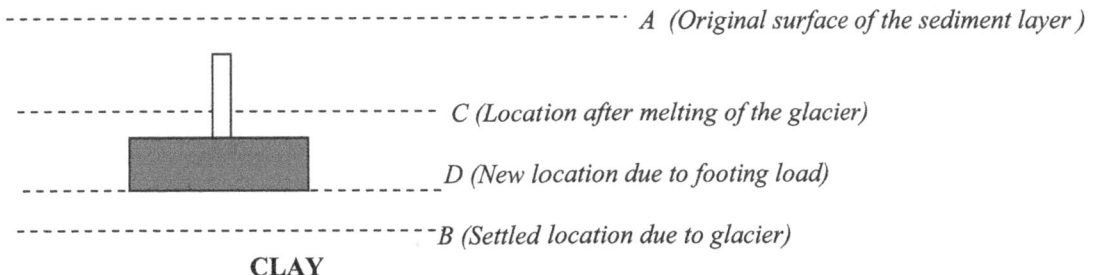

Footing was placed on top of the clay layer   (Footing settlement)

Void ratio "e" vs. log "p" graph is shown below.

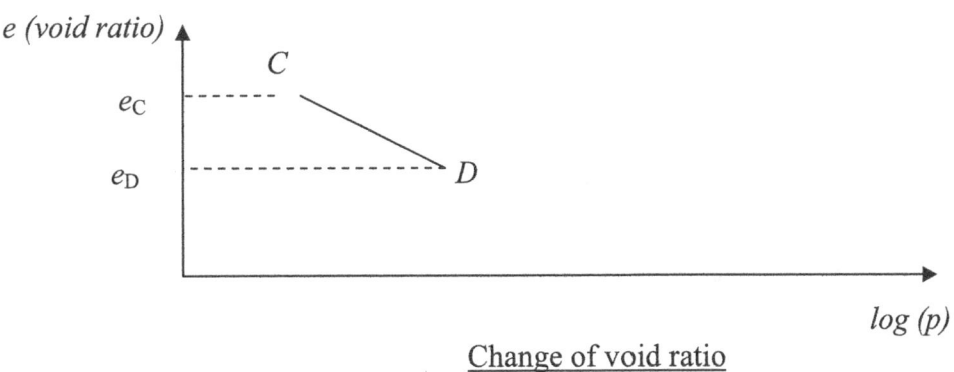

Change of void ratio

e_C = Void ratio of clay after melting of the glacier

e_D = Void ratio of clay after settlement due to foundation load.   Void ratio decreases due to footing load.

Now let us assume that the footing load is increased. When the footing load is increased, the clay layer would further settle. If the load is large enough, the clay layer would settle beyond point "B". Point "B" is the surface of the clay layer during the period glaciers were present.

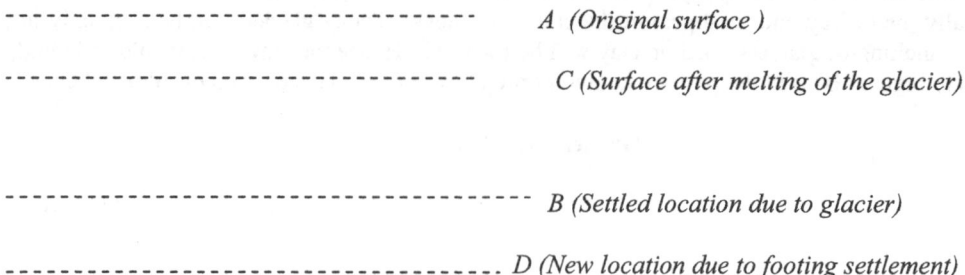

-          *A (Original surface)*

-          *C (Surface after melting of the glacier)*

-          *B (Settled location due to glacier)*

- *D (New location due to footing settlement)*

<u>Settlement of footing beyond point "B":</u>   Point B is the location of the clay layer after been compressed by the glacier. Now it is possible that footing load is high enough that the clay would settle even below point B. This happens when the footing load is very high.
Void ratio "e" vs. log "p" graph is shown below.

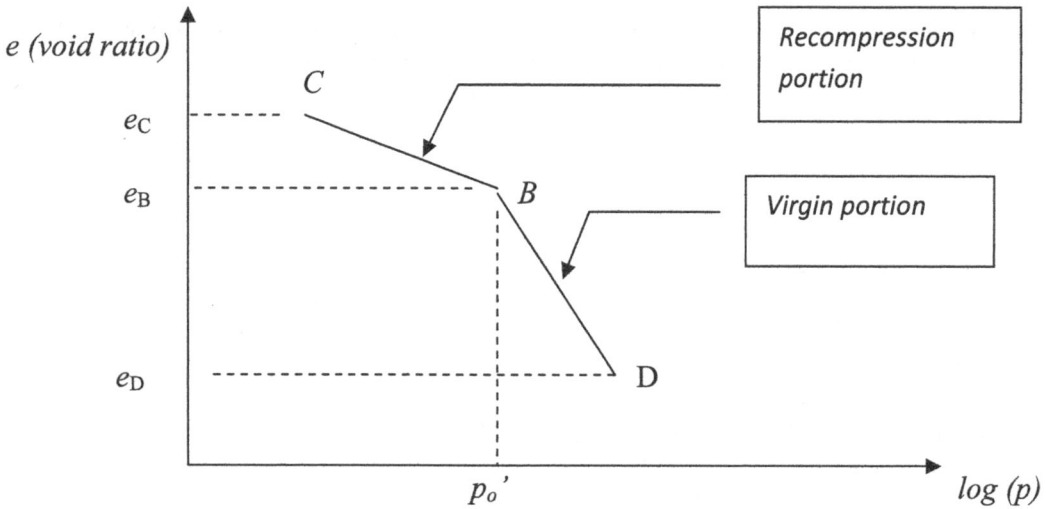

e_B = Void ratio of clay after settlement due to glacier
e_C = Void ratio of clay after melting of the glacier (rebound)
e_D = Void ratio of clay after settlement due to foundation load.

When the load is increased, void ratio of clay would decrease due to settlement.

In the above example, during the glacier was present the void ratio of the clay layer was e_B. When the glacier melted away, the clay layer rebounded. Hence the voids inside clay increased. After the melting of the glacier, void ratio increased from e_B to e_C.
This is the present state of the clay layer until man comes and decided to place a footing.
When the footing load is increased, the void ratio would decrease. As soon as the void ratio goes below e_B (void ratio when the glacier was present) the curve bends.

Top portion of the *"e" vs. log (p)* curve is known as the re-compression curve and the bottom portion is known as the virgin curve.
C to B – Recompression curve
B to D – Virgin curve

The gradient of the virgin curve is steeper than the re-compression curve.

The gradient of the re-compression curve is usually denoted by C_r and the gradient of the virgin curve is denoted by C_c. The settlement rate is high for a given pressure change within the virgin portion of the graph.

The pressure at point "B" is the largest pressure that the clay layer was subjected to prior to placing the foundation. Pressure at point "B" is known as the pre-consolidation pressure and denoted by p_c'. Present stress prior to placement of the footing is usually denoted by p_0'.

3.7.2 Total Primary Consolidation:

When a clay layer is loaded, as we learnt in the previous section it would start to consolidate and settle.

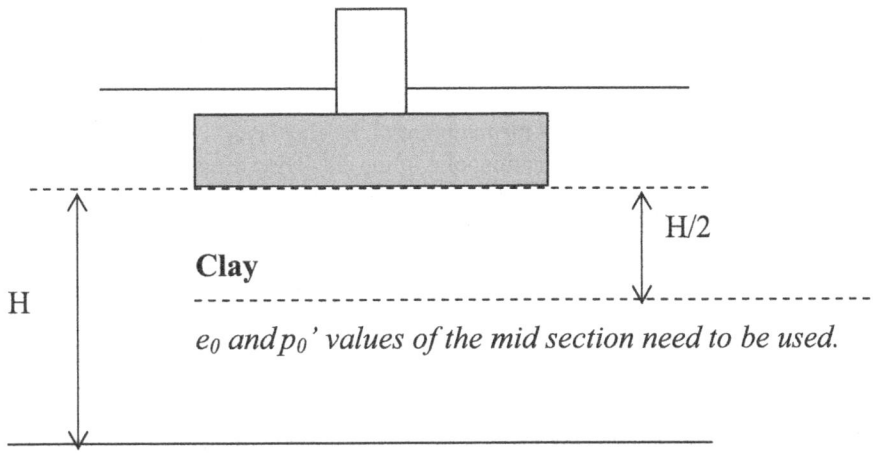

Shallow foundation on clay soil (properties of mid section)

Settlement due to total primary consolidation in normally consolidated clay is given by the following equation. Please note that normally consolidated clays have never been subjected to a higher stress than it enjoy now.

$$\Delta H = H \cdot \frac{C_c}{(1+e_0)} \log \frac{(p_0' + \Delta p)}{p_0'}$$

Description of terms:

ΔH	= Total primary consolidation settlement
H	= Thickness of the clay layer
C_c	= Compression index of the clay layer
e_0	= Void ratio of the clay layer at midpoint of the clay layer prior to loading
p_0'	= Effective stress at the midpoint of the clay layer prior to loading
Δp	= Increase of stress at the midpoint of the clay layer due to the footing.

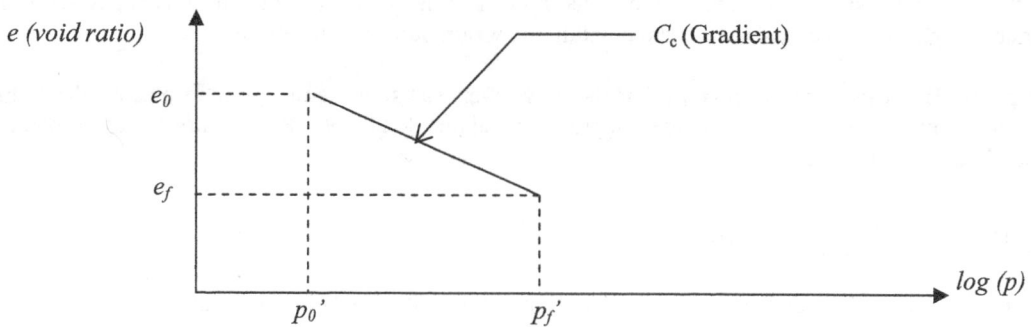

Change of void ratio for normally consolidated clay

e_0 = Present void ratio
e_f = Final void ratio after consolidation due to the footing load
p_0' = Present vertical effective stress at the midpoint of the clay layer
p_f' = Final vertical effective stress at the midpoint of the clay layer after placement of the footing
Δp = $p_f' - p_0'$
C_c = Gradient of the curve (Compression index)

In the following problem, consolidation Settlement in Normally Consolidated Clay is considered. No groundwater is present.

Practice Problem: Find the settlement due to consolidation of the (5 ft x 5 ft) column foundation with a load of 50 kips. The foundation is placed 3 ft below the top surface and the clay layer is 13 ft thick. There is a sand layer underneath the clay layer. Density of the clay layer is 108 lbs /ft3, the compression index (Cc) of the clay layer is 0.3, and initial void ratio (e0) of clay is 0.76. Assume that the pressure is distributed at 2:1 ratio and the clay is normally consolidated.

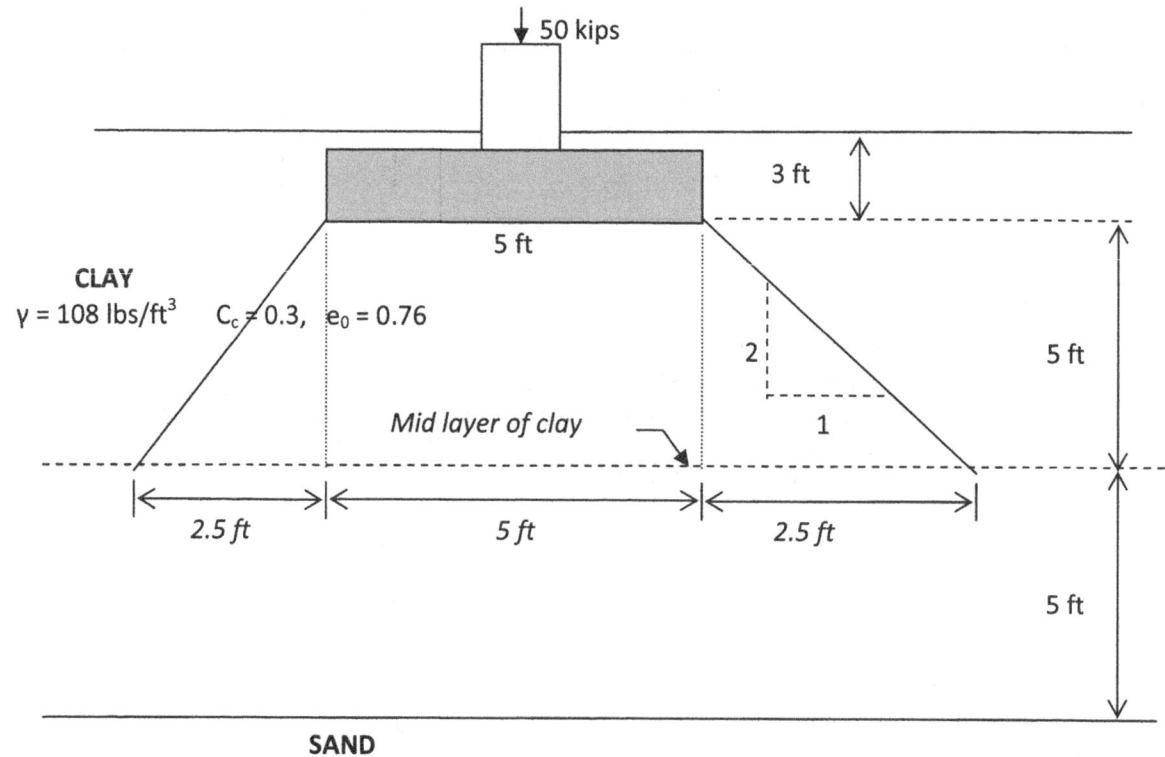

Footing on normally consolidated clay layer

STEP 1:   Write down the consolidation settlement equation for normally consolidated clay;

$$\Delta H = \frac{H. C_c}{(1+e_0)} \log \frac{(p_0' + \Delta p)}{p_0'}$$

ΔH = Total primary consolidation settlement
H = Thickness of the compressible clay layer
C_c = Compression index of the clay layer
e_0 = Void ratio of the clay layer at midpoint of the clay layer prior to loading
p_0' = Effective stress at the midpoint of the clay layer prior to loading

STEP 2:   The clay layer is 13 ft thick and the footing is placed 3 ft below the surface. Top 3 ft is not subjected to consolidation. A clay layer with a thickness of 10 ft below the footing is subjected to consolidation due to footing load.
Find the effective stress at the midpoint of the compressible clay stratum; (p_0');
$$p_0' = \gamma_{caly. x} (5 + 3) = 108 \times 8 = 864 \text{ lbs/ft}^2.$$

STEP 3:   Find Δp
Δp = Increase of stress at the *midpoint* of the clay layer due to the footing.
Total load of 50 kip is distributed at a larger area at the mid section of the clay layer.
Area of the mid section of the clay layer = $(5 + 2.5 + 2.5) \times (5 + 2.5 + 2.5) = 100 \text{ ft}^2$.
$\Delta p = 50/100 = 0.5 \text{ kip/ft}^2. = 500 \text{ lbs/ft}^2$.

STEP 4:   Apply values in the consolidation equation.

$$\Delta H = \frac{H. C_c}{(1+e_0)} \log \frac{(p_0' + \Delta p)}{p_0'}$$

Following parameters are given:

$C_c = 0.3$, $e_0 = 0.76$;   H = Compressible clay thickness = 10 ft

$$\Delta H = 10. \frac{0.3}{(1 + 0.76)} \log \frac{(864 + 500)}{864}$$

$$\Delta H = 0.338 \, ft \qquad (\text{4 inches})$$

Note: "H" thickness of the clay layer should be calculated starting from bottom of the footing. Though the total thickness of the clay layer is 13 ft, first 3 ft of the clay layer is not compressed.

Practice Problem: (Consolidation Settlement in Normally Consolidated Clay. Groundwater is Present)
Find the settlement due to consolidation of the (4 ft x 4 ft) column foundation with a load of 40 kips. Sand layer is 5 ft thick and the clay layer below is 10 ft thick. The bottom of footing is 3 ft below the ground surface. Density of the sand layer is 110 lbs/ft^3 and the density of the clay layer is 108 lbs/ft^3. Groundwater is 4 ft below the surface. Compression index (C_c) of the clay layer is 0.3 and initial void ratio (e_0) of clay is 0.8. Assume that the pressure is distributed at a 2:1 ratio and the clay is normally consolidated.

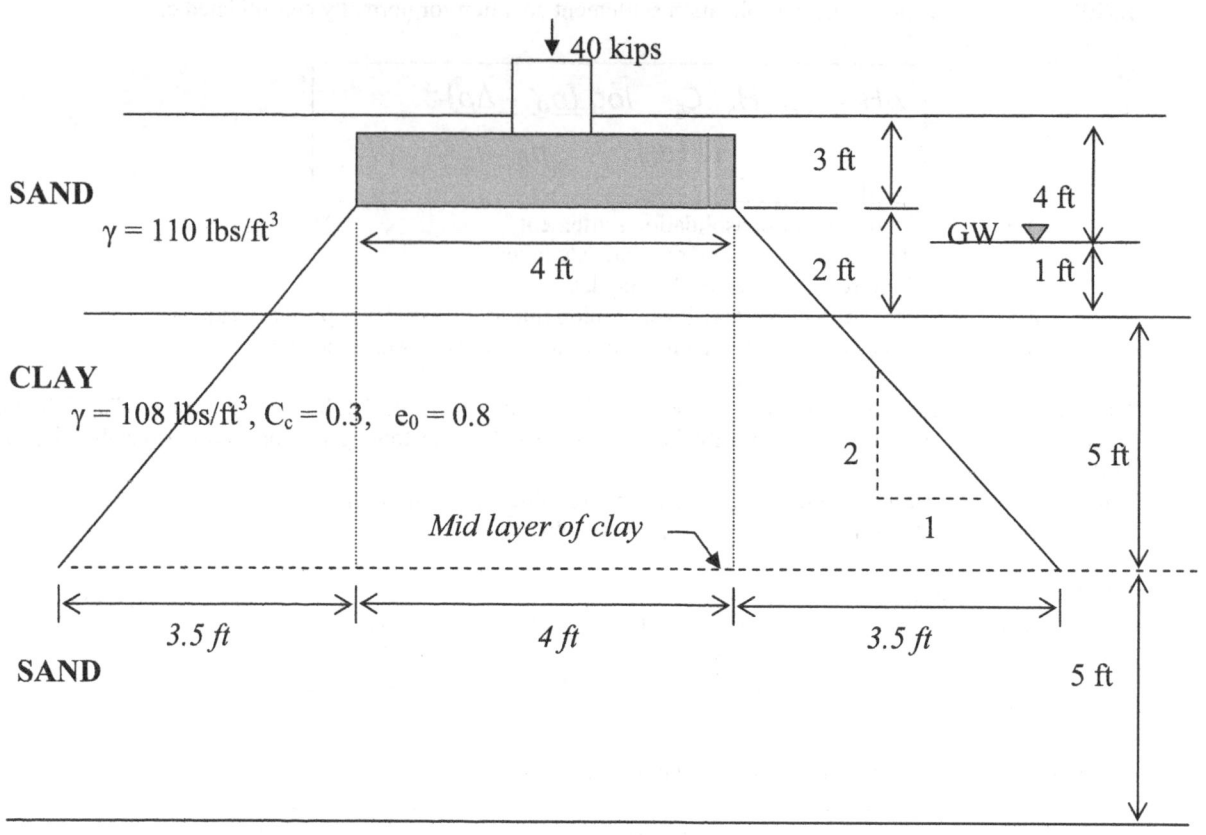

STEP 1:   Write down the consolidation settlement equation for normally consolidated clay;

$$\Delta H = \frac{H.\ C_c}{(1+e_0)}\ log\ \frac{(p_0' + \Delta p)}{p_0'}$$

ΔH = Total primary consolidation settlement
H = Thickness of the clay layer
C_c = Compression index of the clay layer
e_0 = Void ratio of the clay layer at midpoint of the clay layer prior to loading
p_0' = Effective stress at the midpoint of the clay layer prior to loading

STEP 2:   Find the initial effective stress at the mid layer of the clay stratum; (p_0');
        $p_0' = \gamma_{sand}\ x\ 4 + (\gamma_{sand} - \gamma_{water})\ x\ 1 + (\gamma_{clay} - \gamma_{water})\ x\ 5$
        $p_0' = 110\ x\ 4 + (110 - 62.4)\ x\ 1 + (108 - 62.4)\ x\ 5 = 715.6\ lbs/ft^2.$

STEP 3:   Find Δp
        Δp = Increase of stress at the midpoint of the clay layer due to the footing.
        Total load of 40 kips distributed at a larger area at the mid section of the clay layer.
        Area of the mid section of the clay layer = $(4 + 3.5 + 3.5)\ x\ (4 + 3.5 + 3.5) = 121\ ft^2.$

$\Delta p = 40/121 = 0.3306\ kip/ft^2. = 330.6\ lbs/ft^2.$

STEP 4:   Apply values in the consolidation equation.

$$\Delta H = \frac{H \cdot C_c}{(1+e_0)} \log \frac{(p_0' + \Delta p)}{p_0'}$$

H = Thickness of compressible clay layer = 10 ft

$$\Delta H = 10 \cdot \frac{0.3}{(1 + 0.8)} \log \frac{(715.6 + 330.6)}{715.6}$$

$$\Delta H = 0.275 \text{ ft}   (3.3 \text{ inches})$$

Consolidation settlement is very high. Reduce the load per footing or increase the footing area.

Consolidation Settlement in Overconsolidated Clay: In previous examples we solved problems involving normally consolidated clay.   Now it is time to consider problems involving overconsolidated clay.   (Also known as pre consolidated clay).   *"e" vs. log p*, curve for pre consolidated clay is shown below.

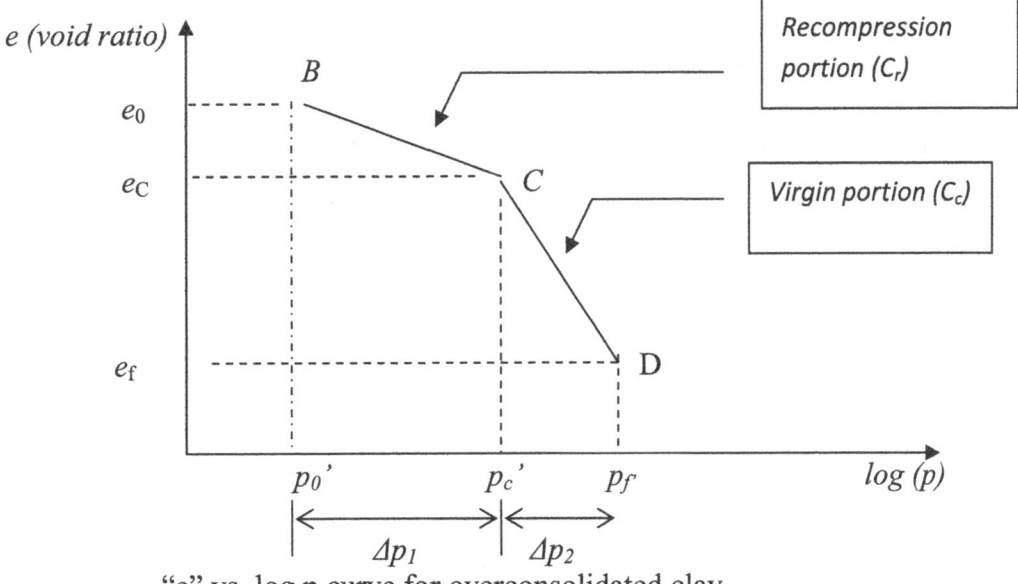

"e" vs. log p curve for overconsolidated clay

e_0 =   Initial void ratio e_c =   Void ratio at pre consolidation pressure
e_f =   Final void ratio at the end of the consolidation process.
p_0' =   Initial pressure prior to the footing load
p_c' =   Overconsolidation pressure (also known as preconsolidation pressure)
p_f'=   Pressure after the footing is placed
C_c =   Compression index C_r =   Recompression index

$$\Delta H = \frac{H \cdot C_r}{(1+e_0)} \log \frac{(p_0' + \Delta p_1)}{p_0'} + \frac{H \cdot C_c}{(1+e_0)} \log \frac{(p_c' + \Delta p_2)}{p_c'}$$

$$\underbrace{\hspace{4cm}}_{\text{Point B to C}} \quad \underbrace{\hspace{4cm}}_{\text{Point C to D}}$$

Point B: Point "B" indicates the present void ratio and existing vertical effective stress
Point C: Point "C" indicates the maximum stress that the clay had been subjected to in the past. Void ratio and the past maximum stress have to be obtained by conducting a laboratory consolidation test. Geotechnical engineer should obtain Shelby tube samples and send to the laboratory to conduct a consolidation test.
Point D: Point "D" indicates the expected stress after the footing is constructed.
Unlike in the normally consolidated soils, in overconsolidated soils, the settlement has to be computed in two parts.
Consolidation settlement from B to C: (C_r should be used)

$$\Delta H_1 = H. \frac{C_r}{(1+e_0)} \log \frac{(p_0' + \Delta p_1)}{p_0'}$$

$$\Delta p_1 = p_c' - p_0' \quad \text{(see the figure above)}$$

Consolidation settlement from C to D: (C_c should be used)

$$\Delta H_2 = H. \frac{C_c}{(1+e_0)} \log \frac{(p_c' + \Delta p_2)}{p_c'}$$

$$\Delta p_2 = p_f' - p_c' \quad \text{(see the figure above)}$$

Hence total settlement is given by
$$\Delta H = \Delta H_1 + \Delta H_2$$
This is better explained using the example below.

Practice Problem: (Consolidation settlement in overconsolidated clay without groundwater)
Find the settlement due to consolidation in the (3 ft x 3 ft) column foundation with a load of 15 kip. The foundation is placed 3 ft below the ground surface and the clay layer is 10 ft thick. There is a sand layer underneath the clay layer. Density of the clay layer is 110 lbs/ft^3 and the compression index (C_c) of the clay layer is 0.32, re-compression index C_r is 0.035, pre consolidation pressure (p_c') is 950 lbs/ft^2 and initial void ratio (e_0) of clay is 0.80. Assume that the pressure is distributed at 2:1 ratio.

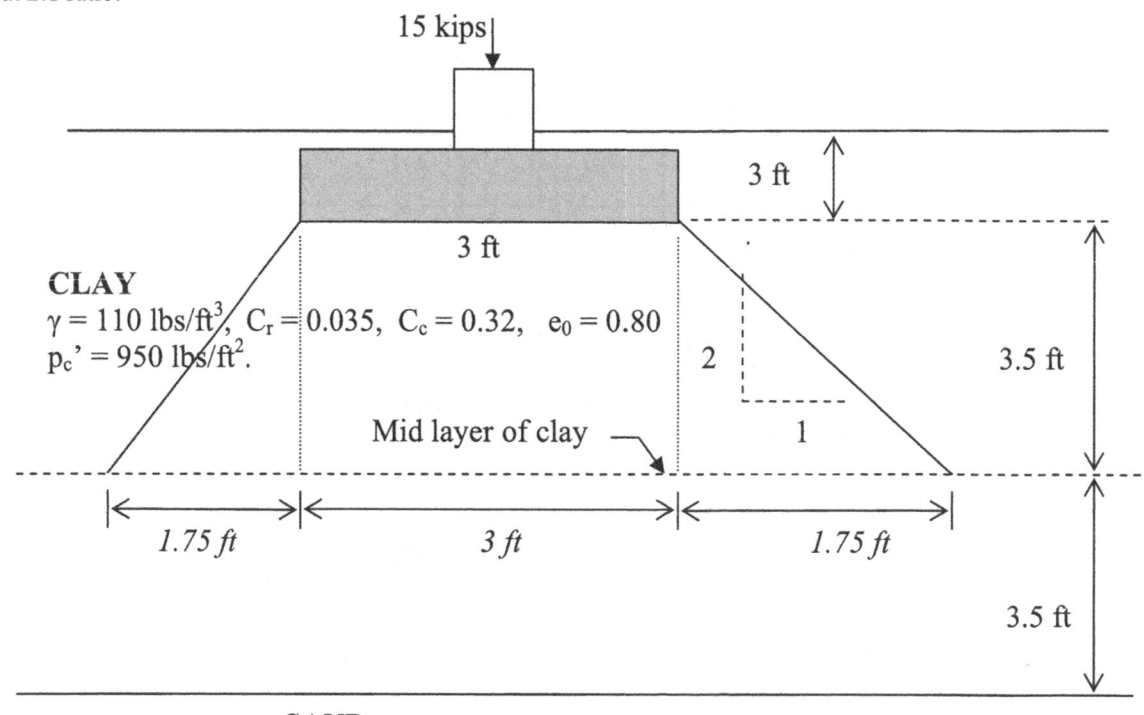

Settlement in overconsolidated clay

STEP 1:   Write down the consolidation settlement equation;

$$\Delta H = H. \frac{C_r}{(1+e_0)} \log \frac{(p_0' + \Delta p_1)}{p_0'} + H. \frac{C_c}{(1+e_0)} \log \frac{(p_c' + \Delta p_2)}{p_c'}$$

Recompression portion   Virgin compression

ΔH	= Total primary consolidation settlement	
H	= Thickness of the compressible clay layer	(7 ft)
C_c	= Compression index of the clay layer	(0.32)
C_r	= Recompression index of the clay layer	(0.035)
e_0	= Void ratio of the clay layer at midpoint of the clay layer prior to loading	(0.8)
p_0'	= Effective stress at the midpoint of the clay layer prior to loading	
p_c'	= Pre consolidation pressure	(950 lbs/ft^2)
p_f'	= Pressure after the footing is placed	

STEP 2:   The clay layer is 10 ft thick and the footing is placed 3 ft below the surface. Top 3 ft is not subjected to consolidation. A clay layer with a thickness of 7 ft below the footing is subjected to consolidation due to footing load. Find the effective stress at the mid layer of the compressible clay stratum; (p_0');

$p_0' = \gamma_{clay.} \times 6.5 = 110 \times 6.5 = 715$ lbs/ft^2.

STEP 3:   Find Δp
     Δp = Increase of stress at the *midpoint* of the clay layer due to the footing.
     Total load of 15 kips is distributed at a larger area at the mid section of the clay layer.

Area of the mid section of the clay layer = $(1.75 + 3 + 1.75) \times (1.75 + 3 + 1.75) = 42.25$ ft^2.
$\Delta p = 15/42.25$ kip/ft^2 = 355 lbs/ft^2.
Existing stress (p_0') = 715 lbs/ft^2.
Preconsolidation stress is given to be = 950 lbs/ft^2
Final pressure after foundation load (p_f') = $p_0' + \Delta p$ = 715 + 355 = 1,070 lbs/ft^2.

STEP 4:   Apply values in the consolidation equation.

$$\Delta H = H. \frac{C_r}{(1+e_0)} \log \frac{(p_0' + \Delta p_1)}{p_0'} + H. \frac{C_c}{(1+e_0)} \log \frac{(p_c' + \Delta p_2)}{p_c'}$$

p_0' = Effective stress at the midpoint of the clay layer prior to loading = 715 lbs/ft^2.
(Δp_1) = Stress increase **from** initial stress (p_0') **to** pre-consolidation pressure p_c'.
Since p_0' is 715 and p_c' was given to be 950,
Δp_1 = (950 – 715) lbs/ft^2 = 235 lbs/ft^2.

(Δp_2) = Stress increase **from** pre-consolidation pressure (p_c') **to** final pressure p_f'.
Since p_c' is 950 and final pressure (p_f') was found to be 1,070, Δp_2 would be (1,070 – 950) lbs/ft^2.
Δp_2 = 120 lbs/ft^2.

$$\Delta H = 7. \frac{0.035}{(1 + 0.8)} \log \frac{(715 + 235)}{715} + 7. \frac{0.32}{(1 + 0.8)} \log \frac{(950 + 120)}{950}$$

$\Delta H = 0.0168 + 0.0643 = 0.0811$ ft = 0.97 in

Line Diagram:

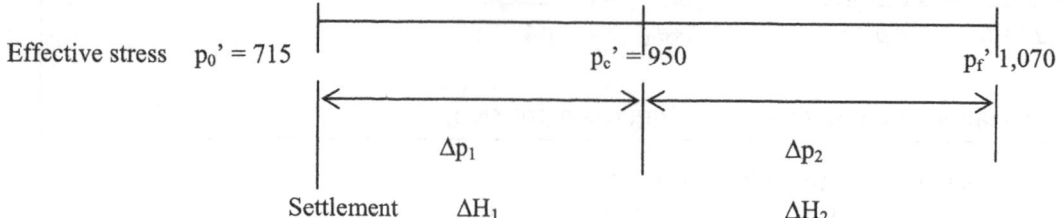

3.7.3 Computation of Time for Consolidation:

In the previous chapter we studied how to compute the total settlement due to primary consolidation. In this chapter, we would discuss how to compute the time taken for the consolidation process.
Time taken for primary consolidation is given by the following equation.

$$t = \frac{H^2 . T_v}{c_v}$$

t   = Time taken for the consolidation process (sec)
H   = Thickness of the drainage layer (in). (see the explanation below)
Tv   = Time coefficient. (No units)
U%   = Percent consolidation
c_v = Coefficient of consolidation (in^2/sec)

T_v can be obtained from the table given below.

U% (Percent Consolidation)	T_v (Time Coefficient)
0	0.00
10	0.048
20	0.090
30	0.115
40	0.207
50	0.281
60	0.371
70	0.488
80	0.652
90	0.933
100	1.0 (Approximately)

U% and T_v

<u>Explanation for the drainage layer (H)</u>:   Thickness of the drainage layer is defined as the **longest** path a water molecule has to take for drainage.

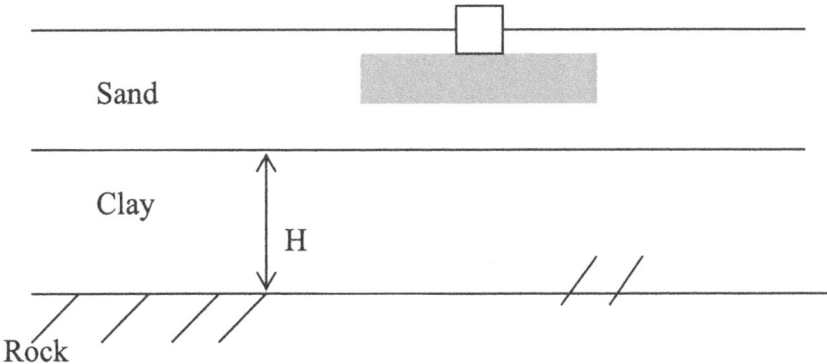

<u>Thickness of the drainage layer (single drainage)</u>

In the case of single drainage, drainage can occur only from one side. In the above figure, water cannot drain through the rock. Water can drain from the sand layer on top. In this case, thickness of the drainage layer = H.
In some situations water can drain from top and bottom.

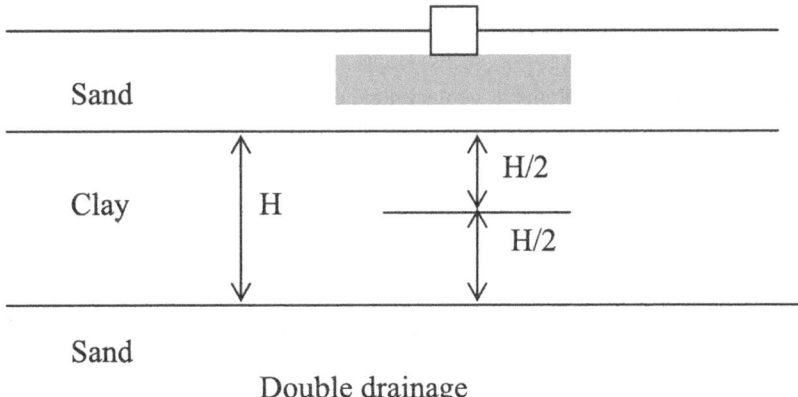

<u>Double drainage</u>

Water molecules in the above figure has the opportunity to drain from either top or bottom. Longest drainage path for a water molecule is only H/2. Hence thickness of the drainage layer for double drainage is H/2.

Practice Problem: A footing is placed 3 ft below the ground. 0 to 3 ft is a coarse sand layer. Clay layer is 9 ft thick. Below the clay layer lies bed rock. Coefficient of consolidation of the clay layer is 0.003 in^2/sec. Find the time taken for 90% primary consolidation.

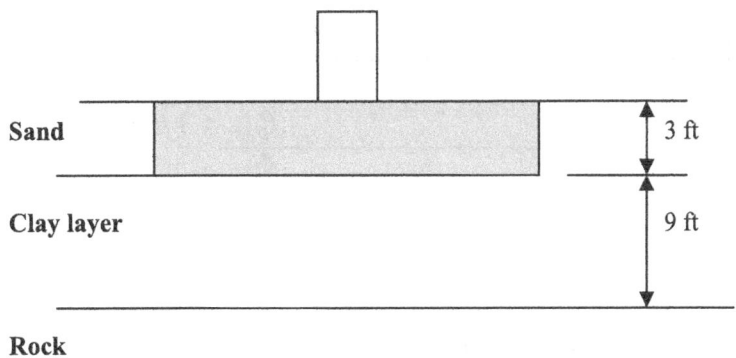

STEP 1: This is a single drainage situation. Water cannot drain thru rock.

STEP 2: Write the time rate consolidation equation;

$$t = \frac{H^2 . T_v}{c_v}$$

H = Thickness of the compressible clay layer = 9 ft = 9 x 12 inches = 108 inches.
c_v = 0.003 in^2/sec
U = 90% (Given)
T_v = 0.933 (From the U vs. T_v table given above)

$$t = \frac{H^2 . T_v}{c_v}$$

$t = \frac{108^2 \times 0.933}{0.003}$ = 3627504 seconds = 41.9 days

3.9 Effective Stress and Total Stress:

Total Density of Soil: Total density of soil is also known as wet density (γ_{wet}), total density (γ_{tot}) or simply density (γ). This is simply mass of soil divided by soil volume.

Buoyant Density: Below groundwater, buoyancy acts on soil.
                             Buoyant density = Total density – density of water = ($\gamma_{wet} - \gamma_w$)
                             γ_w = Density of water

Vertical Effective Stress: Almost all problems in geotechnical engineering require computation of effective stress. Soils also have a lesser effective stress under the water table due to buoyancy.
Rule 1: Above groundwater, use total density to compute effective stress.
Rule 2: Below groundwater, use buoyant density to compute effective stress.

Practice Problem: Find the effective stress at point "A". (no groundwater present). Total density also known as wet density represented with γ.

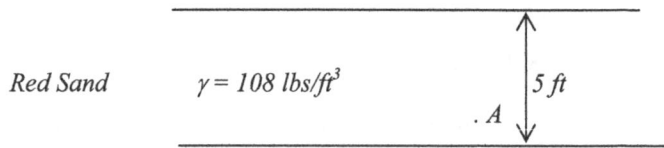

Red Sand $\gamma = 108 \ lbs/ft^3$ 5 ft
                                                        . A

Effective stress

Solution:
Effective stress at point "A" = 108 x 5 = 540 lbs/ft^2
Since there is no groundwater, total density of soil is obtained to compute the effective stress.

Practice Problem (Computing effective stress when groundwater present):
Find the effective stress at point "A". Groundwater is 3 ft below the surface.

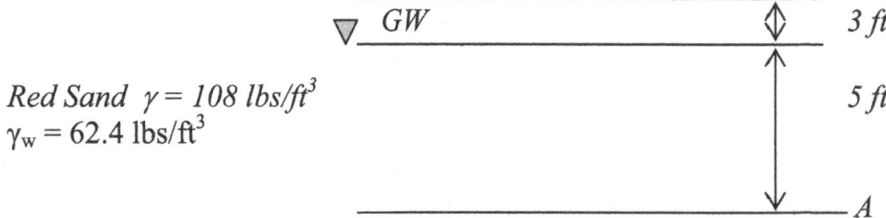

▽ GW   ⇕   3 ft

Red Sand $\gamma = 108 \ lbs/ft^3$ 5 ft
γ_w = 62.4 lbs/ft^3
                                                     A

Effective stress when groundwater present

Effective stress at point "A" = $(108 \times 3) + (108 - \gamma_w) \times 5$ lbs/ft^2
$\qquad\qquad\qquad\qquad = (108 \times 3) + (108 - 62.4) \times 5$ lbs/ft^2.
$\qquad\qquad\qquad\qquad = 552$ lbs/ft^2.

Note: There is no buoyancy acting on first three ft of the soil. Hence density of soil was NOT reduced. However, buoyancy acts on the soil below groundwater level. Hence the density of soil was reduced by 62.4 lbs/ft^3 to account for the buoyancy.

3.10   Slope Stability:

Consider the slope shown below. Soil mass "W" has two components. The component along the slope tries to bring the slope down. The shear stress (τ) acting along the edge resists the slope from failing.

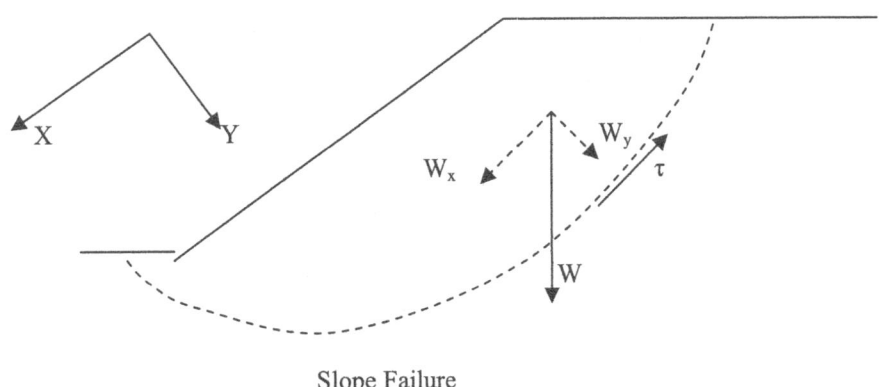

Slope Failure

W   = Weight of the slope
W_x and W_y = Weight of the slope in X and Y directions.
τ = Shear stress in soil

When W_x is greater than shear strength (maximum shear stress that can be accommodated by soil) the slope will fail.

Massive slope failure

3.11 Slab-On-Grade (ACI 360R):   Slab on grade is a situation where a slab is poured on the compacted soil. Construction and design of slab on grade is in done as per ACI 360R. (ACI - American Concrete Institute). A slab on grade is defined in one whose total loading, uniformly distributed, would impart a pressure to the

grade or soil that is less than 50 percent of the allowable bearing capacity. In other words, slab on grade load on soil should be less than 50% of the bearing capacity of the soil. Of course there are exceptions.

<u>Slab on Grade Types</u>:   Following types are identified in ACI 360R.
  a) Plain concrete slab
  b) Slab reinforced for shrinkage and temperature only
  c) Shrinkage-compensating concrete with shrinkage reinforcement
  d) Slab post-tensioned to offset shrinkage
  e) Slab post-tensioned and/or reinforced, with active prestress
  f) Slab reinforced for structural action

ACI 360R provides design methodologies for each type.

<u>Site Preparation for Slabs on Grade</u>; Prior to placing base material, all top soil has to be stripped. In addition, any loose soil and isolated pockets of organic soil need to be removed.

<u>Base and Subbase Material for Slabs on Grade</u>:   Typically base and subbase layers are used below the slab to provide stability.

Slab on grade
Base
Subbase
Existing Ground

Typically cheap sandy fill material is used for the subbase. Base material can be a well graded sand or well graded gravel. Alternatively, a mixture of two. Generally base material is more expensive fill material than subbase material. Both base and subbase should be compacted to 95%.

<u>Loading on Slabs on Grade</u>:

Following loads are identified in ACI 360R.
Live loads
Vehicle wheel loads
Concentrated loads
Line and strip loads
Uniform loads
Construction loads
Environmental effects including expansive soil
Unusual loads, such as forces caused by differential settlement
The slab should be designed for most critical loading combination.

3.12 Shallow Foundations:

3.12.1 Bearing capacity:

Karl Terzaghi (1,883 – 1,963). Terzaghi was the first to come up with a satisfactory theory for bearing capacity of footings. He is known as the father of soil mechanics.

Shallow foundations are the very first choice of foundation engineers. They are first and foremost cheap and easy to construct compared to other alternatives such as pile foundations and mat foundations. Shallow foundations need to be designed for bearing capacity and excessive settlement. There are number of methods available to compute the bearing capacity of shallow foundations. Terzaghi bearing capacity equation, which was developed by Karl Terzaghi still widely used.

What is bearing failure?   When you place a footing on top of soil, the load of the footing has to be supported by the strength of soil. Due to the pressure from the footing, soil particles tend to push down. When the load is too much, soil layer would undergo failure and the footing would sink. This is known as the bearing failure.

Soil strength comes from two parameters. They are

- Cohesion
- Friction

Cohesion is an electro-chemical process that acts between clay particles. Friction is a physical process that acts between particles in sands and silts.

Buildings: Shallow foundations are widely used to support buildings. In some cases doubt is cast upon the ability of a shallow foundations to carry necessary loads due to loose soil underneath. In such situations, engineers use piles to support buildings. Shallow foundations are much more cost effective than piles. Building footings are subjected to wind, earthquake forces and bending moments. Hence shallow foundations need to be designed to withstand all these forces.

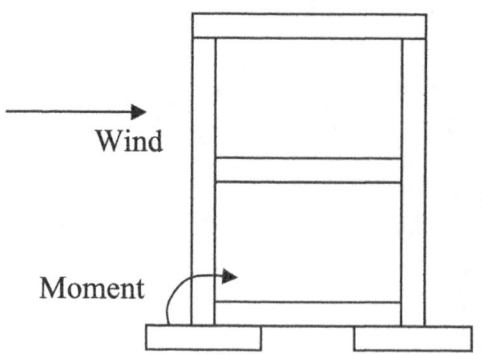

<u>Building footings are subjected to both vertical and lateral forces</u>

<u>Buildings with Basements</u>:   Footings in buildings with basements need to be designed for lateral soil pressure as well. In most cases, the building frame has to support forces due to soil.

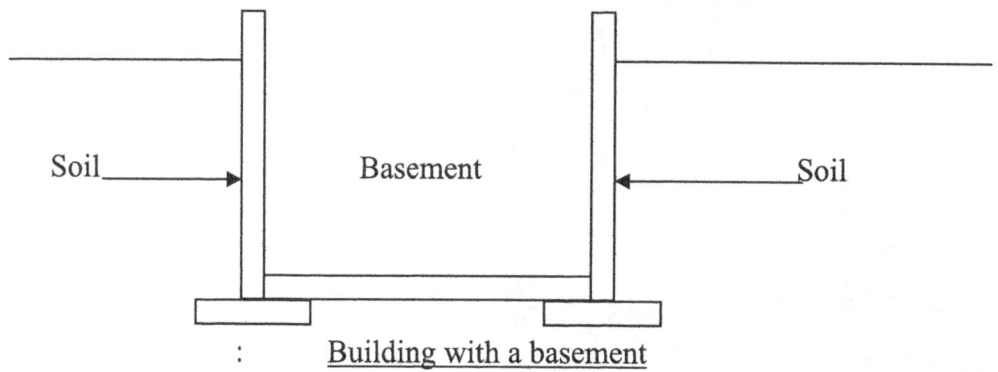

:   <u>Building with a basement</u>

<u>Bridges</u>:   Bridges consist of abutments and piers. Due to large loads many engineers prefer to use piles for bridge abutments and piers. Nevertheless if the site conditions are suitable, shallow foundations can be used for bridges.

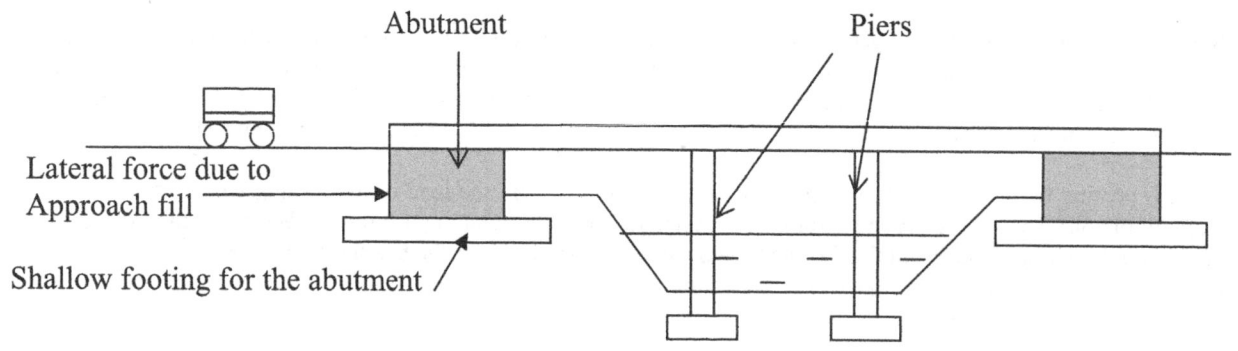

<u>Lateral forces in a bridge abutment due to approach fill</u>

<u>Frost Depth</u>:   Shallow foundations need to be placed below frost level. During winter times, soil at the surface will be frozen. During the summer, frozen soil will melt again. This freezing and thawing of soil generate a change in volume.   If the footing is placed below the frost depth, freezing and melting of soil would generate upward forces and eventually cause cracking in concrete.

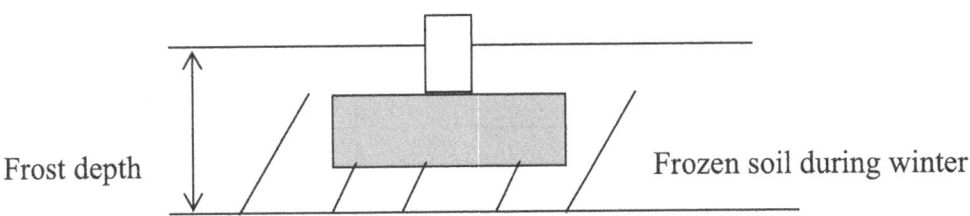

Frost depth   Frozen soil during winter

<u>Soil frozen during winter (Footing placed above frost depth)</u>

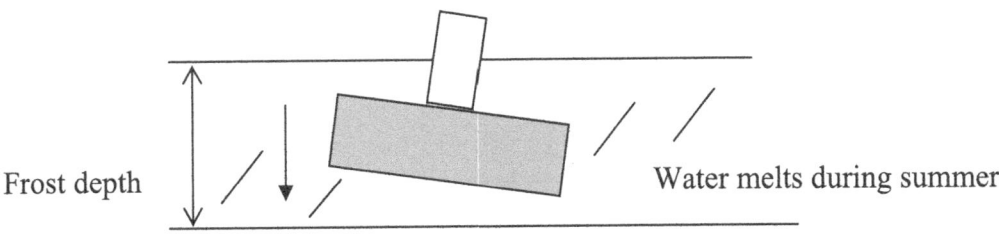

Frost depth   Water melts during summer

<u>Soil during summer (water melts and footing unstable)</u>

During summer, the water in soil will melt and the footing would settle. Due to this reason all footings should be placed below the frost depth in that region.

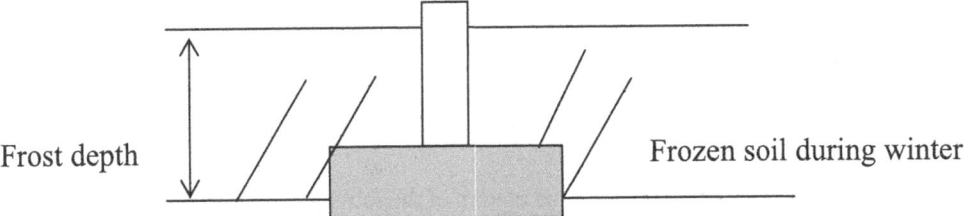

Frost depth   Frozen soil during winter

<u>Footing placed below the frost depth (No effect due to soil freezing) (Correctly placed footing)</u>

Frost depth is dependent upon the region. Frost depth in Siberia and some parts of Canada could be as high as 7 feet while in places like New York it is not more than 4 feet. Since there is no frost in tropical countries, frost depth is not an issue.

Bearing Capacity Computation: (Standard Method, Terzaghi Equation):

Terzaghi bearing capacity equation or variants of it are widely used by geotechnical engineers to compute the bearing capacity. Terzaghi theorized that the soil underneath the footing would generate a triangle of pressure as shown below.

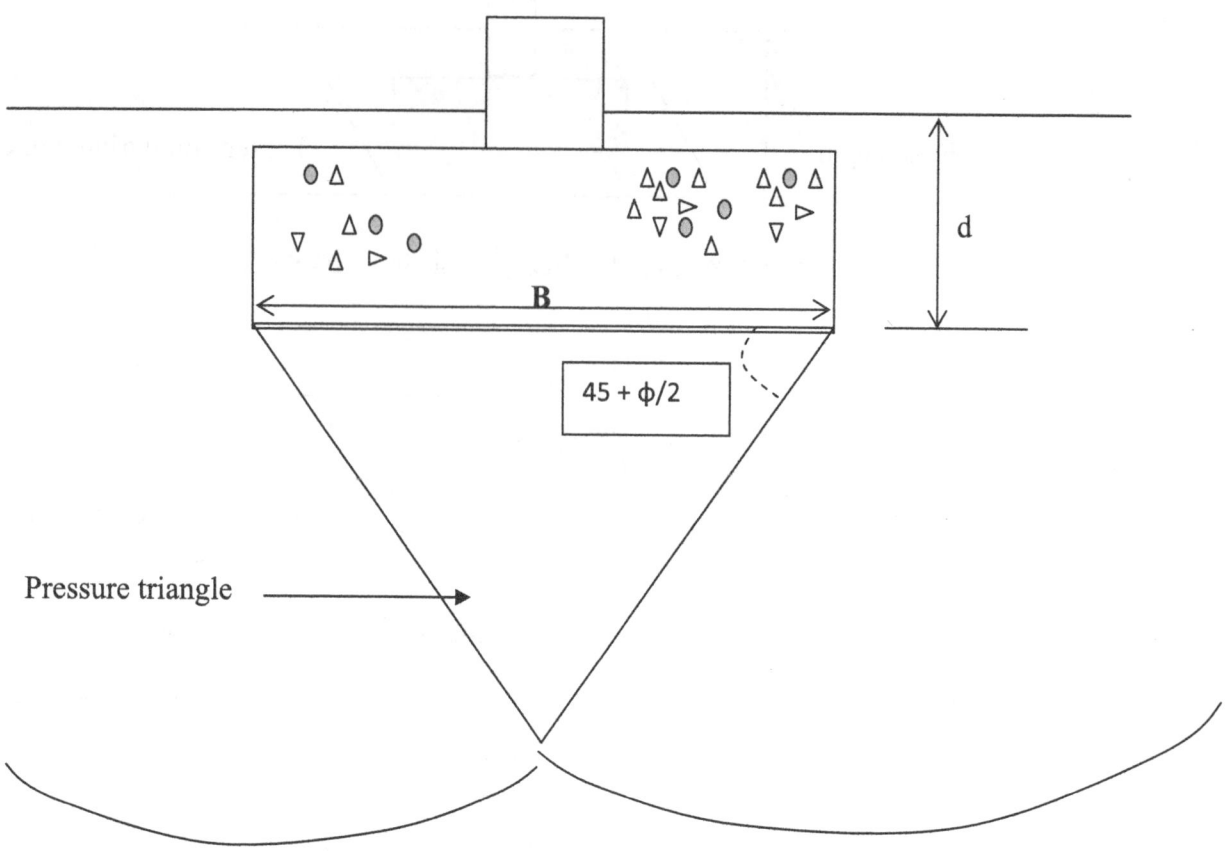

Bearing capacity computation

Terzaghi considered that the angle between the bottom of the footing and the pressure triangle to be $(45 + \varphi/2)$, as shown above. Bearing capacity is developed in a footing due to three properties of soil.

1)   Cohesion
2)   Friction
3)   Density of soil

Terms Used in Terzaghi Bearing Capacity Equation:
<u>Ultimate bearing capacity (q_{ult})</u>:   Ultimate bearing capacity of a foundation is the load that a foundation would fail beyond any usefulness.
<u>Cohesion (C)</u>:   Cohesion is an electro-chemical process. Clay particles tend to adhere to each other due to electrical charges present in clay particles.

Friction: Unlike cohesion, friction is a physical process. Higher the friction between particles, higher the capacity of the soil to carry footing loads. Usually sands and silts inherit friction. For all practical purposes, clays are considered friction free. Friction of a soil is represented by the friction angle. (φ).
B = Width of the footing
$(45 + \varphi/2)$ = The angle of the soil pressure triangle
General Terzaghi bearing capacity equation is given below.

$$q_{ult} \quad = \quad c.N_c.\,s_c.\,d_c \quad + \quad q.N_q.\,s_q \quad + \quad 0.5.\,B.\,N_\gamma.\,\gamma.\,s_\gamma$$

$$\underbrace{}_{\text{Cohesion term}} \qquad \underbrace{}_{\text{Surcharge term}} \qquad \underbrace{}_{\text{Density term}}$$

Description of Terms in the Terzaghi Bearing Capacity Equation:

q_{ult} = Ultimate bearing capacity of the foundation (tsf or psf)

Cohesion Term: $(c\ N_c\ s_c.\ d_c)$:   This term represents the strength due to cohesion. Higher the cohesion, higher the bearing capacity.

c	=	Cohesion of the soil
N_c	=	Terzaghi bearing capacity factor (Obtained from table T-1 given below)
s_c	=	Shape factor obtained from the table T-2 given below.
d_c	=	Depth factor = $1 + 0.4\ d/B$

(d is the depth to bottom of footing and B is the width of the footing. For circular footings, use the diameter instead of B).

Surcharge Term: $(q.\ N_q\ s_q)$:   This term represents the bearing capacity strength developed due to surcharge. Surcharge load is the pressure exerted due to soil above the bottom of footing. It is obvious that if soil surcharge is increased, bearing capacity of the footing also would increase.

"q" is the effective stress at bottom of footing; $q = \gamma.\ d$

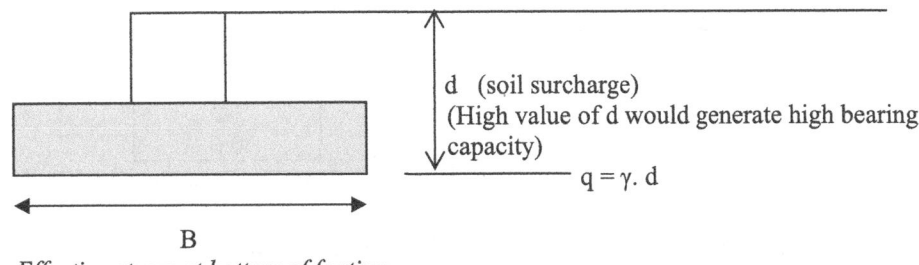

Effective stress at bottom of footing

(d, is the distance from ground surface to the bottom of the footing. γ is the effective density of soil).
Surcharge term = $q \times N_q.s_q$

$q = \gamma.\ d$ (γ = Density of soil, d = depth to bottom of footing)

One can increase the bearing capacity of a foundation by increasing "d".
N_q = Terzaghi bearing capacity factor obtained from table T-1. Bearing capacity factor, N_q depends upon the friction angle of soil. (φ).

s_q = Shape factor = $1 + B/L\ \tan\varphi$
(B is the width of the footing and L is the length of the footing).
For wall footings, length (L) of the footing is very large compared to width (B) hence S_q will approximate to 1.0.

Discussion: If one places the bottom of the footing deeper, "d" term in the equation would increase. Hence the bearing capacity of the footing also would increase. Placing the footing deeper is a good method to increase the bearing capacity of a footing. This may not be a good idea when softer soils are present at deeper elevations.

Density Term $(0.5.\ B.\ N_\gamma.\ \gamma.\ s_\gamma)$: This term represents the strength due to density of soil. If the soil has a higher density, the bearing capacity of that soil would be higher.

B = Width or the shorter dimension of the footing
N_γ = Terzaghi bearing capacity factor obtained from table T-1. Bearing capacity factors depend on the friction angle (φ).
γ = Density of soil s_γ = Shape factor obtained from table T-2.

Terzaghi Bearing Capacity Factors (N_c, N_q and N_γ)

φ (Friction angle)	N_c	N_q	N_γ
0	5.7	1.0	0.0
5	7.3	1.6	0.5
10	9.6	2.7	1.2
15	12.9	4.4	2.5
20	17.7	7.4	5.0
25	25.1	12.7	9.7
30	37.2	22.5	19.7
35	57.8	41.4	42.4
40	95.7	81.3	100.4
45	172.3	173.3	297.5
50	347.5	415.1	415.1

Table T-1   Terzaghi Bearing Capacity Factors (Source, Bowles, J.E)

Shape Factors:

	S_c	S_γ
Square footings	1.3	0.8
Strip footings (wall)	1.0	1.0
Round footings	1.3	0.6

Table T-2   Shape Factors for Terzaghi Bearing Capacity Equation: (Source, Bowles, J.E)

3.12.2 Terzaghi Bearing Capacity Equation (Discussion):

It is important to see how each term in the Terzaghi bearing capacity equation impacts the bearing capacity of a footing.

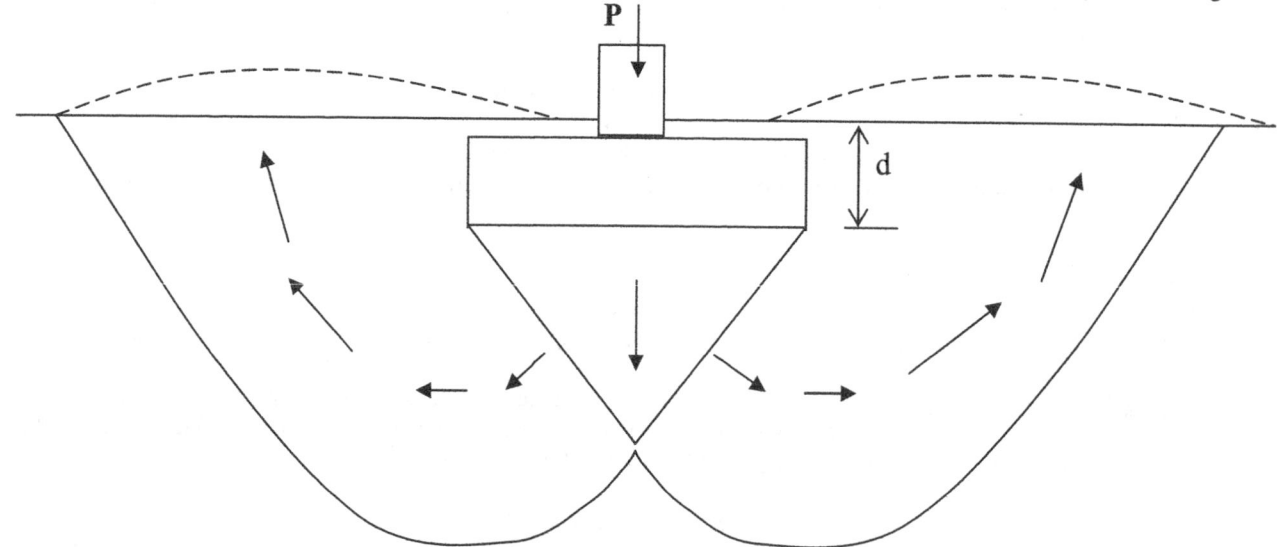

Movement of soil below a shallow footing at failure

When the footing load (P) is increased, the pressure triangle that is below the footing would be pressed down. Soil on sides would get pushed upwards. At failure, soil from the sides of the footing would be heaved. If there is a larger surcharge or the dimension "d" to be increased, soil will be locked in and bearing capacity would increase. It is possible to increase the bearing capacity of a footing by increasing "d".

The bearing capacity increase due to surcharge (q) is given by the second term (qN_q) as discussed earlier.

Increase of bearing capacity when a layer of thickness "X" is added is shown below.

New soil layer added

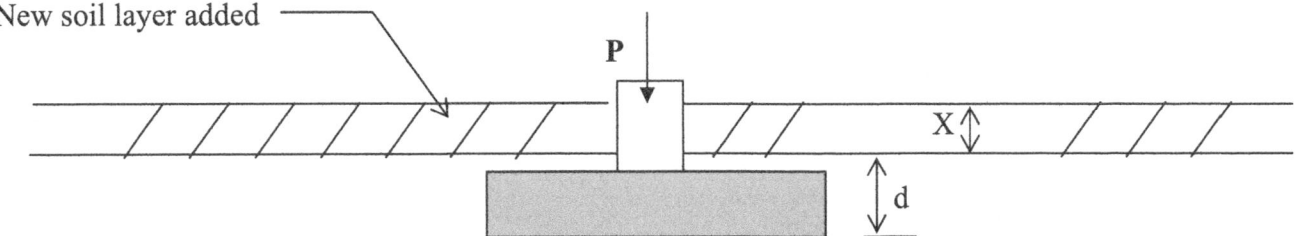

New soil layer added to increase the bearing capacity

In the above figure, additional soil layer of thickness "X" is added. New ultimate bearing capacity will be given by the following equation.

$$q_{ult} = cN_{c.} s_c d_c + qN_q s_q + 0.5. B. N_\gamma. \gamma. s_\gamma$$

First and third terms will not undergo any change due to addition of the new soil layer.
New surcharge "q" in the second term will be $q = \gamma (d + X)$.

$$q_{ult} = cN_{c.} s_c d_c + \gamma (d + X).N_q s_q + 0.5. B. N_\gamma. \gamma. s_\gamma$$

If the layer X was not present, q would be simply ($\gamma.d$).
Foundation engineer can increase the bearing capacity of a footing by increasing "d". This can be done by adding fill or by placing the foundation deeper.

Effect of Density: (0.5. B. N_γ. γ. s_γ):   The effect due to density of soil is represented by the third term (0.5. B. N_γ. γ. s_γ). Higher the density of soil, higher the ultimate bearing capacity. When "γ" increases, ultimate bearing capacity also increases.

Effect of Friction Angle (φ):   One may wonder where is the effect of the friction angle in the Terzaghi bearing capacity equation? All bearing capacity coefficients (N_c, N_q and N_γ) are obtained using the friction angle.
Higher the friction angle, higher the bearing capacity factors. (See the table T-1 for bearing capacity factors). Hence the effect of the friction angle is incorporated into Terzaghi bearing capacity coefficients in an indirect way.

Bearing capacity in sandy soil: Cohesion in sandy soils and non plastic silts are considered to be zero. The following example show the computation of bearing capacity in sandy soils

Practice Problem: (Column footing placed in a homogeneous sand layer)
Find the net bearing capacity of a (3 ft x 3 ft) square footing placed in a sand layer. The density of the soil is found to be 112 lbs/ft³ and friction angle to be 30^0. Footing is placed 3 ft below the surface.

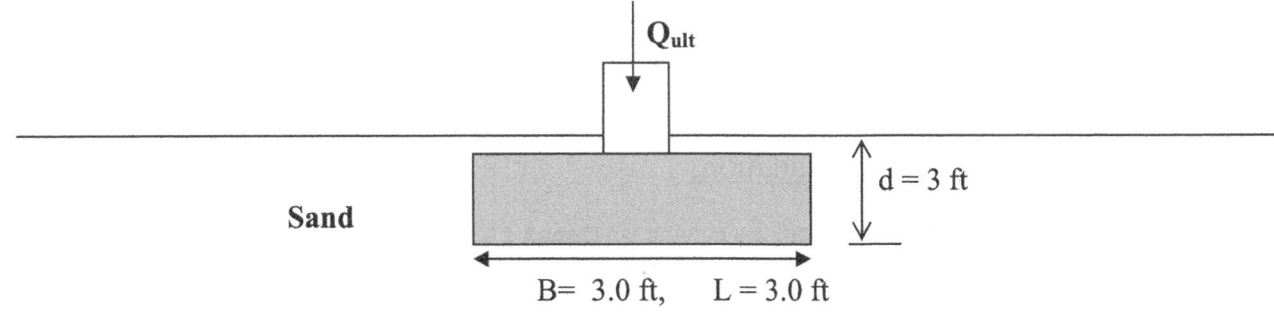

γ = 112 lbs/ft³   c = 0   (Usually cohesion in sandy soils is considered to be zero). $\varphi = 30^0$
Factor of Safety (F.O.S) = 3.0

Column footing in a homogeneous sand layer

Solution:
Write down the Terzaghi bearing capacity equation.

$$q_{ult} = c.N_c.s_c.d_c + q.N_q.s_q + 0.5.B.N_\gamma.\gamma.s_\gamma$$

STEP 1: Find Terzaghi bearing capacity factors from table T-1, using $\varphi = 30^0$.

$N_c = 37.2,$ $N_q = 22.5,$ $N_\gamma = 19.7$

STEP 2:   Find shape factors using the table T-2.
For a square footing $s_c = 1.3$ and $s_\gamma = 0.8$
Equation for shape factor s_q is given below
s_q = Shape factor = $1 + B/L \tan \varphi$ (B is the width and L is the length of the footing)
$s_q = 1 + 3/3 \tan 30 = 1 + 0.577 = 1.577$

STEP 3: Find the surcharge (q).
          $q = \gamma.d = 112 \times 3 = 336$ lbs/ft^2.

Find the depth factor using typical depth factor equation;
          d_c = Depth factor $= 1 + 0.4\ d/B$
          $d_c = 1 + 0.4\ d/B = 1 + 0.4 \times 3/3 = 1.4$

STEP 4: Apply the Terzaghi bearing capacity equation.
          $q_{ult} = c\ N_c.s_c\ d_c +$ $q\ N_q\ s_q$ $+$ $0.5.B.N_\gamma.\gamma.s_\gamma$
          $q_{ult} =$ 0 $+$ $336 \times 22.5 \times 1.577$ $+$ $0.5 \times 3 \times 19.7 \times 112 \times 0.8$
          $q_{ult} = 14,569$ lbs/ft^2

STEP 5: Find the allowable bearing capacity;
$q_{allowable} = q_{ult}\ /F.O.S$
F.O.S = Factor of safety.
Typically, a factor of safety of 3.0 is used.

$q_{allowable} = q_{ult}\ /3.0$
$q_{allowable} = 14,569/3.0 = 4,856.3$ lbs/ft^2

STEP 6: Find the net bearing capacity (q_{net}) ;

$q_{net\ allowable} = q_{allowable} - \gamma.d$

"d" is the depth to bottom of footing.

$q_{net\ allowable} = 4,856.3 - 112 \times 3 = 4,520.3$ psf

Total load ($Q_{net\ allowable}$) that could safely placed on the footing;
          $Q_{net\ allowable} = q_{net\ allowable} \times$ Area of the footing $= 4,520.3 \times (3 \times 3) = 40,683$ lbs.

Net Bearing Capacity Explanation:

Discussion of $q_{net\ allowable}$:   Let us imagine a point in the ground 3 ft below the surface. At this point, at the present time existing pressure is $\gamma.D$, where D = 3 ft. If the soil density is 120 pcf, pressure at a point 3 ft below the surface = 3 x 120 = 360 psf.
The settlement due to this pressure is zero. We assume that soil has settled and attained stability.
Let us ask the following question.
What would happen if we remove a pressure of 360 psf and put back a pressure of 360 psf. In other words what would happen if you withdraw 100 dollars from your bank account and deposit 100 dollars back. The answer to both questions is "Nothing".

In order to construct the footing, we will have to remove 3 ft of soil.

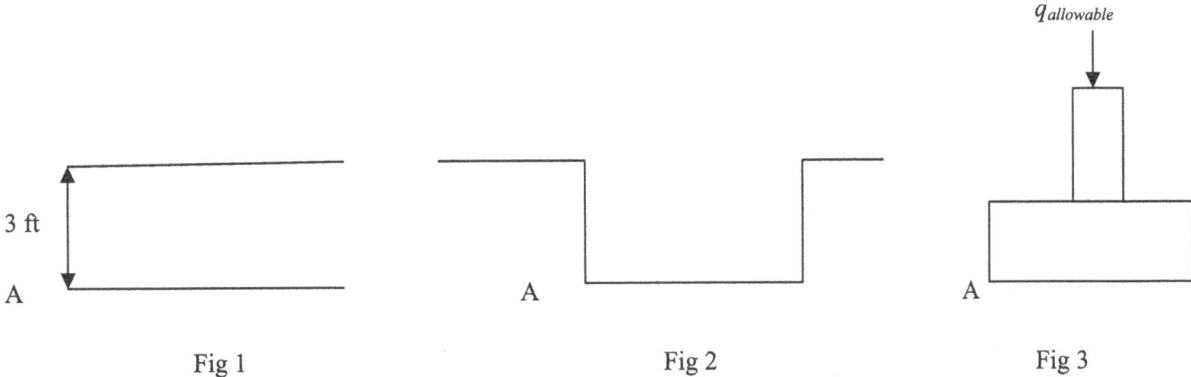

Fig 1   Fig 2   Fig 3

Figure 1 shows a soil profile. Figure 2 shows the excavation of the footing. Figure 3 shows after construction of the footing.

Fig.1: The pressure at point $A = \gamma.D$ (No settlement. Soil is stable).

Fig 2: The pressure at point $A = 0$ (Soil pressure relief or slight upheaval)

Fig 3: The pressure at point A = Load due to footing and column = $q_{ult} /F.O.S = q_{allowable}$

If $q_{allowable} = \gamma.D$, there will not be any settlement. There would be settlement only if the pressure exceeds $\gamma.D$.

$q_{net\ allowable}$ is defined as the pressure at bottom of footing level **in excess** of what was there before.

Hence $q_{net\ allowable} = q_{allowable} - \gamma.D$

Now let us try to see the significance of $q_{net\ allowable}$.

Let us assume $q_{allowable} = \gamma.D$.

What is the settlement if $q_{allowable} = \gamma.D$?

If $q_{allowable} = \gamma.D$, then settlement would be zero. This is due to the fact that there was no settlement in the soil before the excavation. Prior to excavation, the soil was subjected to a pressure of $\gamma.D$ and the soil had attained stability. If you remove a pressure of $\gamma.D$ and put back $\gamma.D$, then soil will not undergo any additional settlement. Settlement occurs only if the pressure due to footing exceeds $\gamma.D$. The pressure in excess of $\gamma.D$ will cause settlement. The pressure in excess of $\gamma.D$ is known as $q_{net\ allowable}$. This value is important for structural engineers since they are interested in soil settlement.

If $q_{allowable} = \gamma.D$, then $q_{net\ allowable} = 0$.

If $q_{net\ allowable}$ is zero, then there will not be any consolidation settlement due to footing.

I should mention here that some text books divide q_{net} by a factor of safety instead of $q_{ultimate}$. I do not think it is correct since $\gamma.D$ is known with good accuracy. In addition, this is not conservative. [Note that $q_{net} = q_{ultimate} - \gamma.D$]

Practice Problem: (Column footing placed in a homogeneous clay layer)
Find the net bearing capacity of a (3 ft x 3 ft) square footing placed in a clay layer. The density of the soil is found to be 100 pcf and the cohesion was found to be 220 psf. Footing is placed 2.5 ft below the surface.

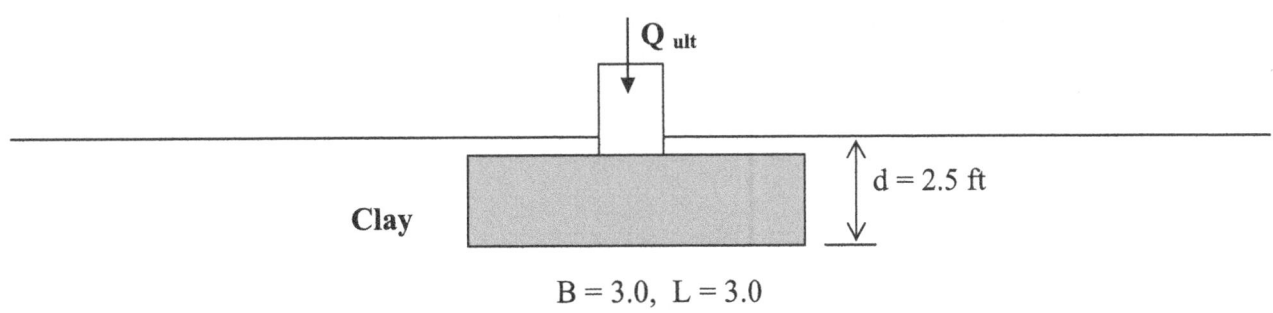

B = 3.0, L = 3.0
Column footing in a homogeneous clay layer

$$\gamma = 100 \text{ pcf} \qquad c = 220 \text{ psf}, \qquad \varphi = 0^0 \qquad \text{Factor of Safety (F.O.S)} = 3.0$$

Solution:

$$q_{ult} = c. N_{c.} s_c \ d_c \ + \qquad q. N_q s_q \qquad + \quad 0.5. B. N_\gamma. \gamma. s_\gamma$$

STEP 1:   Find Terzaghi bearing capacity factors from the table T-1.
For clay soils, friction angle (φ) is considered to be zero.
    $N_c = 5.7$, $N_q = 1.0$, $N_\gamma = 0.0$

STEP 2:   Find shape factors using table T-2.
For a square footing $s_c = 1.3$ and $s_\gamma = 0.8$
s_q = Shape factor = $1 + B/L \tan \varphi = 1 + 3/3 \tan (0) = 1.0$

STEP 3:   Find the surcharge (q).
    $q = \gamma. d = 100 \times 2.5 = 250 \text{ psf}$.

Depth factor $d_c = 1 + 0.4 \ d/B = 1 + 0.4 \times 2.5/3.0 = 1.33$

STEP 4:   Apply the Terzaghi bearing capacity equation.
$q_{ult} = c N_{c.} s_c \ d_c \ + \qquad q. N_q s_q \qquad + \quad 0.5. B. N_\gamma. \gamma. s_\gamma$
$q_{ult} = (220 \times 5.7 \times 1.3 \times 1.33) + (250 \times 1.0 \times 1.0) \quad + \qquad 0 \qquad = 2,418 \text{psf}$

STEP 5: Find the allowable bearing capacity;
$q_{allowable} = q_{ult} /F.O.S = 2,418/3.0 = 806 \text{ psf}$

STEP 6: Find the net bearing capacity ($q_{net \ allowable}$) ;

$q_{net \ allowable} = q_{allowable} - \gamma. d$

$q_{net \ allowable} = 806 - 100 \times 2.5 = 556 \text{ psf}$

Total load ($Q_{net \ allowable}$) that could safely placed on the footing;
$Q_{net \ allowable} = q_{net \ allowable} \times$ Area of the footing = $q_{net \ allowable} \times (3 \times 3) = 5,004$ lbs.

3.13   Lateral Pressure and Earth Retaining Structures:

There are many types of retaining walls. Some of the most famous ones are listed below.

Retaining Wall Types:
- Gravity walls
- Cantilever walls
- Stability analysis
- Mechanically stabilized earth walls
- Braced and anchored excavations
- Soil and rock anchors

Left: *William Rankine (1,820 – 1,872) developed Rankine theory for earth pressure*
Right: *Charles De Coulomb (1,736 – 1,820) –Developed Coulomb theory for earth pressure computations*

3.13.1 Gravity Walls:

Earth retaining structures are important to hold back soil. Gravity retaining walls as the name indicates hold soil through its weight.

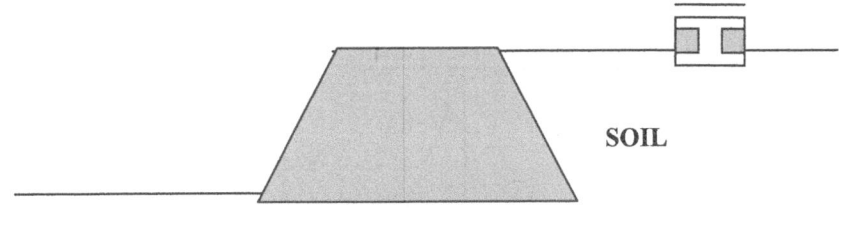

Gravity walls

In the case of gravity walls, pure weight of the retaining wall holds the soil. Gravity walls are made of rock, concrete and masonry. Presently concrete has been the most common material to construct gravity walls.

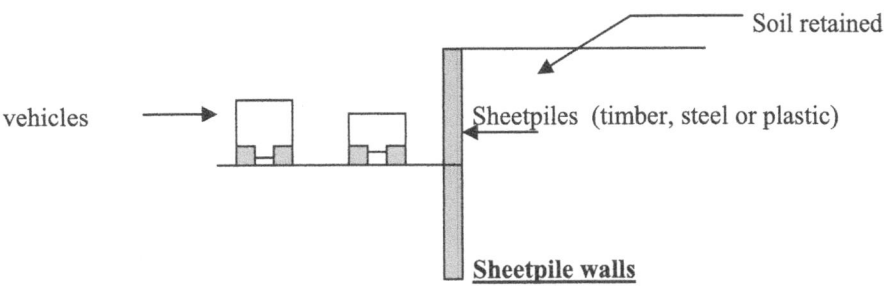

Sheetpile walls

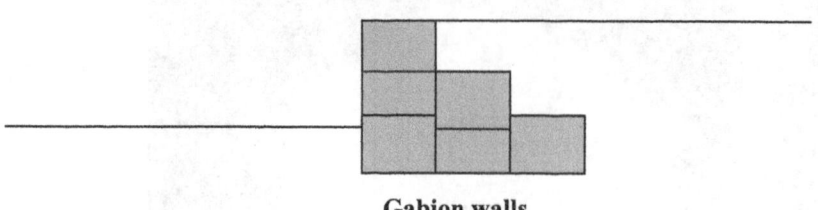

Gabion walls

Gabions are baskets filled with rocks. These rock baskets can be used to construct retaining walls. Gabion walls are designed as gravity walls. Easy drainage through gabions is a major advantage. Gabion walls are in most cases cheaper than concrete gravity walls.

Water Pressure Distribution: Before we discuss the horizontal force due to soil, let us take a look at the horizontal force due to water. Water pressure is same in all directions since it is a liquid.   Vertical stress at a point inside water is same as the horizontal stress at that location.

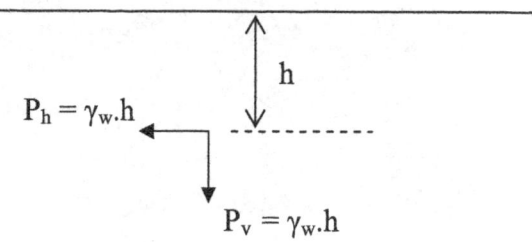

Pressure in water

$P_h = P_v = \gamma_w.h$

P_h = Horizontal pressure P_v = Vertical Pressure

γ_w = Density of water, (Usually taken as 62.4 pcf or 9.81 kN/m^3)

h = Depth to the point of interest

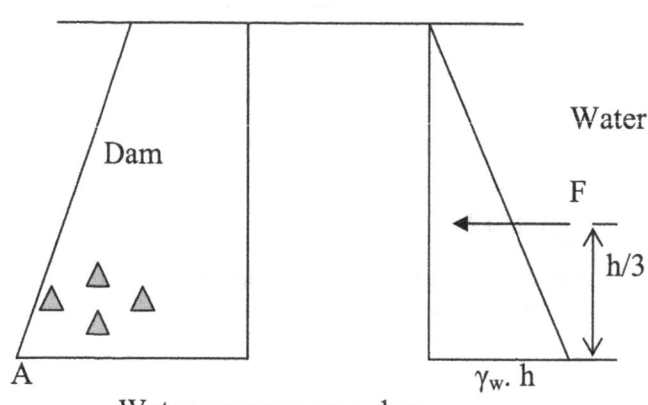

Water pressure on a dam

Water pressure at a depth of "h" = $\gamma_w. h$

Force due to water pressure = Area of the pressure triangle = $F = \gamma_w. h . h/2 = \gamma_w. h^2/2$

The moment around point "A" can be computed.

Total Moment (M) = Force x Distance to the force

$M = \gamma_w. h^2/2 \ x \ h/3 = \gamma_w. h^3/6$

The resultant pressure of the triangle acts h/3 distance from the bottom.

Practice Problem: Find the horizontal force on the dam shown due to water.

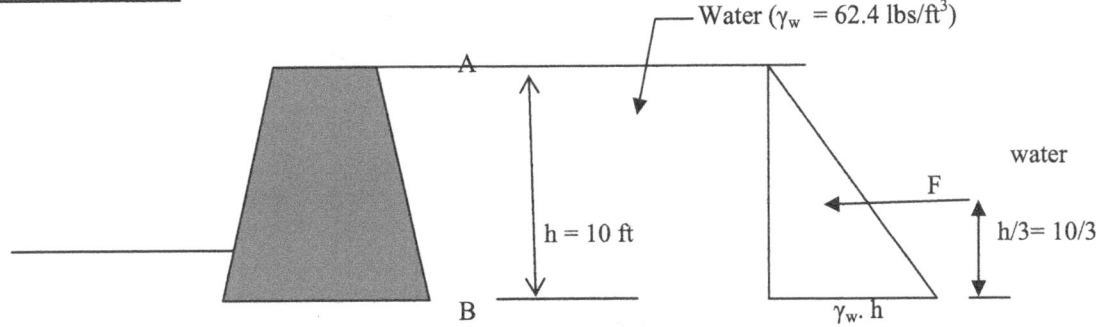

Dam subjected to water pressure

Solution: Find the pressure at point B.

$\gamma_w \cdot h = 62.4 \times 10 = 624 \ lbs/ft^2$.

Total pressure acting on the dam due to water is obtained by computing the area of the pressure triangle.

Total force acting on the dam = Area of the pressure triangle = ½ x 10 x 624 = 3,120 lbs

In the case of water, pressure in all directions is the same.

Overturning moment = Force x Distance to center of gravity

Overturning moment = 3,120 x 10/3 = 10,400 lbs. ft

Computation of Horizontal Pressure in Soil:

Following equations are used to compute the horizontal pressure in soils.

Vertical pressure in soil = density of soil x depth = γ x h

(This is when there is no groundwater).

Horizontal pressure in soil = Lateral earth pressure coefficient (K) x density of soil x depth

$$= K \times \gamma \times h$$

K = Lateral earth pressure coefficient

There are three lateral earth pressure coefficients.

- Active earth pressure coefficient (K_a)
- Passive earth pressure coefficient (K_p)
- Lateral earth pressure coefficient at rest (K_0)

Failure Planes on Active and Passive Sides;

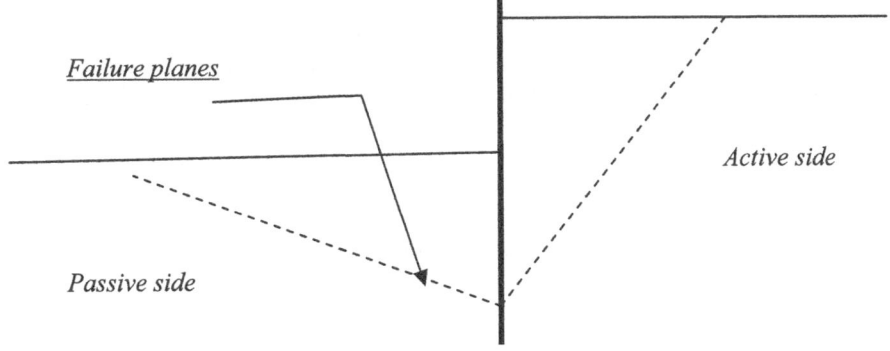

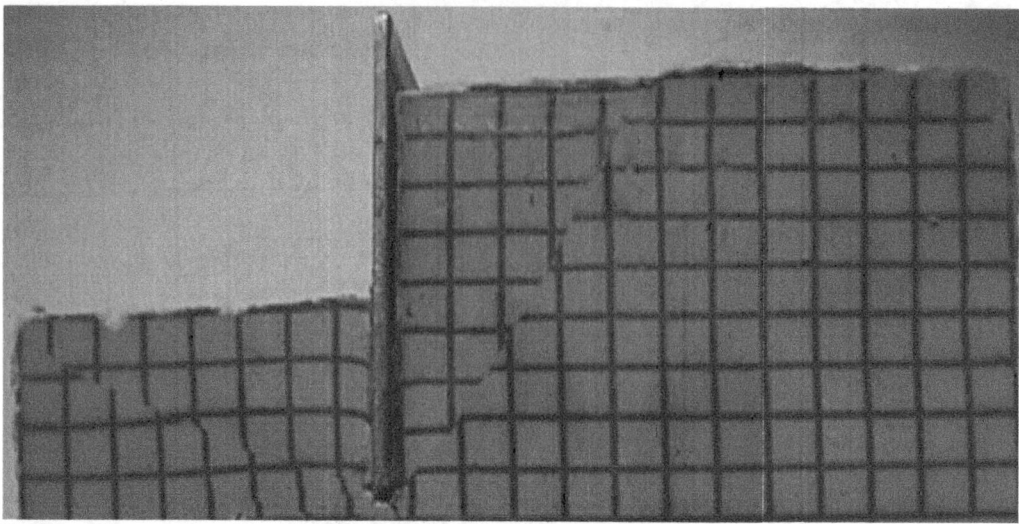

Failure planes on active side and passive side is shown in the figure

Active earth pressure coefficient (K_a):   Active earth pressure coefficient is used when the retaining wall has the freedom to move.

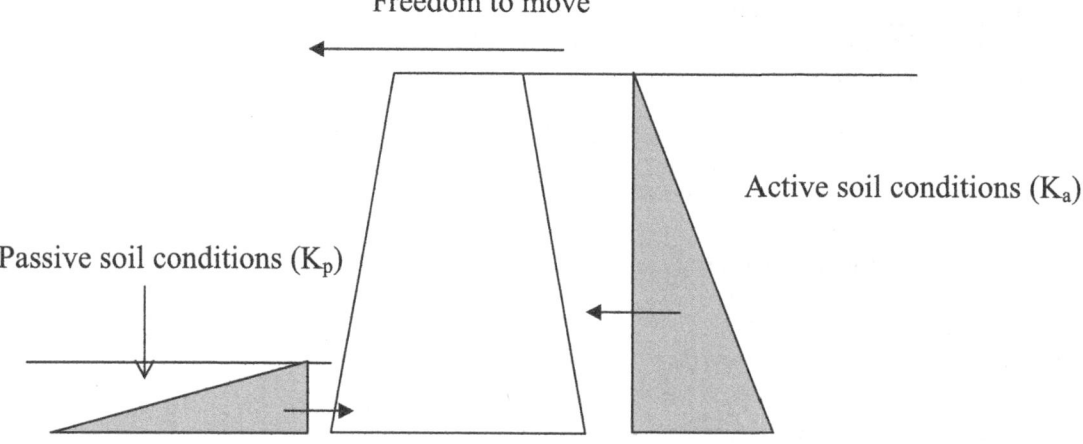

Freedom to move

Active soil conditions (K_a)

Passive soil conditions (K_p)

Freedom of movement of the retaining wall

The retaining wall shown above is free to move to the left. When slight movement in the retaining wall occurs pressure on the right hand side will be reduced. On the other hand, pressure on the left hand side will be increased.
Hence we can see that K_a is smaller than K_p.
Following equations are used to compute K_a and K_p.

$$K_a = \tan^2(45 - \phi/2)$$
$$K_p = \tan^2(45 + \phi/2)$$

Many designers do not consider the passive soil conditions in front of the retaining walls. Some codes require that erosion protection to be provided when passive earth pressure in front of a retaining wall is considered for the design.

Earth pressure coefficient at rest (K_0):   Following equation is used to compute K_0.

$$K_0 = 1 - \sin \varphi$$

Gravity Retaining Walls: (Sand Backfill):   In gravity walls, soil pressure is restrained by the weight of the retaining wall.

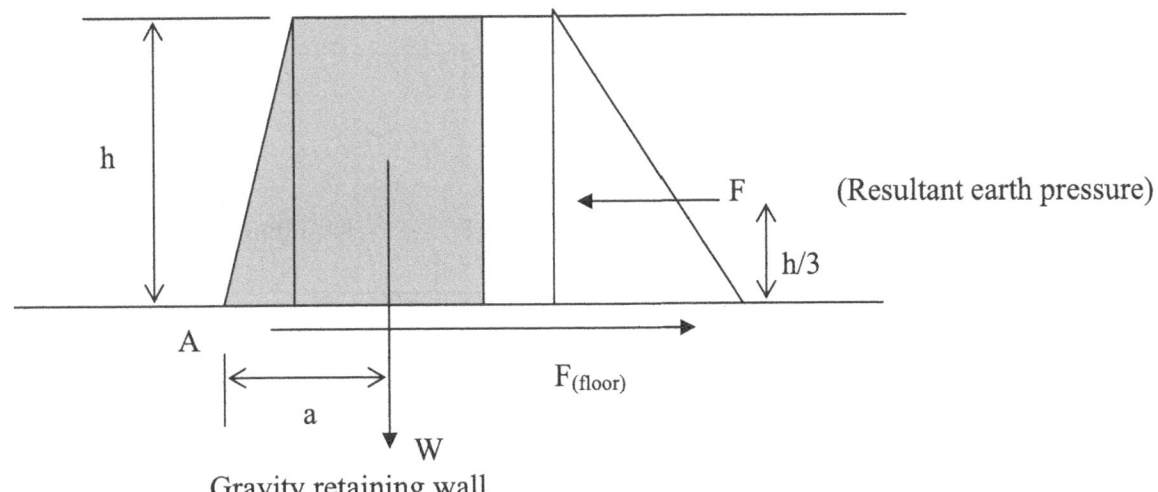

Gravity retaining wall

In the above figure, simple gravity retaining wall is shown. Modern gravity retaining walls are made of concrete. The retaining wall can fail in three different ways.

1)   The retaining wall can slide - Sliding failure
2)   The retaining wall can overturn around the toe (point A)
3)   Bearing failure of the foundation

Resistance against Sliding Failure:

For stability; $F_{(floor)} > F$

F = Resultant earth pressure
$F_{(floor)}$ = Friction between concrete and soil at the bottom face
$F_{(floor)} = \tan \delta \times W$
W = Weight of the retaining wall
δ = Friction angle between concrete and soil.

Factor of safety against sliding failure = $F_{(floor)}/F$

Resistance against Overturning: The retaining wall can overturn around the toe. (point A).
Overturning moment = $F \times h/3$
Resisting moment = $W \times a$
Factor of safety against overturning = Resisting moment/Overturning moment = $W.a/(F.h/3)$ = $3.W.a/(F.h)$

How to find the resultant earth pressure?
Earth pressure at any given point is given by $K_a \gamma h$

K_a = Active earth pressure coefficient = $\tan^2 (45 - \varphi/2)$
γ = Density of soil   h = Height of the soil

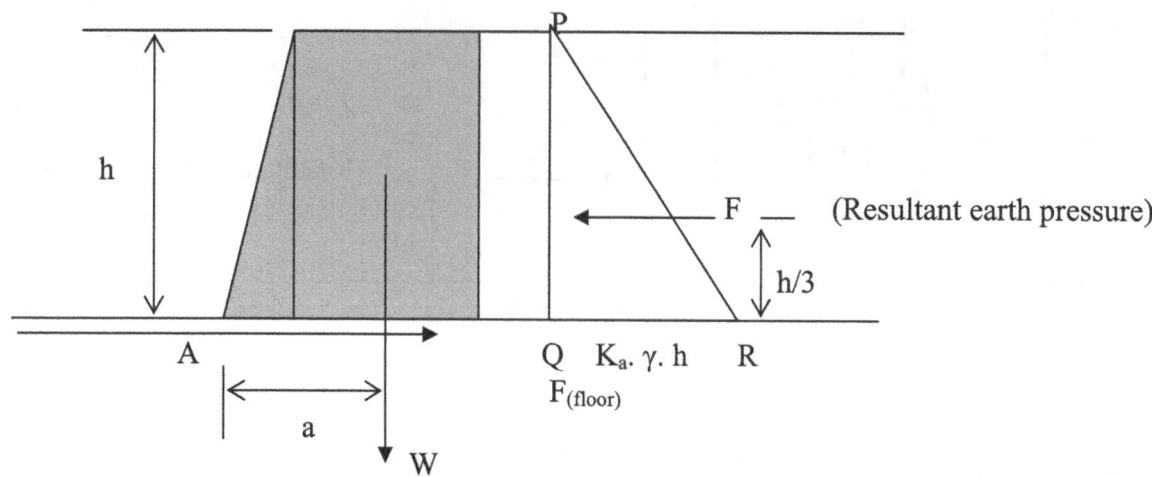

Gravity retaining wall with soil pressure

Resultant earth pressure (F) = Area of the pressure triangle (PQR)
$$= K_a \, \gamma \, h \times (h/2)$$
$$= K_a \, \gamma \, h^2/2$$

The resultant earth pressure acts at the center of gravity of the pressure triangle, h/3 above the bottom.
Total Moment (M) around point A = Force (F) x distance to the force

$$M = (K_a . \gamma . h^2/2) \times h/3 = K_a . \gamma . h^3/6$$

Practice Problem: (Gravity retaining wall with sand backfill – No groundwater)
Find the factor of safety for the retaining wall shown. Height of the retaining wall (H = 10 ft), weight of the retaining wall 7 kip for 1 ft length of the wall. Vertical through center of gravity is at a distance of 5ft from the toe (point X). Friction angle of the soil backfill is 30^0. The soil backfill is mainly consists of sandy soils. Density of the soil is 110 pcf. Resultant earth pressure force acts H/3 distance from the bottom of the wall. Friction angle between soil and earth at the bottom of the retaining wall is 20^0.

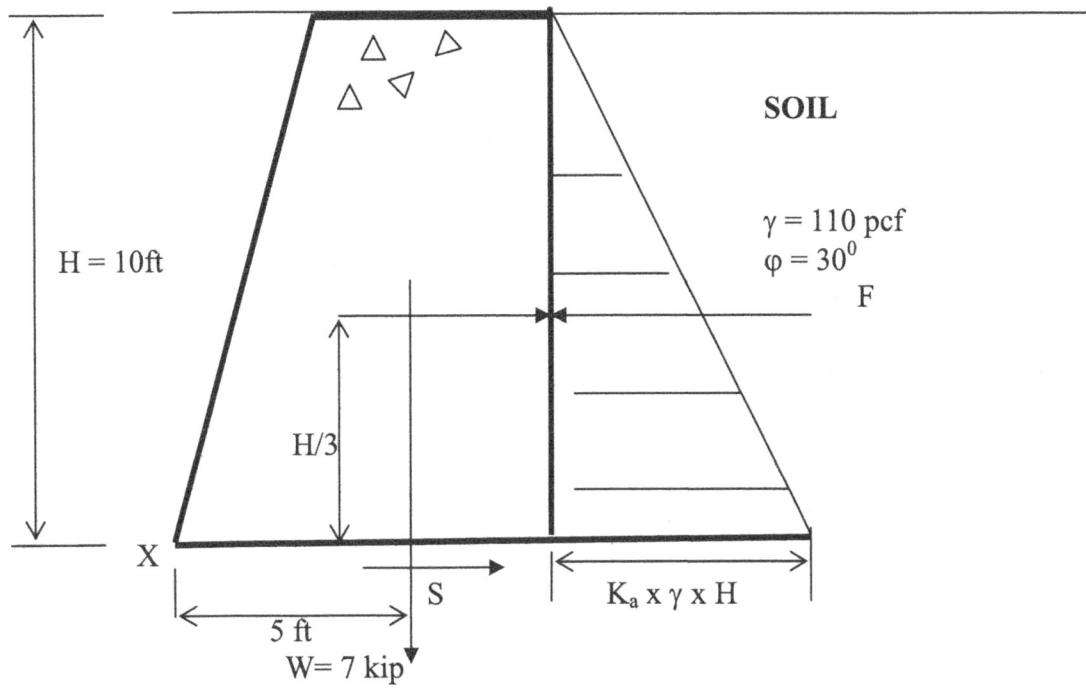

<u>Gravity retaining wall with sand backfill</u>

Solution:
STEP 1: Find the resultant earth pressure
Due to soil pressure, gravity wall will slightly move. Hence K_a is used to compute the lateral earth pressure.
$K_a = \tan^2(45 - \varphi/2) = \tan^2(45–30/2)$
$K_a = \tan^2(30) = 0.33$
Earth pressure at the bottom of the base = $K_a \gamma H$ = 0.33 x 110 x 10 = 363 psf.
Resultant earth pressure (F) = Area of the pressure triangle = 363 x 10/2
        = 1,815 lbs per 1 ft length of the wall = 1.815 kip per 1 ft length of the wall

STEP 2: Find the resistance against sliding at the base (S)
S = Friction between concrete and soil at the base
S = Weight of the wall x tan (δ)
δ = Friction angle between concrete and soil at the bottom of the retaining wall
S = W x tan (δ) = 7 x tan (20^0) = 7 x 0.36397 = 2.548 kip per 1 ft length of the wall
Weight of the retaining wall is given to be 7 kip per 1 ft length of the wall.
Factor of safety against sliding = F.O.S = 2.548/1.815 = 1.4
Factor of safety of 2.5 or more is desirable. Hence it is necessary to increase the weight of the wall.

STEP 3: Find the resistance against overturning (O)

Overturning will occur around point "X" of the retaining wall.
Resistance to overturning is provided by the weight of the retaining wall.

Overturning moment = F x H/3 = 1.815 x 10/3 kip. ft
            = 6.05 kip. ft (per 1 ft. length of the wall)

Resisting moment = W x a = 7 x 5 = 35 kip. ft (per 1 ft length of the wall)
Factor of safety against overturning = Resisting moment/Overturning moment
                = 35/6.05 = 5.78

Retaining Wall Design when Groundwater is Present: Groundwater exerts additional pressure on retaining walls. Following concept needs to be understood when drawing pressure triangles.

- Total density should be used above groundwater level to compute the lateral earth pressure. Below groundwater, **buoyant** density of soil should be used.
Buoyant density = $(\gamma - \gamma_w)$.
Pressure due to water should be computed separately.

Calculation of pressure acting on a gravity retaining wall when groundwater is present needs to be calculated in two parts.

- First draw the pressure diagrams with effective stresses of soil
- Then draw the pressure diagram for groundwater
- Both pressure diagrams are considered when computing the sliding force and the overturning moment.

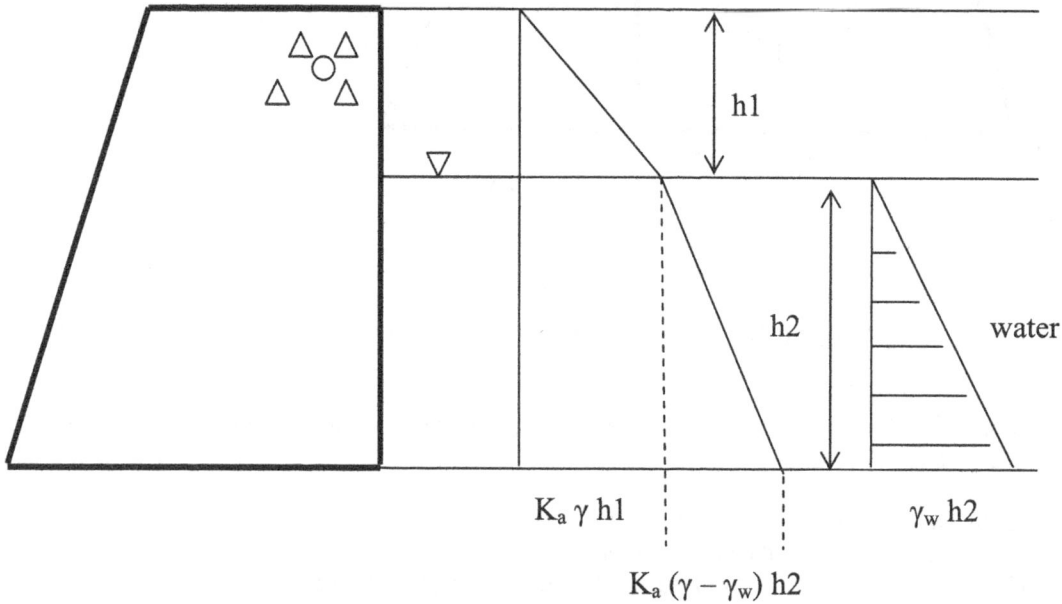

$K_a \gamma h1$

$K_a (\gamma - \gamma_w) h2$

$\gamma_w h2$

Earth pressure   *Pressure due to water*

<u>Earth pressure when water is present</u>

Earth pressure above groundwater = $K_a \times \gamma \times h1$
Earth pressure below groundwater = $K_a \times (\gamma - \gamma_w) \times h2$
Pressure due to water alone = $\gamma_w \times h2$
γ = Total density of soil γ_w = Density of water
$(\gamma - \gamma_w)$ = Buoyant density of soil
Following figure shows how to compute the factor of safety of a retaining wall when groundwater is present.

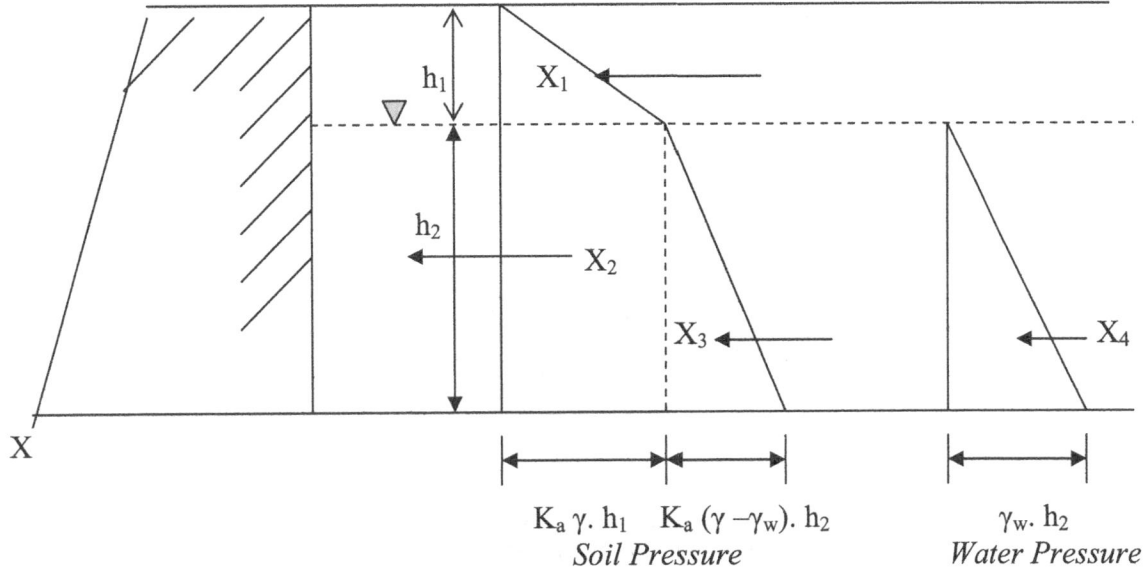

$$K_a\,\gamma.\,h_1 \qquad K_a\,(\gamma-\gamma_w).\,h_2 \qquad\qquad \gamma_w.\,h_2$$
Soil Pressure   *Water Pressure*

Forces acting on the retaining wall

Total Force = Areas of all the triangles and rectangles
Area of the triangle X1 = $K_a.\gamma.\,h_1^2/2$
Area of the rectangle $X_2 = K_a\,.\gamma.\,h_1.\,h_2$
Area of the triangle X3 = $K_a\,.\,(\gamma-\gamma_w).\,h_2^2/2$
Area of the triangle X4 = $\gamma_w\,h_2^2/2$

Total Force   = X1 + X2 + X3 + X4
                  = $K_a.\gamma.\,h_1^2/2 + K_a\,.\gamma.\,h_1.\,h_2 + K_a\,.\,(\gamma-\gamma_w).\,h_2^2/2 + \gamma_w\,h_2^2/2$

Total Moment (M) around point X = Moment of each area

Moment of triangle X1 = $K_a\,.\gamma.\,h_1^2/2 \times (h_2 + h_1/3)$
Moment of rectangle X2 = $K_a\,.\gamma.\,h_1.\,h_2 \times (h_2/2)$
Moment of triangle X3 = $K_a\,.(\gamma-\gamma_w).\,h_2^2/2 \times (h_2/3)$
Moment of triangle X4 = $\gamma_w\,h_2^2/2 \times h_2/3$

Total moment = $K_a\,.\gamma.\,h_1^2/2\,(h_2 + h_1/3) + K_a\,.\gamma.\,h_1.\,h_2\,(h_2/2) + K_a\,.(\gamma-\gamma_w).\,h_2^2/2\,.\,(h_2/3) + \gamma_w\,h_2^2/2.\,h_2/3$

3.13.2 Gabion Walls:

Gabion walls are also a type of gravity walls. Computations involved in Gabion walls are no different from regular gravity earth retaining walls. Earth pressure forces are computed as usual and stability of the wall with respect to rotation and sliding is computed. Gabion baskets are manufactured in different sizes. Typical basket is approximately 3 feet in size. Smaller baskets are easier to handle. At the same time smaller baskets would have more seems to be connected. Not all Gabion baskets are perfect cubes. Some baskets are made with elongated shapes.

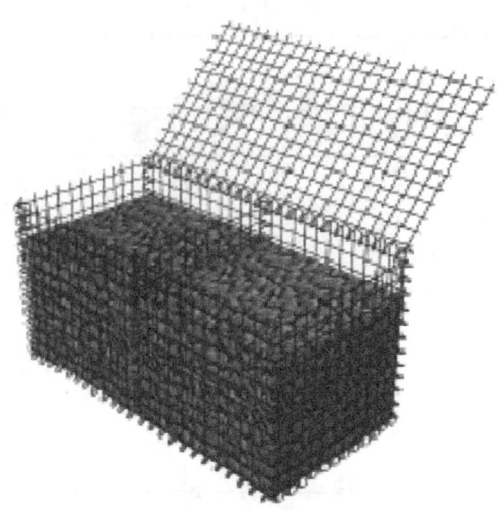

Gabion basket
Gabion baskets are connected to build a wall.

Practice Problem: An embankment of 15 feet high has to be contained. 5 ft Gabion baskets are placed as shown in the diagram has been proposed. Find the factor of safety of the gabion wall. Soil density (γ) = 120 pcf, Density of stones 160 pcf, soil friction angle (φ) = 30^0.   Friction angle between gabion baskets and soil (δ) = 20^0.
Assume all groundwater is drained.

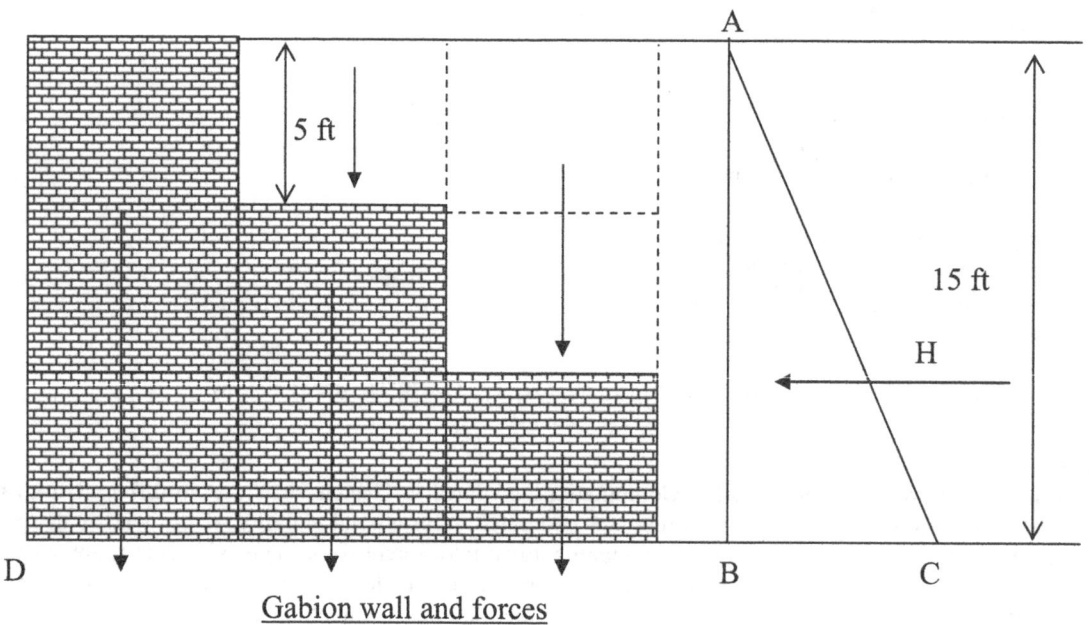

Gabion wall and forces

Solution:

STEP 1:   Find the lateral earth pressure coefficient
$K_a = \tan^2(45 - \varphi/2) = \tan^2(45 - 30/2) = 0.333$

STEP 2:   Compute horizontal force due to earth pressure
Horizontal force (H) = Pressure due to soil
BC = $K_a \times \gamma \times h$ = 0.333 x 15 x (120) = 600 psf
Area (ABC) = ½ x 15 x BC = ½ x 15 x 600 = 4,500 lbs

Total horizontal force (H) = 4,500 lbs/linear ft

STEP 3: Compute the weight of the Gabion wall
Density of stones is given to be 160 pcf. There are 6 Gabion baskets.
W = Weight of gabion baskets = 6 x (5 x 5) x 160 = 24,000 lbs/linear ft
There is soil sitting on gabion baskets.
Weight of soil = 3 x (5 x 5) x 120 = 9,000 lbs
Total weight of the gabion wall = 24,000 + 9,000 = 33,000 lbs/linear ft.

STEP 4: Resistance against sliding
Resistance against sliding = Weight of the gabion wall x tan (δ)

R = 33,000 x tan (20^0) = 12,011 lbs
Factor of safety against sliding = Resistance against sliding/Horizontal force
= 12,011/4,500 = 2.67

STEP 5:   Overturning Moment
Resultant force acts 1/3 of the length of the triangle.
Obtain moments around point "D".
Overturning moment = H x 1/3 x 15 = 4,500 x 5 = 22,500 lbs. ft

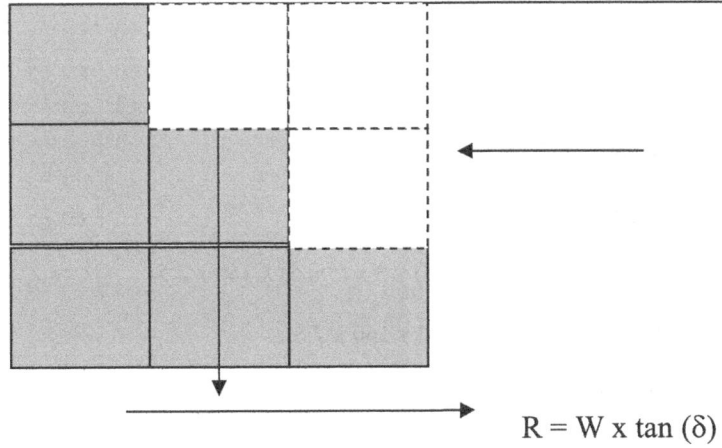

R = W x tan (δ)

R =Resistance against sliding

Forces acting on a gabion wall

STEP 6:   Resisting moment
There are six rock blocks and three soil blocks.
There are three blocks in the far left and one block at far right. There are two blocks in the middle.

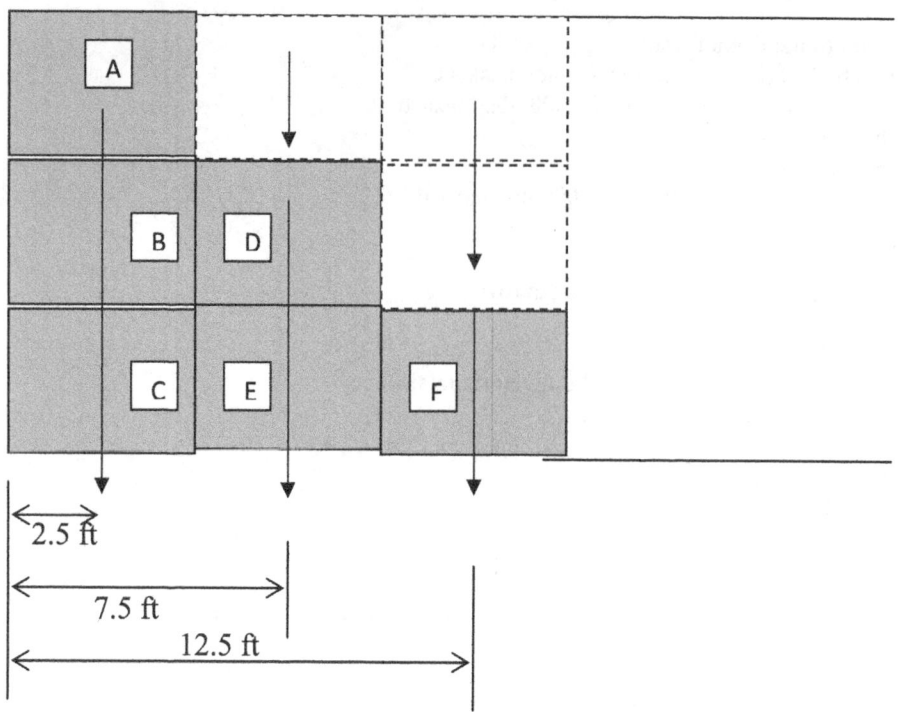

Weight due to gabion baskets

Resisting moment due to gabion baskets:

Resisting moment due to three blocks in the left (ABC) = 3 x (5 x 5) x 160 x 2.5
 = 30,000 lbs. ft
Resisting moment due to two blocks (DE) = 2 x (5 x 5) x 160 x 7.5
= 60,000 lbs. ft
Resisting moment due to one block (F) = 1 x (5 x 5) x 160 x 12.5
= 50,000 lbs. ft
Resisting moment due to soil sitting on top of gabion baskets
= 1 x (5 x 5) x 120 x 7.5 + 2 x (5 x 5) x 120 x 12.5 = 97,500
Total resisting moment = 97,500 + 30,000 + 60,000 + 50,000 = 237,500 lbs. ft

STEP 7: Factor of safety against overturning = Resisting moment/overturning moment
= 237,500/22,500   = 10.55

3.14   Slope Stability and Excavation Support

Deaths from failed excavations is a common occurrence. Excavations need to be conducted in a safe manner.

Unsupported and unsafe excavation:

OSHA recommends following methods.

    1) Steps
    2) Sloping
    3) Timber Shoring
    4) Trench boxes:
    5) Excavation bracing
    6) Sheetpiles
    7) Tied anchors
    8) Soldier piles and lagging wall
    9) Raker Bracing
    10) Secant and Tangent walls

<u>Steps</u>:

Step Excavation:

Step excavation is safer and easy to construct. Only drawback is that this method needs a large area. In some urban settings, this may not be possible.

Sloped Excavation:

Sloped Excavation:

Excavations can be sloped so that the slopes are stable.

Timber Shoring: Sloping and steps require room. In some cases this is not possible. Imagine a trench in the side of the road. There is no room to slope the excavation. In such situations, timber shoring can be used.

Timber Shoring:

Trench Boxes: Trench boxes can be used in place of timber shoring. Trench boxes can be rented and placed in the excavation. This is much faster than providing timber shoring. Also in some cases, trench boxes may be the cheaper option.

Trench boxes:

<u>Excavation Bracing:</u>   Large excavations need to be braced. Figure below shows horizontal braces spanning from side to side.

Excavation Braces:

Sheetpiles: Sheetpiles are widely used in construction sites.

Sheetpiles:

Sheetpiles are used for slope stability, retaining walls and coffer dams.

Tied Anchors:

Tall retaining walls need to be tied. Holes are drilled and anchors are placed and concreted.

Soldier Pile and Lagging Walls: Soldier piles and lagging walls are suitable for large excavations. I-beams are installed and timber planks known as timber lagging is placed in between I-beams.

Timber lagging ———   ┌—— I- Beams (Soldier piles)

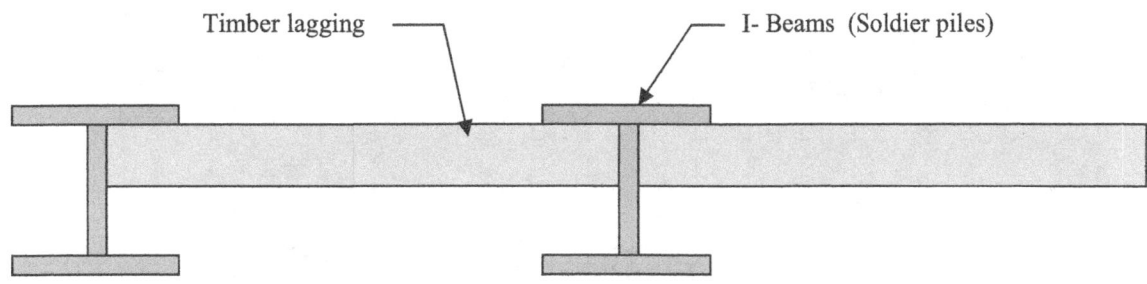

Soldier piles and timber lagging retaining wall:

Raker Bracing:

 Rakers can be used to brace excavations as shown in the above figure:

Tangent Pile Walls;   Tangent pile walls are built by drilling holes and filling with concrete. Generally to achieve higher strength, steel beams or rebar cages are inserted.

Tangent Pile Walls:

General construction procedure of tangent pile walls is;

> Drill holes
> Insert a rebar cage or steel I beams
> Concrete the hole.

Left: *Augering for tangent pile wall*
Right: *Tangent pile wall constructed by concreting the holes*

Secant Pile Walls:

Tangent pile walls had one major problem. They are not good in keeping the excavations dry. When the water table is high, water would leak thru the wall. This can be remediated by constructing a secant pile wall. Secant pile walls cut each other and creates a water tight wall.

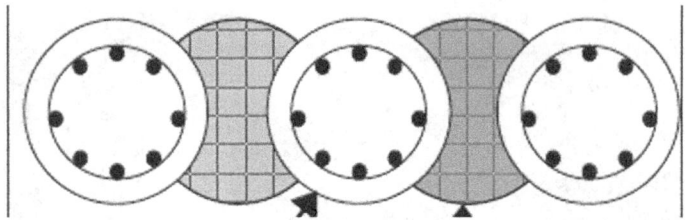

Secant pile wall seen from top

4.0   Structural Mechanics   -   6 Questions

4.1 Loadings:

Following loads are identified by ASCE 7.0. (ASCE 7.0 is used for the PE exam).

Dead Loads (D) - Dead loads are the loads due to weight of slabs, columns, windows etc.

Live Loads (L) - Live loads are due to people and furniture

Earthquake Loads (E) - Loads during earthquakes

Wind Loads (W) - Loads due to wind

Loads due to Fluids (F) - Some structures have fluid storage tanks. Loads due to fluids that are known and well defined.

Flood loads (F_a) - Loads due to floods

Horizontal Loads (H) - Horizontal loads due to soil or groundwater pressure. Also some solids could exert horizontal pressure

Roof live load (L_r) – Live loads in roofs

Ice Load (D_i) - Load due to ice accumulation

Snow Load (S) – Load due to snow accumulation

Rain Load R) – Load due to rain water.

Self straining force (T) – Loads due to self strain of a member due to temperature change or force changes. For an instance a beam can expand due to high temperature and create additional loads

4.1.1. Dead loads:

Permanent loads such as columns, beams, slabs, windows, masonry walls, mechanical piping, lighting and ducts are considered to be dead loads. Let us look at a problem involving a slab.

Practice Problem: A 6 inch concrete slab is supported by four beams as shown. Find the dead load of beam AB in lbs/ft. Concrete density is 145 pcf. Also find the total dead load on beam AB due to slab.

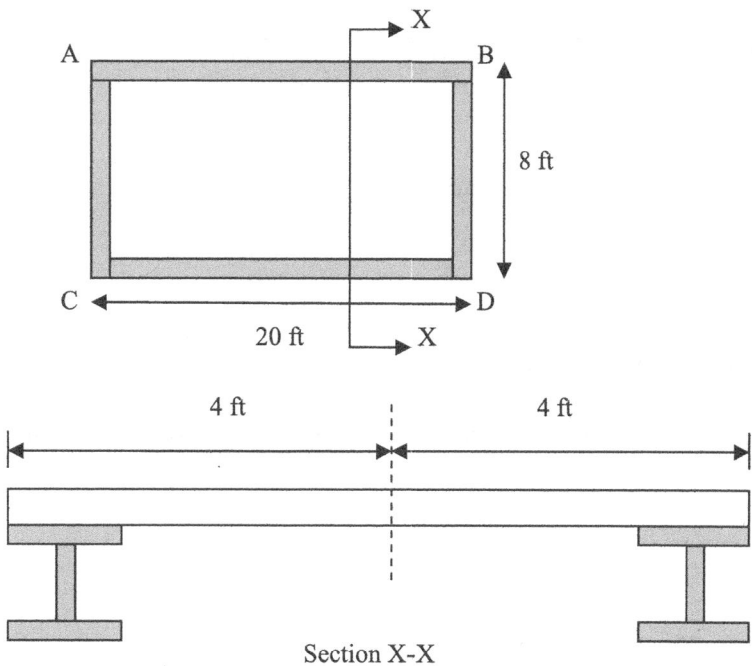

Section X-X

Solution: 4 ft section of the slab will rest on beam AB, while other 4 ft rest on beam CD.

Slab load on beam AB per foot = (4 x 1) x 0.5 x 145 lbs/ft = 290 lbs/ft
Total slab load on beam AB = 290 x 20 = 5,800 lbs

One way slabs and two way slabs;

When the length of a slab is more than twice the width of the slab it is known as a one way slab. If the length is less than twice the width, then it is a two way slab.

L/B > 2 ------>   One way slab
L/B < 2 ------>   Two way slab
L = Length;   B = Width
Load transfer to beams in one way slabs is different than two way slabs.

Practice Problem: A slab has a width of 10 ft and a length of 25 ft. Is this a one way slab or a two way slab?

Solution: L = 25; W = 10
L/B = 25/10 = 2.5 > 2
This is a one way slab.

Load Transfer in One Way Slabs;

In the case of one way slabs, load transfer occurs only to the longitudinal beams, AB and CD. (See the figure below).

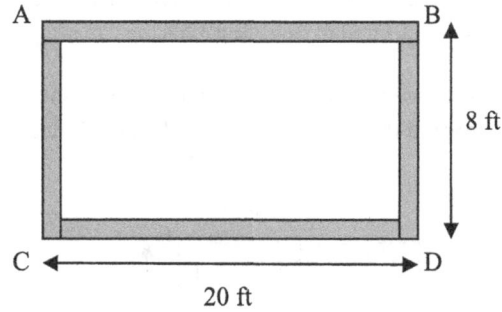

Above slab is a one way slab since L/W is greater than 2.

Load on beam AC = 0
Load on beam BD = 0
Load on beam AB = Load on beam CD
Loads on beams AB and CD are not zero.

Practice Problem: Assume the slab load is 15 psf in the figure shown above. What are the total loads on four beams, AB, BD, DC, CA?

Solution:
Load on beam AC = 0
Load on beam BD = 0
Load on beam AB = 4 x 20 x 15 = 1,200 lbs
Load on beam CD = 4 x 20 x 15 = 1,200 lbs

Flexural Reinforcements in One Way Slabs;

Bending in one way slabs occur only in the shorter span. Hence flexural reinforcements are provided only across the shorter span. (See the figure below). Flexural reinforcements are also known as main reinforcements.

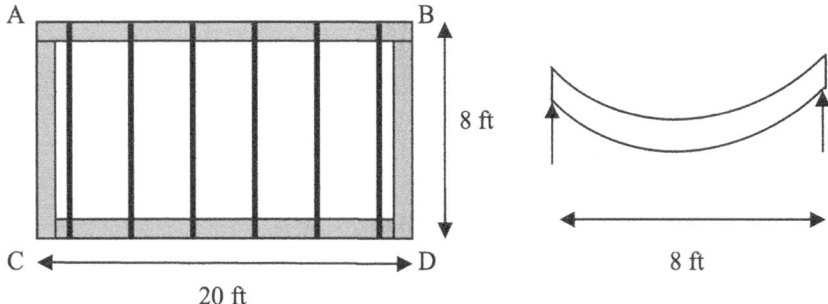

Load Transfer in Two Way Slabs;

Load transfer in two way slabs is different than one way slabs.

Two way slabs   ----------->   Length/Width < 2

Following example shows the load transfer in two way slabs;

Practice Problem: A slab is supported by four beams as shown. The thickness of the slab is 6 inches and the density of concrete is 145 pcf. Find the loading on beams AB, BC, CD and DA.

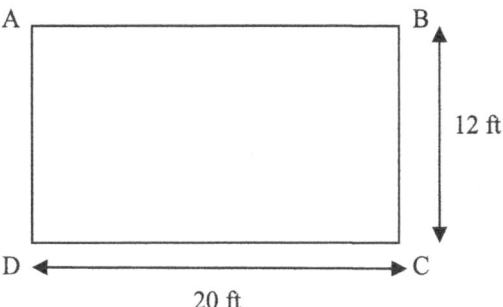

Solution: Length/Width = 20/12 < 2,   Hence this is a two way slab.

STEP 1:   Draw 45^0 lines as shown.

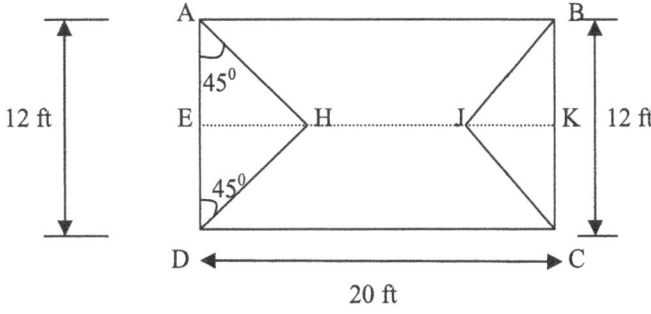

STEP 2:   Load on triangle AHD will be transferred to beam AD. Load on triangle BJC will be transferred to BC. Load on trapezoid ABJH will be transferred to beam AB. Load on trapezoid DHJC will be transferred to beam DC.

STEP 3:   Find the load on triangle AHD;

AE = 6 since AD = 12 (given).

From symmetry, AE = EH = 6
Area of triangle ADH = 1/2 x AD x EH = 1/2 x 12 x 6 = 36 sq. ft

Concrete slab is 6 inches (1/2 ft) thick and has a density of 145 pcf.
Load on triangle ADH = 1/2 x 145 x 36 lbs = 2,610 lbs
Hence the load on beam AD = 2,610 lbs.
From symmetry, load on beam BC = 2,610 lbs

STEP 4:   Find the load on trapezoid AHJB;

HJ = DC - EH - JK = 20 - 6 - 6 = 8 ft

Area of trapezoid AHJB = 0.5 x (AB + JH) x AE = 1/2 x (20 + 8) x 6 = 84 sq. ft
Load on trapezoid AHJB = 0.5 x 145 x 84 = 6,090 lbs   (0.5 here represents thickness of the slab)
Load on beam AB = 6,090 lbs
By symmetry, load on beam DC = 6,090 lbs

4.1.2   Live Loads:

Live load is defined as the loads due to occupancy of the building.
Live loads are due to;
- People
- Partitions
- Exercise equipment
- Furniture
- Computer equipment, refrigerators, ovens, TVs, book racks

Typical live loads as per ASCE 7;
- Office areas without computers = 50 psf
- Office partitions = 15 psf
- Office areas with computers = 100 psf
- Lobbies and balconies = 100 psf
- Single family dwellings = 40 psf
- Library reading rooms = 60 psf
- Library book stack areas = 150 psf
- School classrooms = 40 psf

Practice Problem: Column layout of an office area is as shown in the figure. Columns are located 20 ft apart.

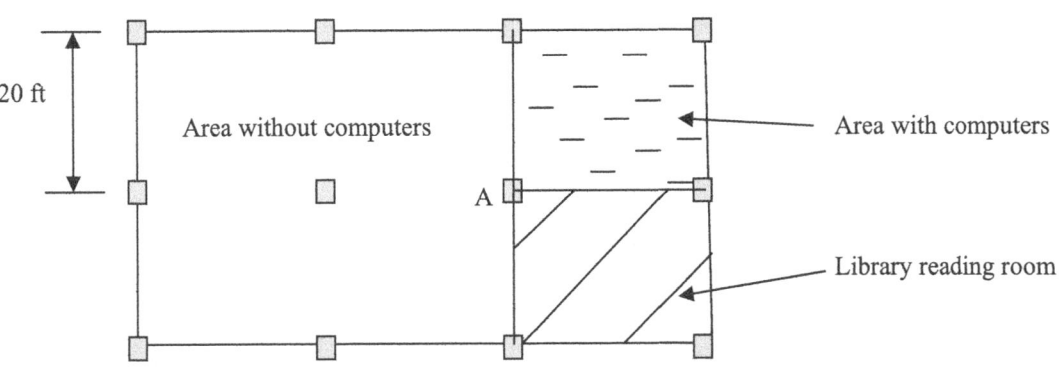

Loading;
Area without computers (50 psf), Area with computers (110 psf), Library reading room (60 psf).
Find the total live load on column A

A) 27,000 lbs   B) 32,000 lbs   C) 21,780 lbs   D) 43,900 lbs

Solution: Draw the influence area for column A

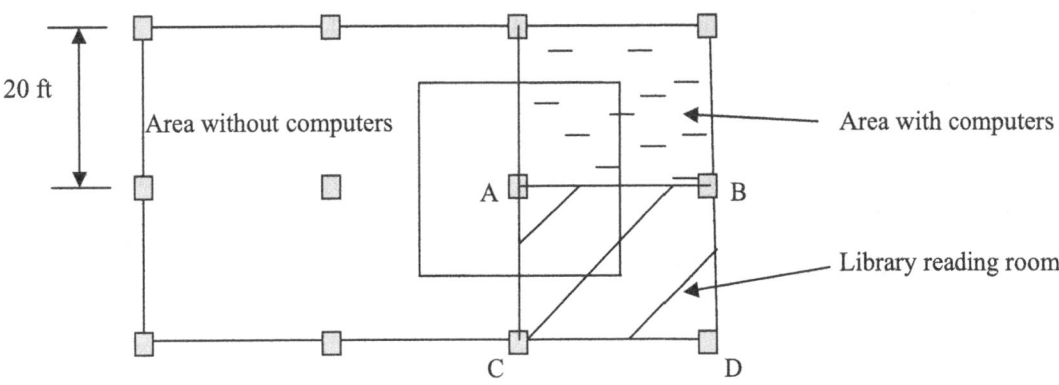

Total influence area of column A = 20 x 20 = 400 sq. ft
Half of the influence area of column A is in the "area without computers" section.

Area of "area without computers" inside the influence area = 400/2 = 200 sq. ft
Load on "area without computers" = 200 x 50 = 10,000 lbs

Area of "library reading room" inside the influence area = 400/4 = 100 sq. ft
Load on " library reading room " = 100 x 60 = 6,000 lbs

Area of "area with computers" = 400/4 = 100 sq. ft
Load on "area with computers" = 110 x 100 = 11,000 lbs

Total live load on column A = 10,000 + 6000 + 11,000 = 27,000 lbs   (Ans A)

4.1.3 Construction Loads:

Loading during construction is different from permanent loading on a slab. During construction, workers will be working with their tools. In addition, concrete buggies, finishing equipment, hoses and vibrators also will be on the slab. ASCE Standard, SEI/ASCE 37-02, 2002, defines following loads during construction. Note that load combinations and definitions are different during construction compared to permanent situation.

Dead load (D) – Dead load due to permanent structure constructed at a given time. Shoring and other dead loads that are not part of the permanent structure are not considered for (D).

Construction dead load (C_D) - Dead load due to shoring, scaffolds and other construction related dead loads. These loads will be gone after the construction is completed.

Live load (L) – Live load due to occupants during construction. In many cases this would be zero. However, occasionally occupants would like to use the structure during construction. Construction workers and their equipment are not part of this load.

Construction personal and equipment Load (C_P) – Construction workers working on the structure, concrete buggies, concrete pumping hoses, equipment for formwork and shoring, generators, gang boxes and compressors.

Horizontal Construction Load (C_H) – Construction activities can create horizontal loads. Moving vehicles, people, vibrating machines, compressors can generate horizontal loads.

Wind Load (W) – Load due to wind acting on the structure during construction. Wind load is mostly lateral and could uplift the structure as well. Lateral stability and resistance to uplift needs to be assessed. Wind load is a dynamic load. However, for computation ease, wind load is considered to be a static horizontal load.

Snow load (S) and Ice Load (I): Accumulation of snow on a structure during construction needs to be addressed. In some areas snow could turn into ice in a short period of time.

Earthquake Load (E): Earthquake load is a dynamic load. However, for computation ease, earthquake load is considered to be a static load acting horizontally.

Material Loads: Load due to material during construction is divided into two. Fixed material loads (C_{FML}) and Variable material load (C_{VML}).

(C_{FML}) Fixed Material Loads : During construction, material has to be stockpiled in the structure that has been constructed. If the magnitude of the material load is fixed, then it is considered as C_{FML}. Load due to fuel and various other materials required during construction may be relatively fixed if they are replenished.

(C_{VML}) Variable Material Load: Load due to some material may be variable. Steel, nuts and bolts may be stored. Once they are constructed it becomes part of the permanent dead load (D).

Horizontal construction loads (C_H): Moving wheelbarrows, moving personnel generates lateral loads on the structure. Wind load is not considered in this item since wind load is considered separately. It is assumed moving person would exert 50 lbs lateral load on the structure.

4.2.0   Structural Analysis:

Structures consist of members. Members are subjected to axial forces, shear forces and bending moments. To size the members one has to find these forces and bending moments. If all the forces (including bending moments) can be found in a structure, we call these structures to be statically determinate structures.

Types of Connections:
There are four main types of connections. They are fixed, pinned, rollers and sliders.

Pinned Support or Connection:

Pinned support

In the case of pinned supports, bending moment is not transferred to the members.

Left Photo: *Sydney Harbor Bridge: Pinned support (bearing pin is 14·5 inches in diameter and 13 ft 8 in long.*
Right Photo: *Column pinned support*

Fixed Support or Connection:

In the case of fixed supports or connections, two reactions and a bending moment is imparted to the member.

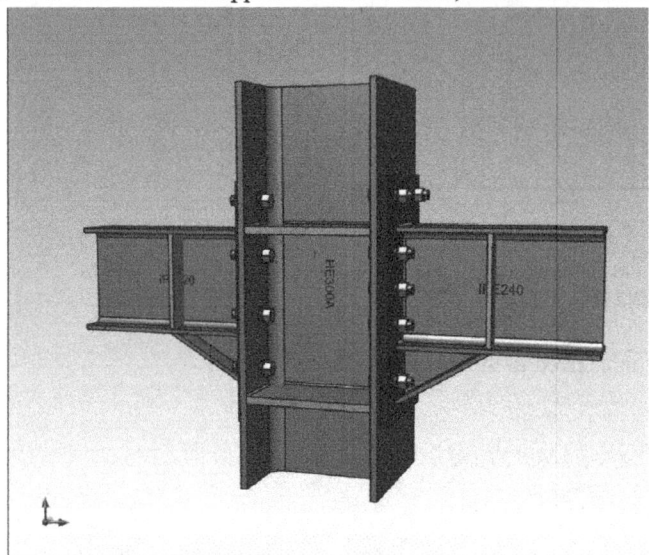

Fixed connection - In this case, reactions and bending moments are transferred from beams to columns

Roller Support:

In the case of roller supports, only one reaction is imparted to the structure. In pinned supports, two reactions were imparted to the structure.

Roller support in a bridge

Sliding Support:

In the case of sliding support, one reaction and a bending moment is imparted to the structure.

4.2.1. Determinate Analysis:

Some structures can be solved for member reactions. Let us look at a simple beam.

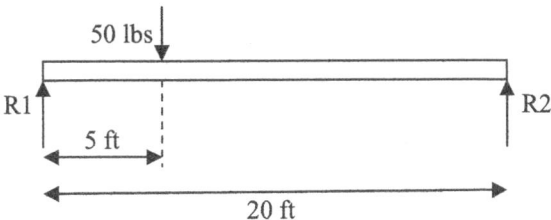

In this case, there is only one beam or one member. Each member can generate 3 equations. They are

- Resolving forces in horizontal direction

- Resolving forces in vertical direction
- Obtain moments

Resolve forces vertically; $R1 + R2 = 50$
Take moments around R1; $20. R2 = 50 \times 5 = 250$
          $R2 = 12.5$ and $R1 = 37.5$

Since there are no horizontal forces, there is no benefit in resolving the forces in the horizontal direction. Hence we have two equations and two unknowns. The above beam is a determinate structure.

Now let us look at the following beam;

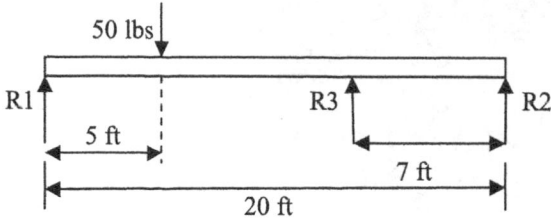

There are three unknowns. R1, R2 and R3. Nevertheless, we can generate only two equations. Resolving forces in horizontal direction would not yield any equation. Three unknowns cannot be found from two equations. Hence the above beam is indeterminate.

4.2.2   External Indeterminacy and Internal Indeterminacy;

A structure could be indeterminate externally or internally.

If all support reactions can be found, then that structure is externally determinate. If all member forces can be found, then that structure is internally determinate.

External Determinacy:

When support reactions are less than or equal to 3, the structure is *externally* determinate.

$S < 3$ The structure is externally determinate but unstable
$S = 3$ The structure is externally determinate
$S > 4$ The structure is externally indeterminate

S = Support Reactions

Example:

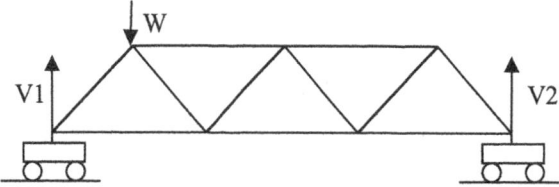

In the above structure there are two support reactions. Roller supports will have only one vertical reaction.

Support reactions = $S = 2$ (V1 and V2)

S < 3   The structure is *externally* determinate but unstable. Slightest horizontal force would make the structure move horizontally. Stability of the structure has to be found by observation.

Is the structure internally determinate? If we can find all member forces, then the structure is internally determinate. We will discuss internal determinacy in the next chapter.

Example:

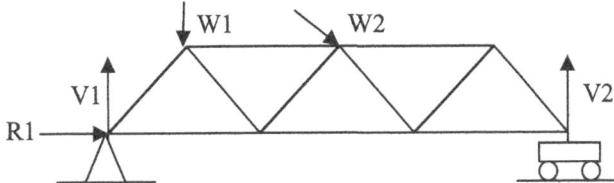

In the above structure one end is pinned and the other end is on a roller support. There are three support reactions.

Support reactions = S = 3   (V1, V2, R1)

S = 3   The structure is *externally* determinate and stable. As mentioned earlier, stability of a structure has to be found thru observation. The above structure will not move horizontally due to the horizontal reaction at the pinned support.

Example:

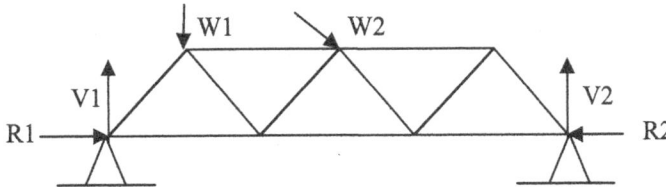

In the above structure, both ends are pinned. There are four support reactions

Support reactions = S = 4

S > 3   The structure is *externally* indeterminate and stable.

It is not possible to find all support reactions. Only three equations can be generated. They are

- Resolving forces horizontally
- Resolving forces vertically
- Obtaining moments around a point

There are three equations and four unknowns. (R1, V1, R2, V2). Hence support reactions cannot be found. The structure is externally indeterminate.

Example:

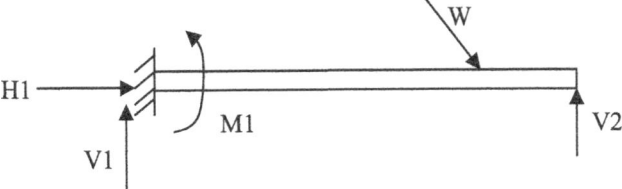

Above beam has three support reactions at the fixed end. In addition, there is one support reaction at the other end.

S = Support reactions = 4   (H1, V1, M1, V2)

S > 3   The structure is *externally* indeterminate

Practice Problem: The 50 lb load is acting 30^0 to the horizontal.   Is the structure statically determinate externally?

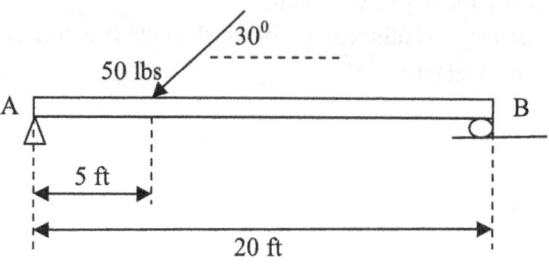

A) Yes   B) No   C) Can't say

Solution: Support at A is a pinned support. The support at B is a roller support.

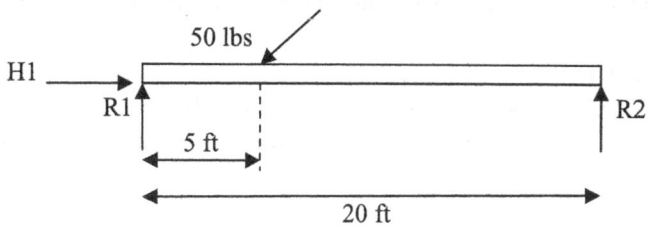

At pinned supports, there are two reactions. (R1 and H1). There is only one reaction at point B. (R2)

Number of members = 1   (N = 1)
Number of support reactions = 3   (R1, R2 and H1).
S = 3
The structure is statically determinate externally.   (Ans A)

Internal Determinacy: A structure is *internally* determinate when all member forces can be found.
Following equation is used to find the internal determinacy.

A structure is statically *determinate* internally if

$$R = 3. N$$

R = Number of unknown forces or moments.
N = Number of members.

If R > 3. N, the structure is statically *indeterminate* internally. In such situations, other methods such as moment distribution, virtual work, portal method and moment area method is used.
One has to be careful because some equations may not yield any value.

Practice Problem: 100 lb load is applied to the center of the horizontal beam and the slanted beam. Is the structure statically determinate internally?

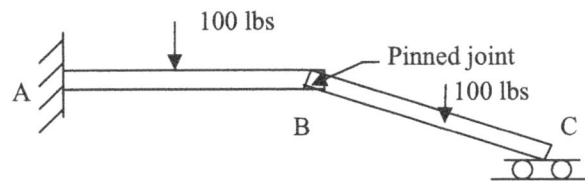

STEP 1:   Draw the forces;

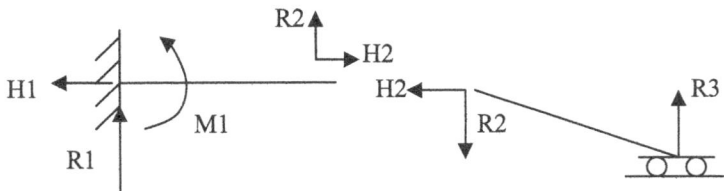

At the fixed joint there are three forces. (H1, R1 and M1). At the pinned joint there are two forces. (H2 and R2). At the roller joint there is only one force (R3).

A structure is statically determinate if

R = 3. N

R = Number of unknown forces or moments.

N = Number of members.

R = Number of forces (including bending moments) = 6   (H1, R1, M1, H2, R2, R3)

N = Number of members = 2

R = 3N

The structure is statically determinate internally.

Practice Problem: 100 lb load is applied to the center of the horizontal beam and the slanted beam. Is the structure statically determinate?

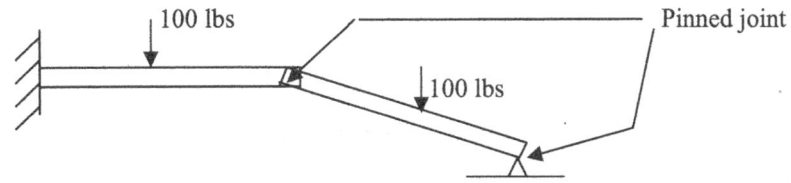

STEP 1:   Draw the forces;

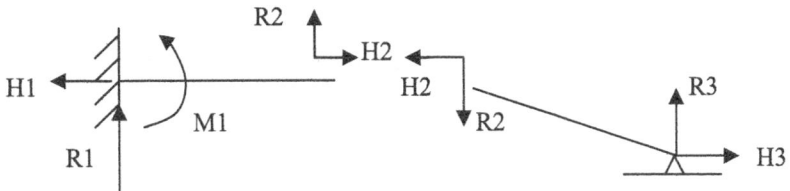

R = Number of forces (including bending moments) = 7   (H1, R1, M1, H2, R2, R3, H3)

N = Number of members = 2

R > 3N

The structure is statically **indeterminate** internally.

4.2.3   Degree of Indeterminacy;

Degree of indeterminacy is given by the following equation;

Degree of indeterminacy = R - 3N

If R = 7 and N = 2, then degree of indeterminacy is 1.

If R = 8 and N = 2, then degree of indeterminacy is 2.

If R = 17 and N = 5, then degree of indeterminacy is 2.

Practice Problem: What is the degree of indeterminacy of the structure shown below;

Pinned joint

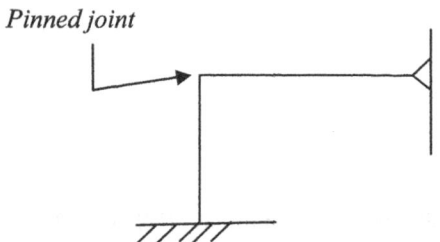

Solution:

Draw the forces;

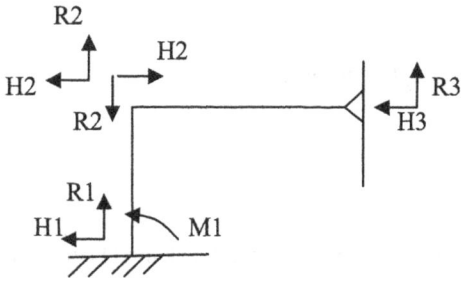

Number of forces = 7   (H1, R1, M1, H2, R2, H3, R3)

Number of members = 2

Degree of indeterminacy = R - 3.N = 7 - 3 x 2 = 1

Practice Problem: What is the degree of indeterminacy of the structure shown below;

Pinned joints

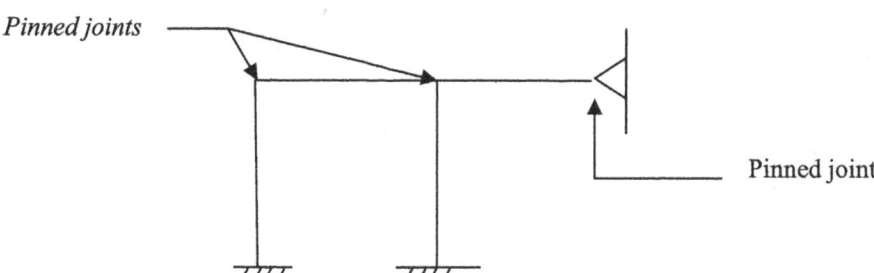

Pinned joint

Solution: Draw the forces;

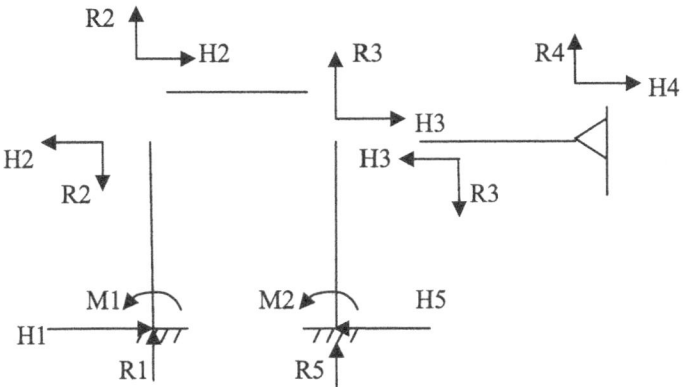

Total number of forces =   12 (H1, R1, M1, H2, R2, H3, R3, H4, R4, H5, R5, M2)
Number of members = 4

Degree of indeterminacy = R - 3.N = 12 - 3 x 4 = 0
The structure is determinate.

Practice Problem: What is the degree of indeterminacy of the structure shown below;

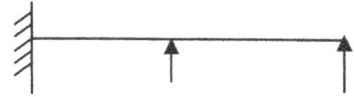

Solution: Draw the forces;

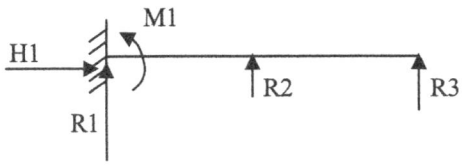

Total number of forces = 5
Number of members = 1

Degree of indeterminacy = R - 3.N = 5 - 3 x 1 = 2
The structure is indeterminate to the second degree.

4.2.4   Stability Analysis:

First you need to find out whether the structure is determinate. If so then stability analysis can be done. If the structure is indeterminate, then stability analysis cannot be performed.

Let us look at some problems.

Practice Problem: Two beams are connected using a pin joint as shown. 200 lbs and 100 lbs loads are acting at center of each beam. Find all the reactions. Assume a roller support at point C.

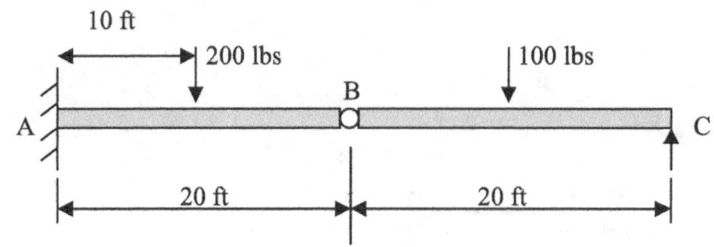

Solution:

STEP 1:   Draw the force diagram;

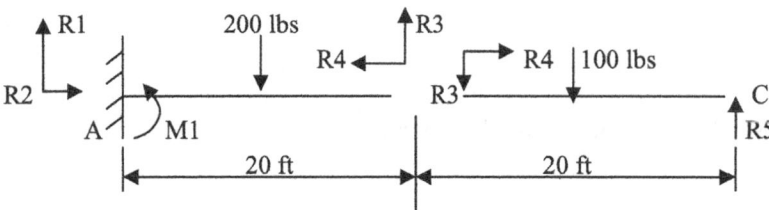

STEP 2: Is the structure statically determinate?

Number of forces = 6 (R1, R2, M1, R3, R4, R5)
Number of members = 2
R - 3. N = 6 - 3 x 2 = 0
The structure is statically determinate.

STEP 3:   Conduct stability analysis for beam AB:
Resolve forces vertically for beam AB;
R1 + R3 = 200   ---------------------------------(1)
Resolve forces horizontally;
R2 = R4   --------------------------------(2)
Take moments around point A;
M1 + R3 x 20 = 200 x 10 --------------------(3)

STEP 4:   Conduct stability analysis for beam BC:
Resolve forces vertically;
R5 = R3 + 100 ----------------------------------(4)
Resolve forces horizontally;
R4 = 0   -------------------------------------(5)
Hence from equation (2);   R2 = 0

Take moments around point C;
R3 x 20 + 100 x 10 = 0
R3 = -1,000/20 = -50 lbs
Negative value indicates that the actual direction of the force is opposite to the assumed direction.

Now we can find R5 using equation (4).
R5 = -50 + 100 = 50 lbs

Now we can find M1 using equation (3).

M1 + R3 x 20 = 200 x 10 --------------------(3)
M1 + (-50 x 20) = 2000
M1 = 3,000 lbs. ft
Note that when R3 is negative for beam BC, R3 is negative for beam AB as well.
Use equation (1) to find R1.
R1 + R3 = 200
R1 + (-50) = 200
R1 = 250 lbs.

Practice Problem: Two beams are connected using a pin joint as shown. 200 lb load is acting 60^0 to the horizontal. 100 lb load is acting vertically. Both loads are acting at the center of each beam. Find all the reactions.

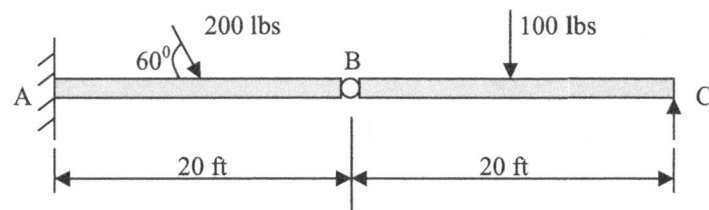

Solution:
STEP 1:   Draw the force diagram;

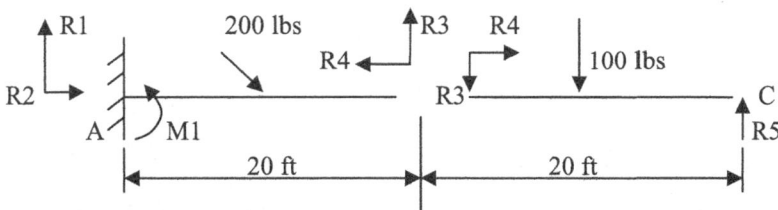

STEP 2: Is the structure statically determinate?

Number of forces = 6 (R1, R2, M1, R3, R4, R5)
Number of members = 2
R - 3. N = 6 - 3 x 2 = 0
The structure is statically determinate.

STEP 3: Conduct stability analysis for beam AB:
Resolve forces vertically for beam AB;
R1 + R3 = 200 sin 60 = 173.2   --------------------------------(1)
Resolve forces horizontally;
R2 + 200 cos 60 = R4
R2 + 100 = R4   ---(2)

Take moments around point A;
M1 + R3 x 20 = 200 sin 60 x 10
M1 + R3 x 20 = 1,732.1 --(3)

STEP 4: Conduct stability analysis for beam BC:
Resolve forces vertically;

R5 = R3 + 100 -----------------------------------(4)
Resolve forces horizontally;
R4 = 0   --------------------------------------(5)

Take moments around point C;
R3 x 20 + 100 x 10 = 0
R3 = -1,000/20 = -50 lbs

Now we can find R5 using equation (4).
R5 = -50 + 100 = 50 lbs

Now we can find M1 using equation (3).
M1 + R3 x 20 = 1,732.1 ---------------------(3)
M1 + (-50 x 20) = 1,732.1
M1 = 2,732.1 lbs. ft
Note that when R3 is negative for beam BC, R3 is negative for beam AB as well.
Use equation (1) to find R1.
R1 + R3 = 173.2
R1 + (-50) = 173.2
R1 = 223.2 lbs.

4.2.5   Determinacy of Trusses:
Trusses are common structures.   Trusses are mostly used for bridges and roofs.

Bridge truss;

Following equation is used to find the internal and external determinacy of trusses.

$$b + r = 2.j \quad \text{(Truss is statically determinate internally and externally)}$$

$$b + r > 2.j \quad \text{(Truss is statically indeterminate)}$$

b = Number of members
r = Total number of external support reactions
j = Number of joints

Example 1:   Is the truss shown statically determinate internally and externally?

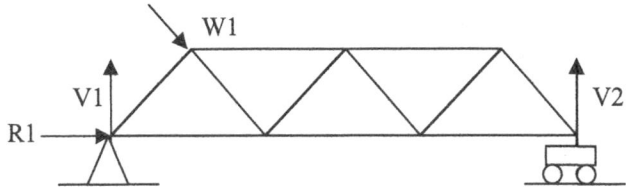

b = Number of members = 11
There are 3 horizontal members at the bottom and 2 horizontal members at the top. In addition, there are 6 angled members.
r = Total number of external support reactions = 3   (R1, V1, V2)
j = Number of joints = 7

b + r = 11 + 3 = 14
2.j = 2. 7 = 14
Hence;   b + r = 2.j
The truss is statically determinate.

Example 2:   Is the truss shown below statically determinate internally and externally?

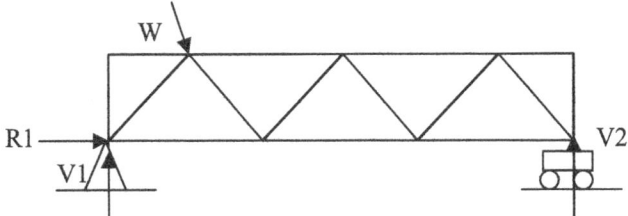

b = Number of members = 15
There are 3 horizontal members at the bottom and 4 horizontal members at the top. There are 6 angled members and 2 vertical members.
r = Total number of external support reactions = 3   (R1, V1, V2)
j = Number of joints = 9

b + r = 15 + 3 = 18
2.j = 2. 9 = 18
Hence;

b + r = 2.j

The truss is statically determinate internally and externally.

4.3   Analysis of Trusses

4.3.1   Method of Joints:

There are two methods to find member forces in trusses. They are

- Method of joints
- Method of sections

First let us study the method of joints. If a structure is stable then all its joints are stable. The forces at the joint should add to zero. The method is better explained using examples.

Practice Problem:

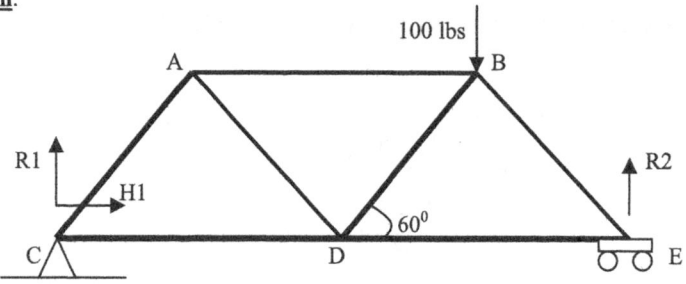

Truss triangles are equilateral triangles. (All legs of the triangles are the same).

CD = DE = 20 ft

Find the member forces of the truss shown.

Solution: STEP 1:   Find external reactions (R1, H1 and R2).

Resolve all external forces horizontally;   H1 = 0   -----------------(1)

Resolve all external forces vertically;   R1 + R2 = 100   -----------------(2)

Take moments around point C;   R2 x CE = 100 x (CD + DB Cos 60) ---------(3)

CD = DE = 20.   Hence CE =40

All triangles have equal length legs. Hence DB = 20 ft

Equation (3) can be rewritten.

R2 x 40 = 100 x (20 + 20x Cos 60) ----------(3)

R2 = 100 x 30/40 = 75

R1 = 25

STEP 2:   Assume member forces as shown below. If the force is in the opposite direction, the value would turn out to be negative.

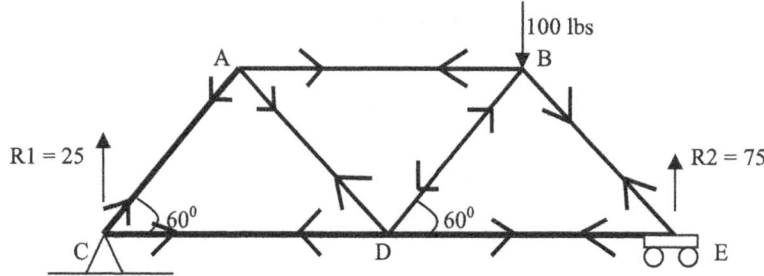

Above member force directions are assumed.

STEP 3:   Resolve forces at joint C;
Resolve forces at joint C, vertically;   25 + CA sin 60 = 0
                                           CA = -25/sin 60 = -28.9

Above minus sign indicates that the real force direction is opposite of the assumed direction.

Resolve forces at joint C, horizontally;   CA cos 60 + CD = 0
                                           -28.9 x 0.5 + CD = 0
                                           CD = 14.5
CD came out to be positive. Hence assumed direction is correct.
Member CD is in tension and member AC is in compression.

Tension   (If the member is pulling the node, then the member is in tension)

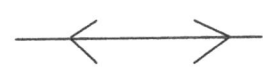

Compression   (If the member is pushing the node, then the member is in
                 compression)

STEP 4:   Resolve forces at joint A;
Resolve forces at joint A, vertically;   AC sin 60 + AD sin60 = 0
Sin 60 cancels off and we get   AC = -AD
                                           CA = AC = - 28.9   (Found in step 3)
                                           Hence AD = 28.9
Resolve forces at joint A, horizontally;   AD cos 60 + AB = AC cos 60
                                           28,9 x 0.5 + AB = -28.9 x 0.5
                                           AB = -28.9

The direction of AB member force is opposite of what is assumed. AB is in compression.

STEP 5:   Resolve forces at joint B;
Resolve forces at joint B, vertically;   BE cos 30 - BD cos 30 + 100 = 0 --------------(A)

Resolve forces at joint B, horizontally;   BA - BD cos 60 - BE cos 60 = 0
                                                   AB = BA = -28.9   (Found in step 4)
Hence;   -28.9 - 0.5 BD - 0.5 BE = 0   -------------------(B)
                                                   BE = 2 x (-28.9 - 0.5 BD)
                                                   BE = -57.8 - BD --------------------------------(C)
Include the BE value in equation (A)

                                                   [-57.8 - BD] cos 30 - BD cos 30 + 100 = 0
                                                   -50.1 – BD cos 30 – BD. cos 30 + 100 = 0
                                                   -50.1 – 2. BD cos 30 + 100 = 0
                                                   -50.1 + 100 = 2 x BD cos 30
                                                   BD = 28.9
Hence from above equation (C)   BE = -57.8 - 28.9 = -86.7
                                                   (Note that BE is compressive)

STEP 6:   Resolve forces at joint D;
Resolve forces at joint D, vertically;   DB sin 60 - DA sin 60 = 0
                                                   DB = DA
                                                   AD = DA = 28.9   (Step 4)
                                                   Hence DB = 28.9

Resolve forces at joint D, horizontally;   DE - DC - DB cos 60 - DA cos 60 = 0
CD = DC = 14.5   (Step 3)
BD = DB = 28.9   (Step 6)
DA = 28.9 (Step 6)
Hence we can write;   DE - DC - DB cos 60 - DA cos 60 = 0
                                                  DE - 14.5 - 28.9 cos 60 - 28.9 cos 60 = 0
                                                  DE = 14.5 + 28.9 = 43.4

STEP 7:   Resolve forces at joint E;
Resolve forces at joint E, vertically;   EB sin 60 + 75 = 0
                                                  BE = EB = -86.7   (Step 5)
                                                  -86.7 x sin 60 + 75 = 0
                                                  -86.7 x 0.866 + 75 = 0   (True)
All the forces at joint E are known. Hence we can use joint E as a check for the computations.

Resolve forces at joint E, horizontally;   ED + EB cos 60 = 0
                                                  DE = ED = 43.4   (Step 6)
                                                  BE = EB = -86.7   (Step 5)
Hence;   43.4 - 86.7 cos 30 = 0
                                                  43.4 - 86.7 x 0.5 = 0   (True)
The values checks out. Hence the calculations are correct.

4.3.2 Method of Sections:

Method of sections is another way to analyze trusses. In this method, the truss is cut thru and stability equations are applied. Let us do the previous problem using method of sections.

Example: Find the force in member BE using method of sections. Length of all members are 20 ft.

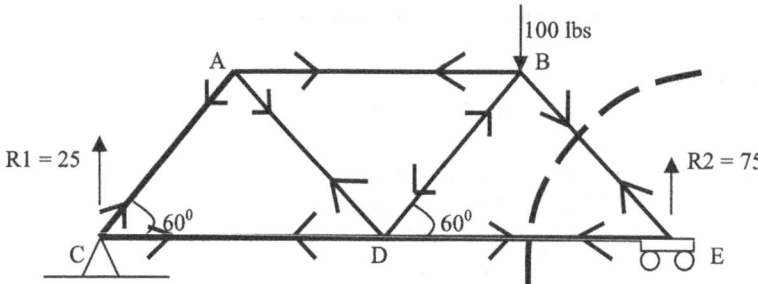

Solution: You are required to find the force in member BE. Hence the truss has to be cut thru BE.

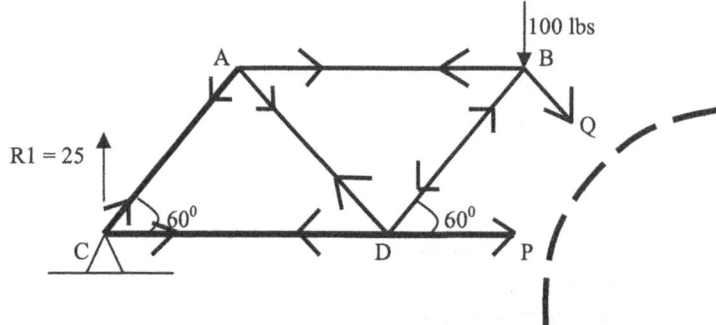

When you cut thru a section as shown, two member forces will be revealed. Let us call them P and Q. (See the figure above). Now let us take moments around point C. To make computations easier, let us substitute member force Q with its horizontal component and vertical component as shown below. Force "Q" is replaced with X and Y.

.

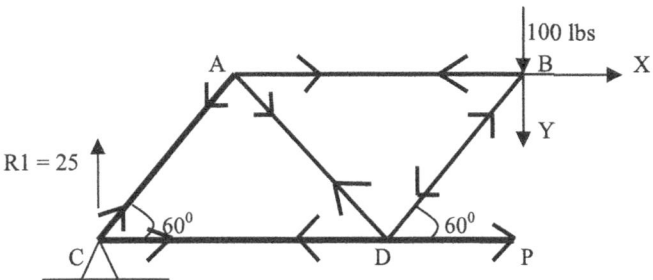

Take moments around point C;

X . AC sin 60 + Y (CD + DB cos 60) + 100 (CD + DB cos 60) = 0

AC = CD = DB = 20 ft (Given)
X. 20 sin 60 + Y (20 x 20 cos 60) + 100 (20 + 20 cos 60) = 0
17.32 X + 30 Y + 3000 = 0 ----------------------------(1)

Resolve forces vertically
25 - 100 - Y = 0
Y = -75
Substitute Y in above equation (1)
17.32 X + 30 x (-75) + 3,000 = 0
X = -43.3
X and Y are negative. X and Y act opposite of what is assumed. Assumed member force was tension. (Look at the first figure of this example). Hence actual member force of BE is compressive.

Next we need to find the value of the member force BE

Member force BE = $[X^2 + Y^2]^{1/2}$
Member force BE = $[-43.3^2 + -75^2]^{1/2}$
BE = 86.7 (compressive)
Compare the answer with the answer we got using method of joints.

4.4 Mechanics of Materials

Introduction: Mechanics of materials can be divided into two parts. They are

- Statics – Deals with static objects and forces
- Dynamics – Deals with moving objects and forces

Statics: It may have been a while that you have performed any force computations. Hence it is important to look at some of the fundamentals in force computations.

Forces:

Consider the force shown below.

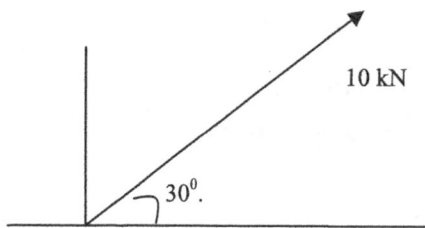

Find the components in horizontal and vertical directions.

Horizontal component = 10 x cos (30^0) = 10 x 0.866 = 8.66 kN
Vertical component = 10 x sin (30^0) = 10 x 0.5 = 5 kN

Resultant Forces: When two or more forces acting on a body, these forces can be resolved into one resultant force.

Parallelogram of Forces: When two forces are acting at an angle θ, they can be resolved into one force using parallelogram.

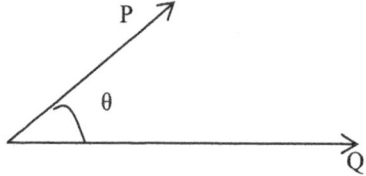

Draw a parallelogram as shown.

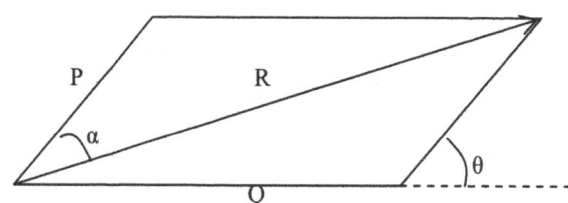

The angle θ is the angle between two forces P and Q.
Resultant R can be found by drawing the parallelogram as shown above. Resultant force R is given by the diagonal. From geometry it can be shown that

$$R^2 = P^2 + Q^2 + 2PQ \, \text{Cos} \, \theta$$

Hence

$$R = (P^2 + Q^2 + 2PQ \, \text{Cos} \, \theta)^{1/2}$$

The angle α is given by

$$\text{Tan} \, \alpha = Q \, \text{Sin} \, \theta / (P + Q \, \text{Cos} \, \theta)$$

Practice Problem: Two forces P and Q are found to be 10 lbs and 22 lbs. The angle between them is 35^0. Find the resultant force and the angle of the resultant force.

Solution:

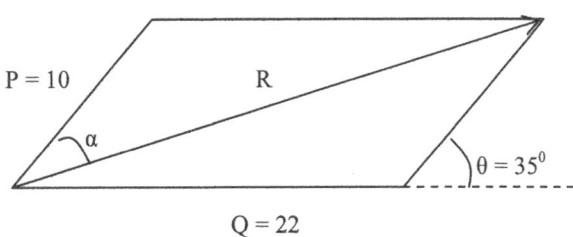

$R = (P^2 + Q^2 + 2PQ \cos\theta)^{1/2} = (10^2 + 22^2 + 2 \times 10 \times 22 \times \cos 35)^{1/2}$
$R = 30.7$ lbs

$\tan\alpha = Q \sin\theta/(P + Q \cos\theta) = 22 \sin 35/(10 + 22 \cos 35) = 0.45$
Angle $\alpha = \tan^{-1}(0.45) = 24.5^0$

Equilibrium of Forces: When there are more than one force, all forces can be resolved in two perpendicular directions. Typically forces are resolved in horizontal and vertical directions.

For equilibrium: $\Sigma H = 0$ and $\Sigma V = 0$

If the resolved forces are not zero, there would be a resultant force R.
R can be found using following equation.

$$R = (\Sigma H^2 + \Sigma V^2)^{1/2}$$

$$\tan\theta = \Sigma V/ \Sigma H$$

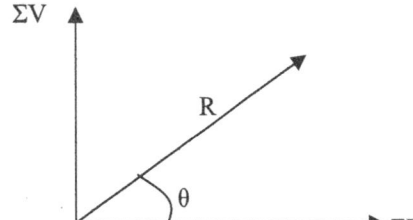

Practice Problem: Find the resultant force of given forces and the resultant angle to the horizontal.

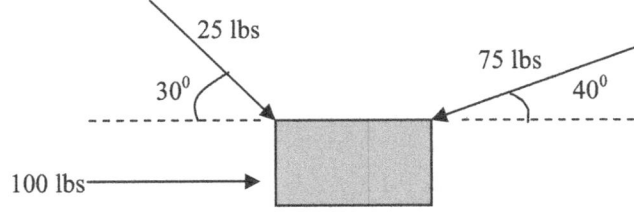

Solution: STEP 1:   Resolve forces in the horizontal direction.
$\Sigma H = 100 + 25 \cos 30 - 75 \cos 40 = 100 + 25 \times 0.866 - 75 \times 0.766 = 64.2$ lbs
$\Sigma V = 25 \times \sin 30 + 75 \sin 40 = 25 \times 0.5 + 75 \times 0.643 = 60.7$
$R = (\Sigma H^2 + \Sigma V^2)^{1/2}$
$R = (64.2^2 + 60.7^2)^{1/2} = 88.4$ lbs

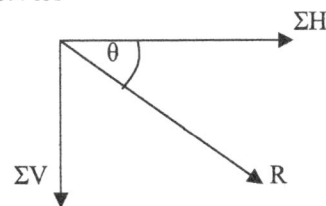

Tan θ = ΣV/ ΣH = 60.7/64.2 = 0.95

Angle θ = 43.5^0

Center of Gravity: Knowing the center of gravity is important for construction engineers. There is plenty of lifting, moving, loading and unloading work and construction engineers should know where the weight is concentrated. Center of gravity of some objects shown below.

Triangle:

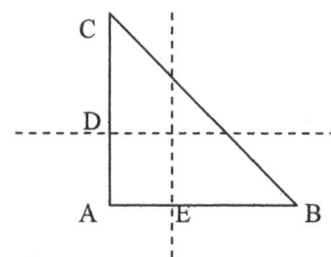

AD = AC/3 and AE = AB/3

CD = 2AC/3 and EB = 2AB/3

Semi Circle:

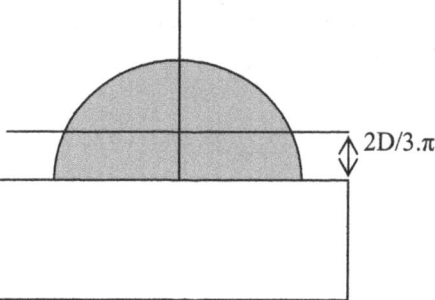

2D/3.π

Curve:

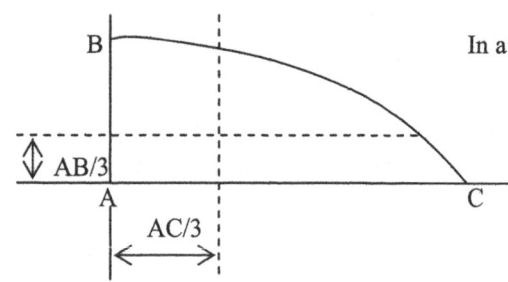

In a curve, center of gravity acts 1/3 of the leg.

Area of a curve = 2/3 x AB x AC

Cone:

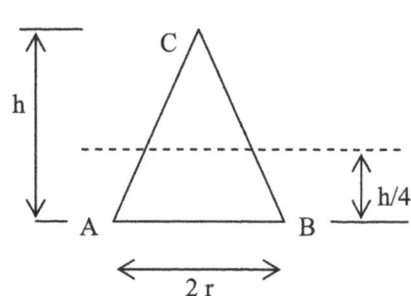

Area of a cone = 1/3. π r^2 h

Trapezium:

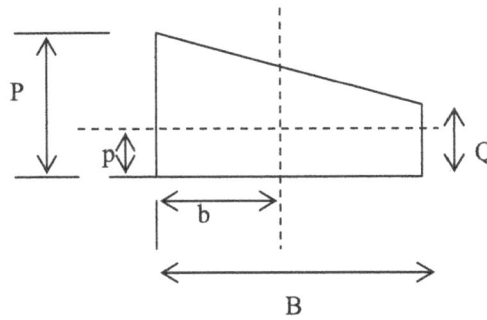

Area = ½ (P + Q).B

$p = (P^2 + PQ + Q^2)/3(P + Q)$

$b = B/3 \times (2Q + P)/(P + Q)$

Q is the shorter side and P is the longer side.
b is measured from the longer side P.

Center of Gravity of Multiple Figures: Center of gravity of many objects can be found using the following equation.

$X = (a1.x1 + a2.x2 + a3.x3…..)/(a1 + a2 + a3……)$

Similarly

$Y = (a.y1 + a2.y2 + a3.y3…….)/(a1 + a2 + a3……)$

a1, a2, a3 are areas of figures. x1, x2, x3 are distances to center of gravity of each object.

Practice Problem: Find the distances X and Y to the center of gravity of the composite figure.

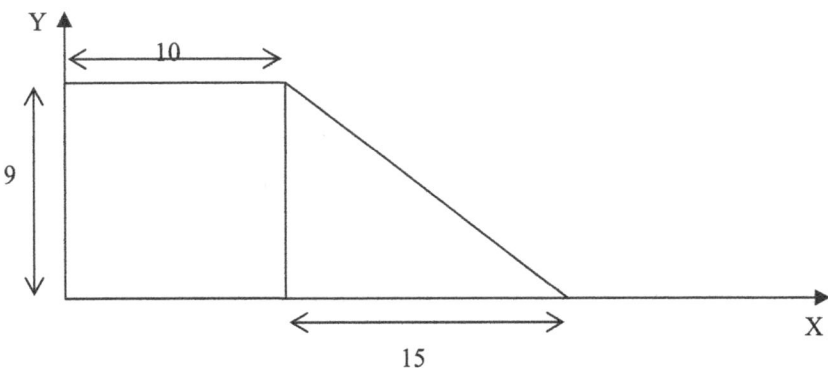

Area of the rectangle = a1 = 90
Area of the triangle = a2 = 15 x 9/2 = 67.5
Next we have to find x1 and x2 or distances to center of gravity of each object.

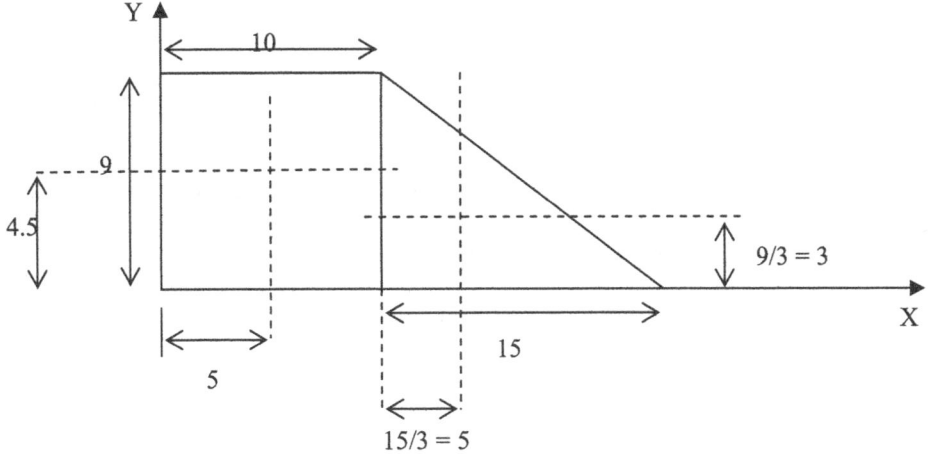

Distance to center of gravity from Y axis = X = (a1.x1 + a2.x2)/(a1 + a2) = [90 x 5 + 67.5 x (10 + 5)]/[90 + 67.5]
          X = 9.3

Distance to center of gravity from X axis = Y = (a1.y1 + a2.y2)/(a1 + a2) = [90 x 4.5 + 67.5 x 3]/[90 + 67.5]
                    Y = 3.86

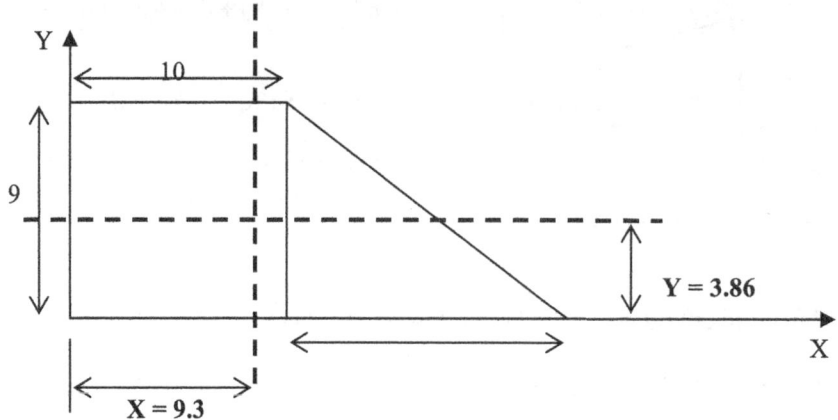

Note that X value is measured from the Y axis and Y value is measured from the X axis.

Moment: Moment of a force about a given point is obtained by multiplying the force with the perpendicular distance to that force.

Practice Problem: Find the moment of the force given from point A.

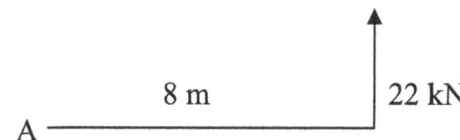

Moment around point A = 8 x 22 = 176 kN. m

Practice Problem: Similarly find the moment of forces around point A of the forces given below.

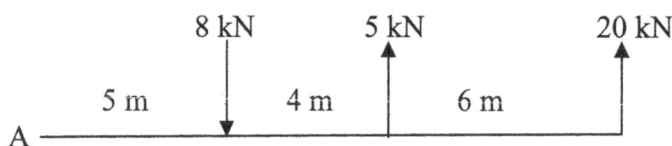

In the above figure, 8 kN acts downward while 5 kN and 20 kN act upwards. Assume downward moment to be positive and upward moment to be negative.
Moment around point A = (8 x 5) – 5 x (5 + 4) - 20 x (5 + 4 + 6) = -305 kN. m
Above negative sign indicates that the moment is acting upwards since we considered downward to be positive.

Uniform Loads: Weight of a beam is a good example for a uniform load.

Practice Problem: Assume the weight of the beam shown is 20 lbs per linear ft. Length of the beam is 15 ft.
Find
a)   Total weight of the beam
b)   Two reactions (P and Q)
c)   Moment around point A.

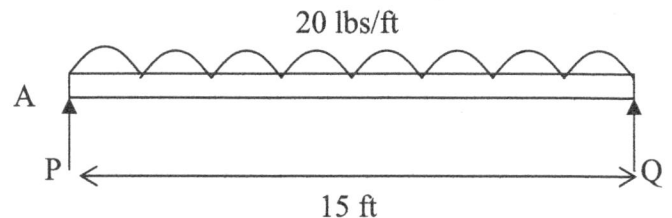

Solution:

a)   Total weight of the beam = 15 x 20 = 300 lbs.

b)   Two reactions:

Due to symmetry, two reactions P and Q has to be equal.

P + Q = 300 lbs.

Since P and Q are equal;   P = 150 and Q = 150

c)   Moment around point A;

The center of gravity of the uniform load act at the center of the beam.

Moment due to uniform load = 20 x 15 x (15/2) = 2,250 lbs. ft

Moment due to Q = -150 x 15 = -2,250

Moment due to Q is minus since it acts opposite to the uniform load.

Net moment around point A = 0

Practice Problem: Find the moment due to force T around point A. Force T is given to be 120 kN.

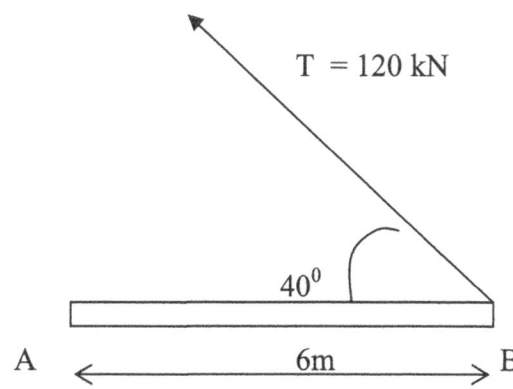

Solution: Draw a perpendicular to force T from point A.

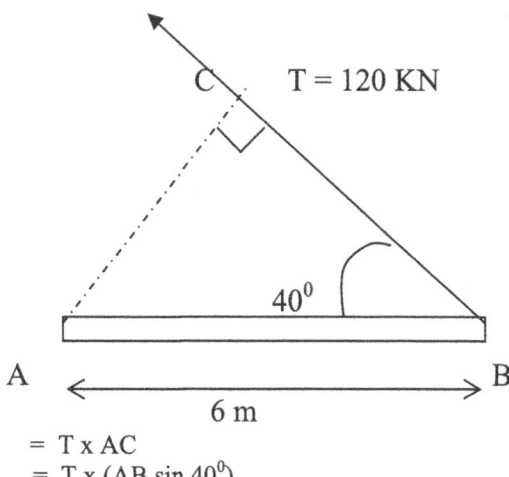

Moment around point A   = T x AC

          = T x (AB sin 40^0)

          = 120 x 6 x sin 40^0 = 462.8 kN. m

Practice Problem: A post is stabilized by guy wires as shown. Following information is given.

P = 1,200 lbs, $\alpha = 60^0$, $\beta = 100^0$,   Find horizontal components of Q and R.

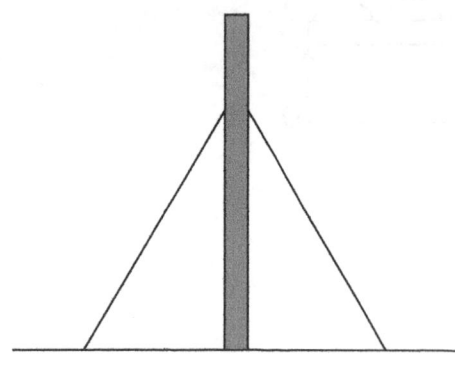
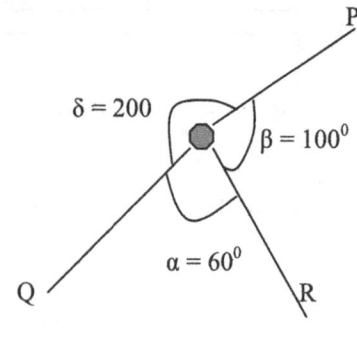

$\delta = 360^0 - (60 + 100) = 200^0$

Following equation can be used. This equation is known as the **"Sine Rule"** and widely used.

$$\frac{P}{\sin \alpha} = \frac{Q}{\sin \beta} = \frac{R}{\sin \delta}$$

$$\frac{1,200}{\sin 60} = \frac{Q}{\sin 100} = \frac{R}{\sin 200}$$

$$\frac{1,200}{\sin 60} = \frac{Q}{\sin 100}$$

Q = 1,200 x Sin 100/Sin 60 = 1,365 lbs

$$\frac{1,200}{\sin 60} = \frac{R}{\sin 200}$$

R = 1,200 x Sin 200/Sin 60 = - 474 lbs

4.5 Bending Moment and Shear:

Shear stress and bending stress are two different types of stresses. Let us imagine a simply supported beam.

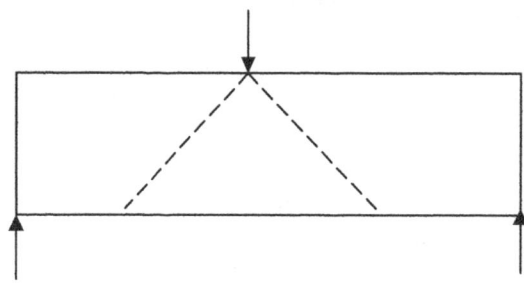

When a beam is loaded, the load is transferred to the beam as shown above figure. When the load is increased, cracks start to appear along the load path.

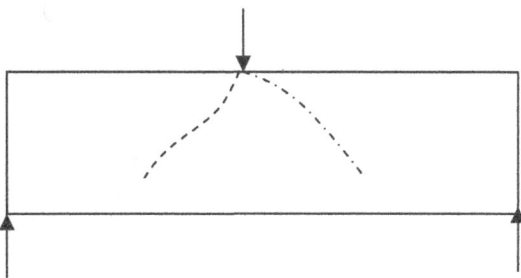

These types of cracks are known as shear cracks. Shear cracks also appear near the supports, since support reaction has to be transferred to the beam.

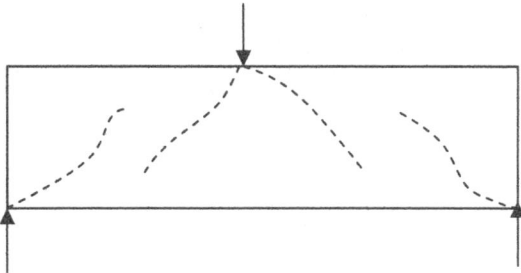

Stirrups are provided to protect beams from failing due to shear. Stirrups are vertical rebars.

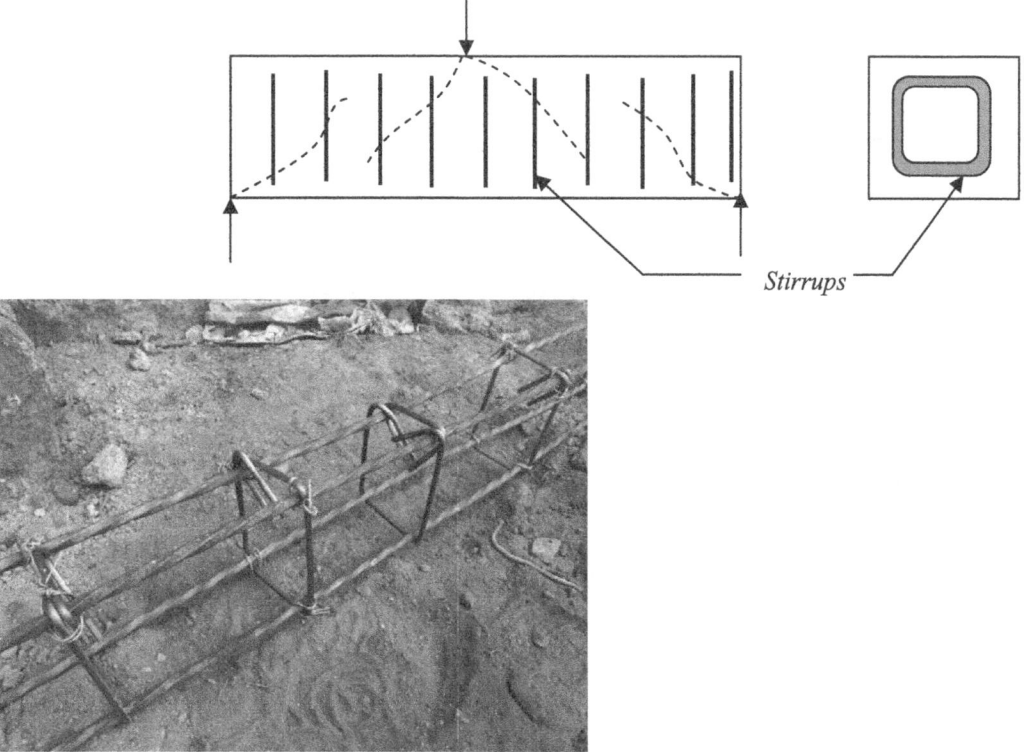

Stirrups

Beam stirrups

Vertical stirrups would increase the resistance against shear failure.

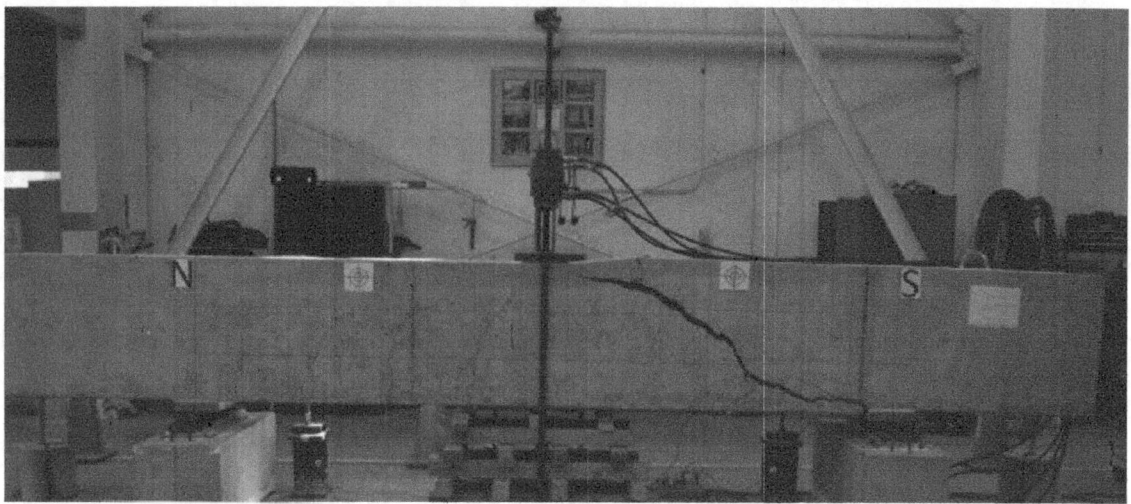

Shear crack of a beam

I would like to point out that shear stress is different from bending stress. Let us look at a beam again. When a load is applied, beams tend to bend. When a beam bends, top side would be under compression and the bottom side would be under tension.

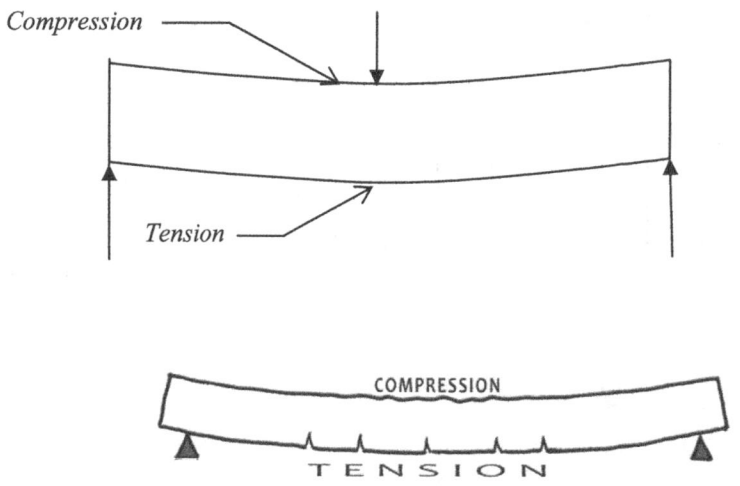

Due to tension, cracks would open up as shown.

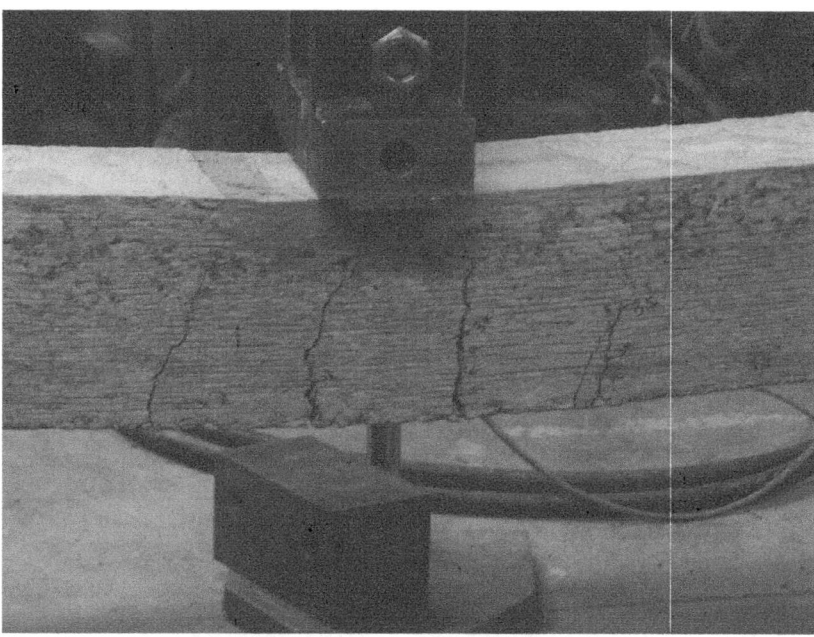

Cracks due to bending stress

Cracks due to bending stress (Note the difference between shear cracks and bending cracks)

Longitudinal reinforcements are provided to resist bending failure. Bottom bars are known as tension reinforcements. When compression is too high, rebars are provided on top of the beam as well. Top bars would resist against compression failure and known as compression reinforcements.

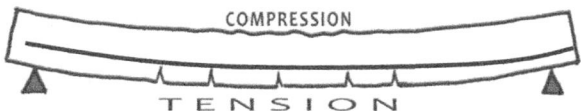

Bottom rebars are installed to resist bending stresses

Top bars, bottom bars and stirrups

Above beam shows bottom tension bars, top compression bars and shear reinforcements or stirrups.

4.6 Shear Diagrams:

To draw shear force diagrams, shear force along the beam need to be calculated. Let us look at a simply supported beam loaded at the middle. If the load is 30 lbs, support reactions would be 15 and 15.

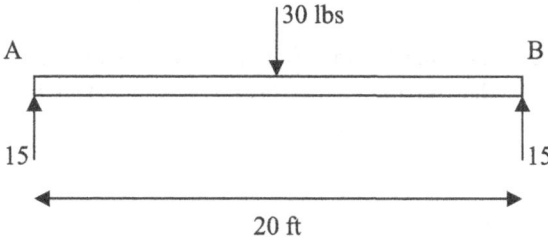

Now let us cut a section at a point 7 ft away from the support.

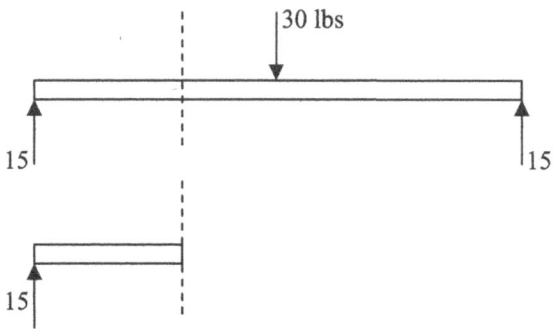

There is a 15 lb support reaction. This support reaction has to be balanced by a shear force at the cut section. Let us call the shear force at the section "V".

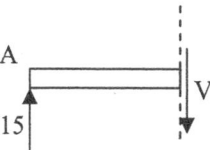

By resolving forces vertically we get;

15 - V = 0;   Hence V = 15   (Acting downward. Assumed direction is correct)

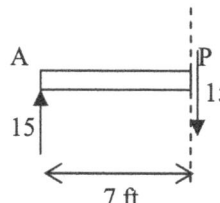

Still the section is not balanced. If you take moments around point P, there is an unbalanced moment.
Moments around point P = 7 x 15 = 105 ft. lbs

This moment has to be balanced. Hence a bending moment has to be included at point P.

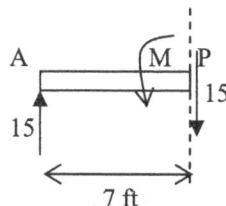

Bending moment at point P = 105 ft. lbs
Now we have found the bending moment and shear force at point P.

Now let us find the shear force and bending moment at another point. Let us find the bending moment and shear force at a point 12 ft away from point A.

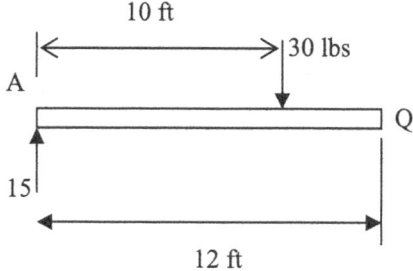

Now we know that when you cut a section, there is a shear force and a bending moment. Let us call the shear force V and bending moment M.

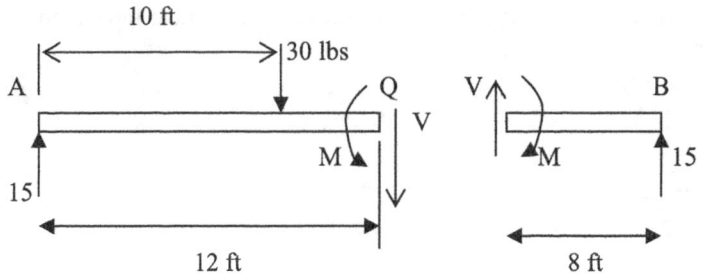

Left section of the beam;

By resolving forces vertically we get;

15 - 30 - V = 0

V = -15

Hence the shear force is acting opposite of assumed direction. The shear force is acting *upward* at this point. Note that at point P, the shear force was acting *downward*. (Computed earlier).

Now let us take moment around point Q;

15 x 12 - 30 x 2 - M = 0

M = 120 ft. lbs

Right section of the beam: Now we can do the same to the right section of the beam. You will find answers that you get for V and M is the same for the section on right of the cut location.

Resolve forces vertically;

15 + V = 0

V = -15

Take moments around point Q;

15 x 8 - M = 0

M = 120 ft. lbs

As you could see both sections give the same answer. It is important to pick the easier section to do the computations.

Easier Way to Draw Shear Force Diagrams: One can find the shear force in few places of a beam as shown above and draw a shear force diagram. There is a much simpler way.

Now let us draw the shear force diagram for the beam shown below;

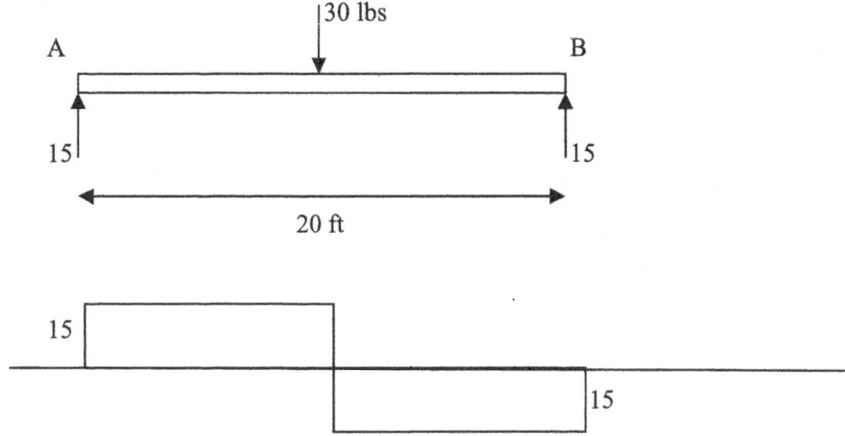

- At point A, the shear force goes up by 15 lbs.
- Then nothing happens till midpoint. Hence the shear force stays same.
- At midpoint shear force goes down by 30 lbs.

- Then nothing happens till the next end.
- At point B, the shear force goes up by 15.

Example: Draw the shear force diagram for the beam shown below;

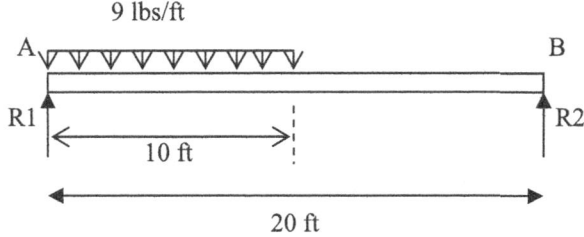

Solution: STEP 1:   Find support reactions, R1 and R2

Resolve forces vertically;   R1 + R2 = 9 x 10 = 90

Take moments around point A;   R2 x 20 = (9 x 10) x 5
The center of gravity of the uniformly distributed load is 5 ft from point A;
R2 = 22.5;   R1 = 67.5

STEP 2:   Draw the shear force diagram;

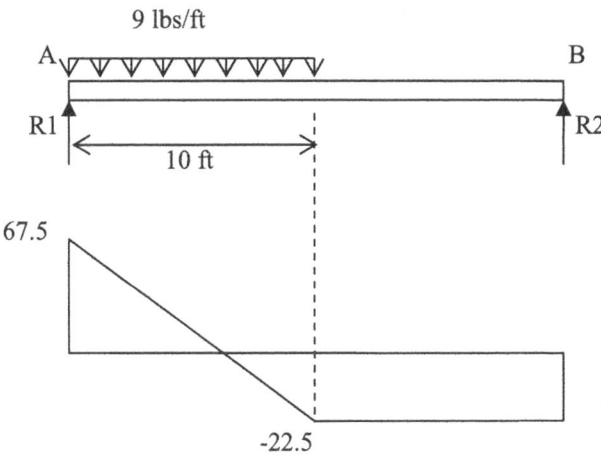

At point A, support reaction is 67.5. Hence shear force goes up to 67.5.
Then due to uniform load, shear force comes down. Total uniform load is 90 lbs. (9 x 10).
Hence the shear force at mid point is 67.5 - 90 = -22.5.
After the midpoint, there are no loads. Hence the shear force stays the same.
At other corner, shear force goes up by 22.5 due to support reaction.

Example: Draw the shear force diagram for the cantilevered beam shown below;

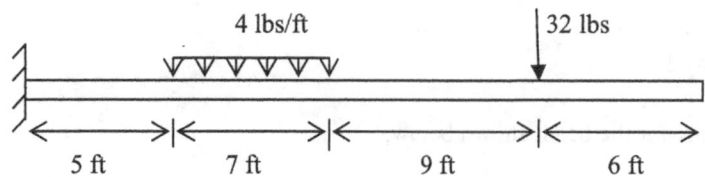

Solution:

STEP 1:   Find support reactions;

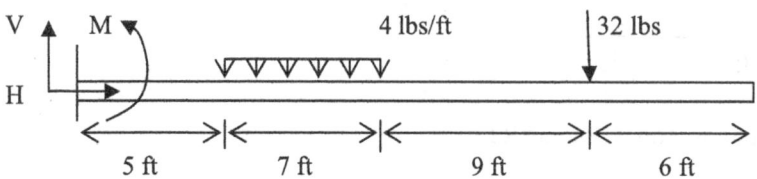

There are three support reactions at the fixed end;

They are V, H and M.

Since there are no horizontal forces, H = 0

By resolving forces vertically;

$\qquad$ V = (4 x 7) + 32 = 60 lbs

By taking moment around fixed end;

$\qquad$ M = (4 x 7) x (5 + 7/2) + 32 x (5 + 7 + 9)

Above (4 x 7) is the total load due to uniform load of 4 lbs/ft. (5 + 7/2) is the distance to the center of gravity of the uniform load.

$\qquad$ M = 238 + 672 = 910 ft. lbs

Note that we do not need the moment to draw the shear force diagram.

STEP 2:   Draw the shear force diagram;

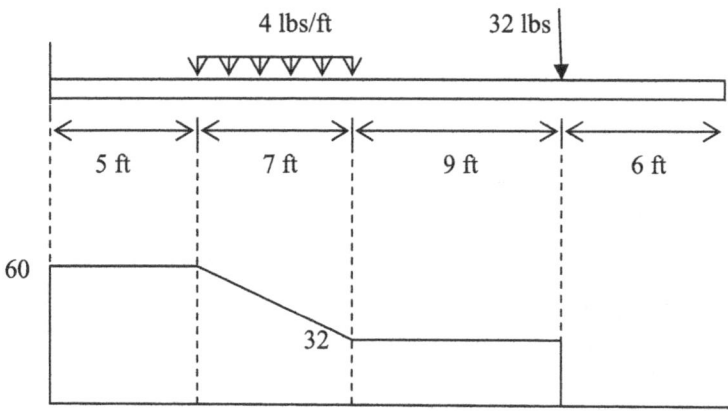

- At the fixed end, the shear force goes up by 60 lbs due to support reaction.
- Then the shear force stays same till the uniform load.
- At the uniform load, shear force would start to come down at a rate of 4 lbs/ft. The uniform load continues to a length of 7 ft. Total uniform load is 28 lbs.
- At the end of the uniform load, the shear force is 32 lbs. (60 - 7 x 4 = 32)

- Then the shear force stays the same till the concentrated load of 32 lbs. At the concentrated load, the shear force drops down to zero.

Example: Draw the shear force diagram for the beam shown below;

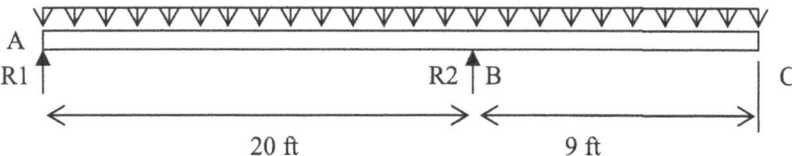

Solution:

STEP 1:   Find support reactions;

Resolve forces vertically;   R1 + R2 = 30 x 29 = 870

Take moments around point A;   R2 x 20 = (30 x 29) x 29/2

Above (30 x 29) is the total load and 29/2 is the distance to the center of gravity from point A.

R2 = 630.75;   R1 = 239.25

STEP 2:   Draw the shear force diagram;

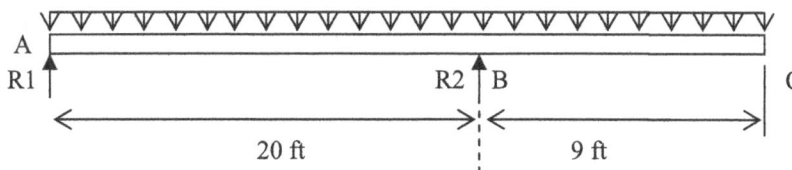

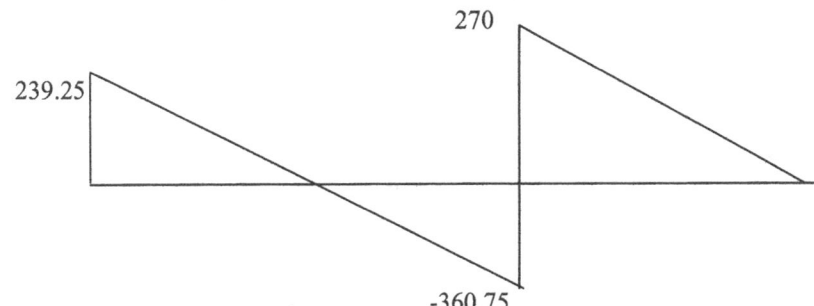

- Let us start from support A. Support reaction at A is 239.25. Hence shear force goes up by 239.25
- Then due to uniform load, shear force start to come down at a rate of 30 lbs/ft.
- Shear force at point B = 239.25 - 30 x 20 = -360.75
- At point B there is a support. At the support shear force goes up by the value of the support reaction.
- Shear force at point B = -360.75 + R2 = -360.75 + 630.75 = 270
- After the support at point B, shear force start to come down.

4.7 Bending Moment Diagrams:

The sign of shear force is not important. There is no reason to know whether the shear is acting upwards or downwards. Typical stirrups would resist both shears. This is not true in the case of bending moment. Look at the two beams shown below.

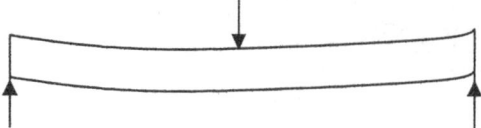

In the above case, the bottom of the beam is under tension. Hence tension rebars have to be placed at the bottom.

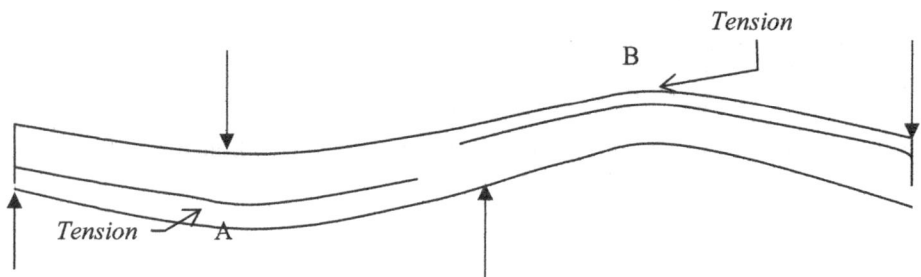

In the above beam, bottom fibers at point A is under tension. Hence tension rebars should be placed at the bottom of the beam. On the other hand, top fibers at point B is under tension. Hence tension rebars should be placed at the top. At point B, bottom fibers are under compression.

Due to this reason, it is important to know the sign of the bending moment.

Sign Convention: Different text books use different sign conventions. In this book I have used sagging moment to be positive. Moment that would create sagging is considered to be positive. When there is positive moment, bottom fibers would be under tension.

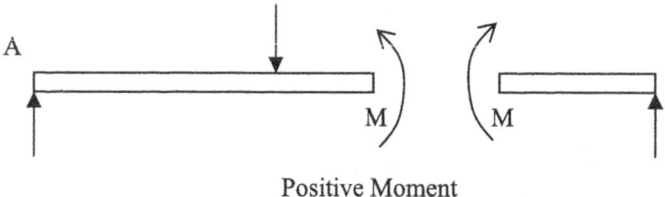

Positive Moment

Above shown moment would put the bottom fibers under tension and top fibers under compression.
Let us do an example;

Example: Find the bending moment at point C, 10 ft from point A. If this is a concrete beam, tension bars should be placed on top or bottom?

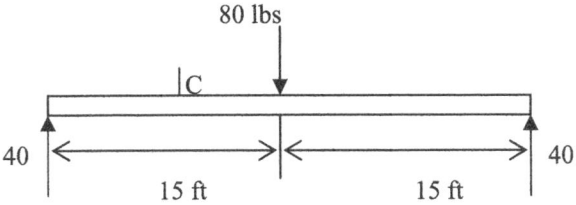

Solution: Support reactions are 40 lbs on each ends.

STEP 1:   Cut a section at point C;

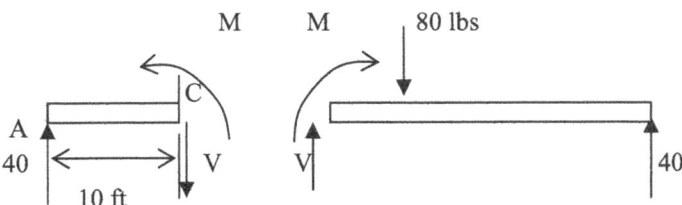

Now you can take moments around point A or C.

Let us take moments around point C, so that shear force V, do not get involved.

40 x 10 = M
M = 400   (This value is positive).
Hence at point C, tension bars should be placed at the bottom of the beam.
Now I would take moments around point A just to illustrate that both points would give same moment values.

Take moments around point A;

        V x 10 = M;
By resolving forces, you can see V = 40.
Hence M = 400   (positive)

Example: Draw the bending moment diagram for the beam given in the previous problem;

Solution:
STEP 1:   Cut a section at point "x" distance from point A;

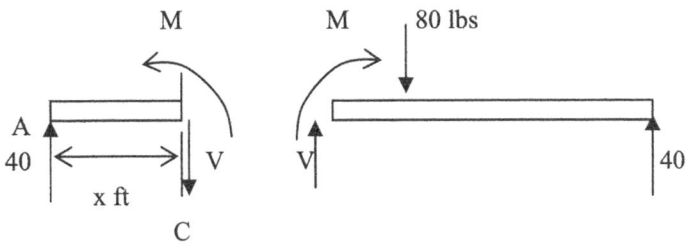

Take moments around point C;

M = 40. x
When x = 0; M = 0

When x = 15, M = 600

Note that this equation is valid only till the 80 lb load. We can draw the bending moment diagram till the 80 lb load.

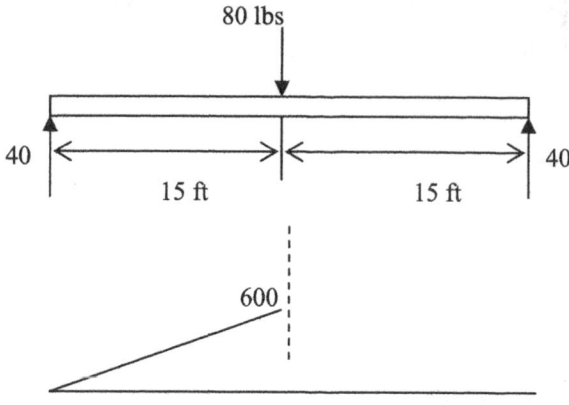

STEP 2:   Cut a section beyond 80 lb load;

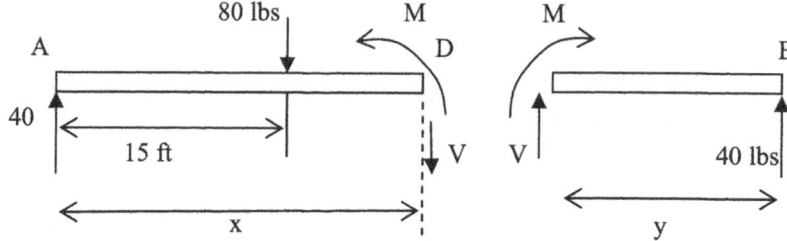

Many options available;
1) You can take moments around point A
2) You can take moments around point D for the section on left hand side
3) You can take moments around point D for the section on right hand side
Calculations would be easier if you take moments around point D for the section on right.
M = 40. y
When y = 0;   M = 0
When y = 15;   M = 600   (Note that M is positive.)

STEP 3:   Draw the bending moment diagram;

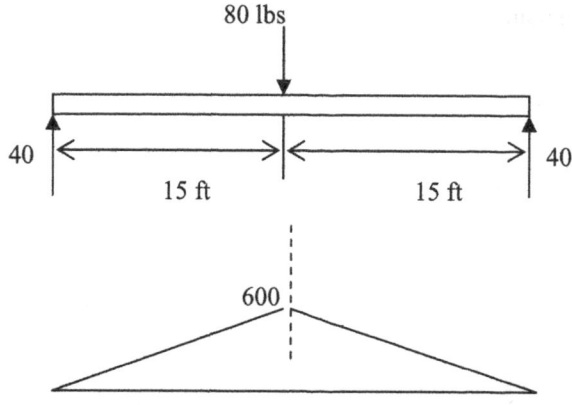

Example: Draw the bending moment diagram for the beam shown;

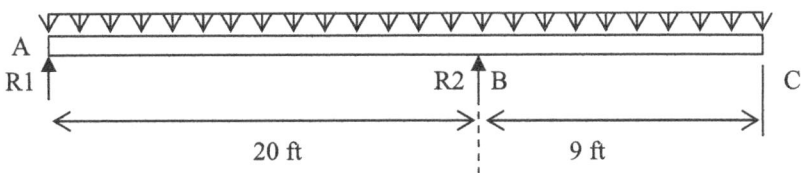

STEP 1:   Find support reactions;
Resolve forces vertically;   R1 + R2 = 30 x 29 = 870
Take moments around point A;   R2 x 20 = (30 x 29) x 29/2
Above (30 x 29) is the total load and 29/2 is the distance to the center of gravity from point A.
R2 = 630.75;   R1 = 239.25

STEP 2:   Cut a section at "x" distance from point A;

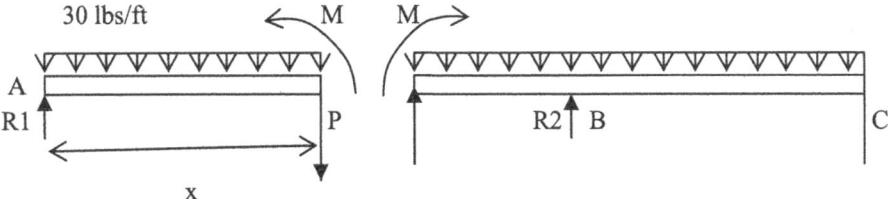

Take moments around point P;   M - R1. x - (30. x). (x/2) = 0
We found R1 = 239.25;
$M = 239.25. x - 30.x^2/2$
$M = 239.25. x - 15.x^2$ -------------------------------------(1)
Above equation is valid from point A to point B.
As per above equation, when   x = 0;   M = 0
                                        x = 5;   M = 821.25
                                        x = 10;   M = 892.5
                                        x = 15;   M = 213.75
                                        x = 20;   M = -1,215
This is a parabolic curve. It would look like as shown below.

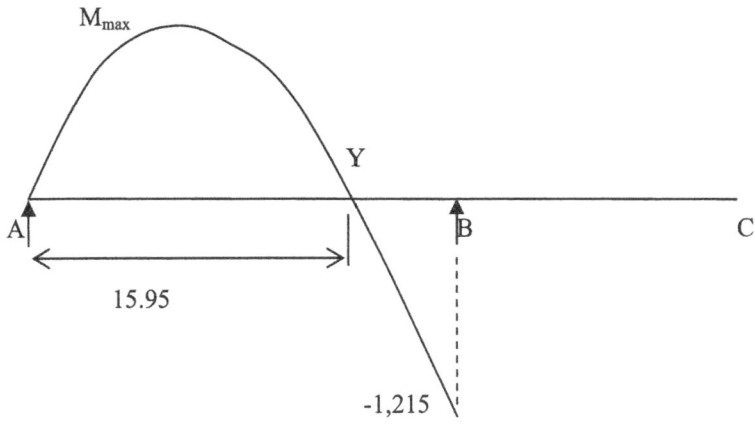

STEP 3:   How to find point Y, where bending moment becomes zero?

$M = 239.25. x - 15.x^2$

Make M = 0 and find "x".

$0 = 239.25 - 15. x$

$x = (239.25/15) = 15.95$

At 15.95 ft from point A, the bending moment becomes zero.

In addition, you may notice that beyond 15.95, bending moment becomes negative. That means according to our sign convention, tension occurs on the top of the beam.

STEP 4:   Find the maximum bending moment (M_{max});

This can be done by differentiating the equation (1) and making it zero.

$M = 239.25. x - 15.x^2$

$dM/dx = 239.25 - 30. x$

$0 = 239.25 - 30. x$

$x = 7.975$

The bending moment becomes a maximum at a point 7.975 ft from point A.

To find the maximum bending moment; insert 7.975 in equation (1)

$M = 239.25 \times 7.975 - 15 \times 7.975^2$

$M_{max} = 953.6$

STEP 5:   Find the bending moment from point B to C;

Cut a section between B and C;

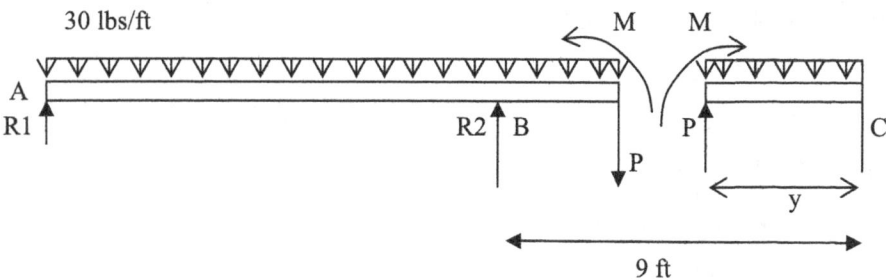

Take moments for the right hand section of the beam;

$M + 30. y \times (y/2) = 0$

$M = -15. y^2$

At point C,   y = 0 and M = 0

At point B,   y = 9 and M = -1,215

Now you can draw rest of the bending moment diagram.

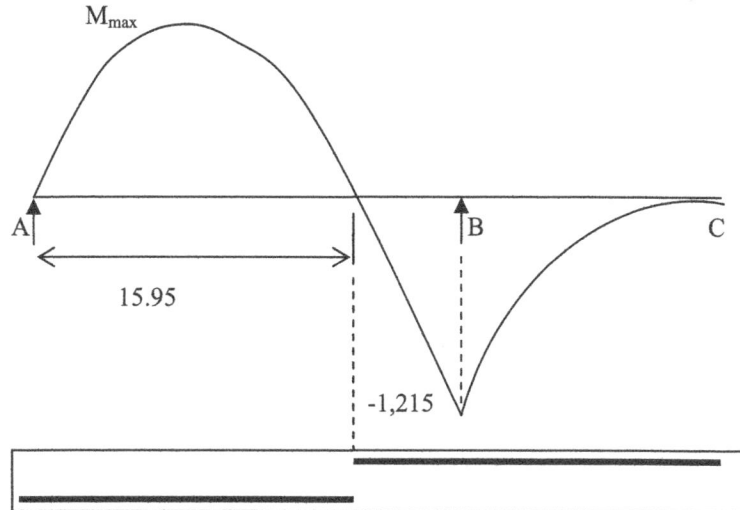

Note: As per our sign convention, rebars should be placed at the bottom of the beam where bending moment is positive. Rebars should be placed top of the beam where bending moment is negative. (Development length of rebars is not shown for clarity)

If you find this problem too difficult to understand, I would suggest to complete the rest of the chapter and come back to it later.

Example: Draw the bending moment diagram for the cantilevered beam shown;

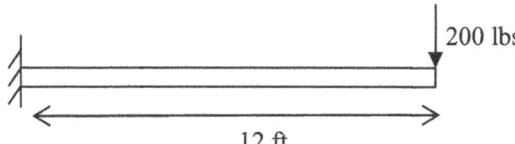

Solution: Cut a section "x" distance from the edge.

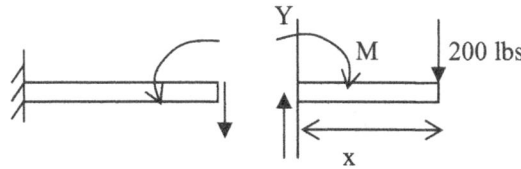

Take moments for the right hand section of the beam from point Y;

$M + 200. x = 0$

M is assumed to be positive. M shown would create tension in bottom of the beam.

$M = - 200. x$

Maximum M would occur at the fixed end. $M_{max} = 200 \times 12 = 2,400$

Hence the bending moment diagram will be as shown below.

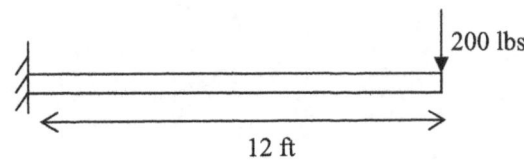

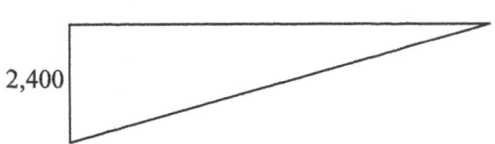

Since the bending moment is negative, rebars has to be placed on top of the beam.

Example: Draw the bending moment diagram for the cantilevered beam shown;

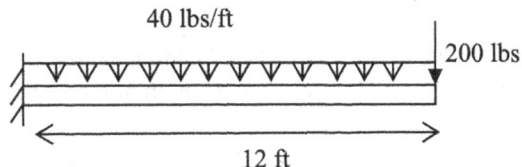

Solution: Cut a section "x" distance from the edge.

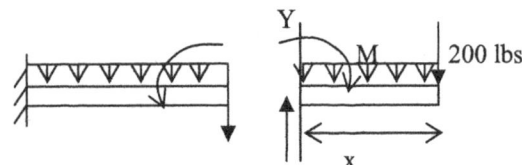

Take moments for the right hand section of the beam from point Y;

M + 200. x + 40.x (x/2) = 0
M + 200.x + 20. x^2 = 0
M is assumed to be positive. M shown would create tension in bottom of the beam.
M = - 200. x - 20. x^2
Maximum M would occur at the fixed end. M_{max} = -200 x 12 - 20. 12^2 = 5,280

Hence the bending moment diagram will be as shown below.

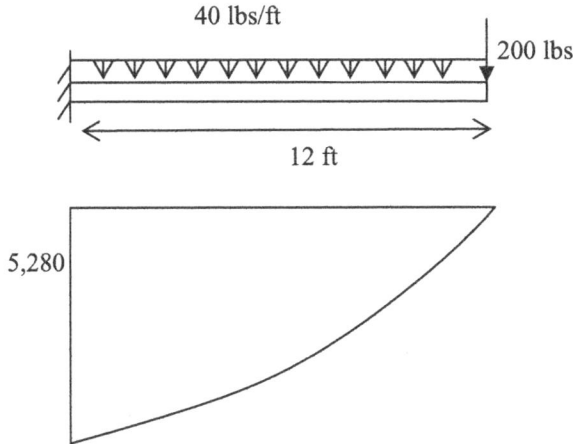

Since the bending moment is negative, rebars has to be placed on top of the beam.

4.8   Compression:

Columns are the main compression elements in a structure.

General equation for column buckling.

$$P_{cr} = \pi^2 . E.I/(K.L)^2 \text{------------------------}(1)$$

P_{cr} = Euler buckling load in lbs
E = Young's modulus in lbs/in^2
L = Length of the column in inches
I = Moment of inertia in in^4.
K = Effective length factor. (no units)
(K depends on end conditions. Following table for K is obtained from AISC Table C-C2.2.)

Buckled shape of column is shown by dashed line	(a)	(b)	(c)	(d)	(e)	(f)
Theoretical K value	0.5	0.7	1.0	1.0	2.0	2.0
Recommended design value when ideal conditions are approximated	0.65	0.80	1.2	1.0	2.10	2.0

End condition code		
		Rotation fixed and translation fixed
		Rotation free and translation fixed
		Rotation fixed and translation free
		Rotation free and translation free

	Theoretical Effective length factor (K)	Recommended design (K) factor
a) Both ends fixed	0.5	0.65
b) One end pinned, other end fixed	0.7	0.8
c) One end fixed other end rotation fixed but translation free	1.0	1.2
d) Both ends pinned	1.0	1.0
e) One end fixed other end free	2.0	2.10
f) One end pinned other end rotation fixed but	2.0	2.0

Moment of inertia (I) also can be expressed in terms of radius of gyration.

$I = A.r^2$

A = Cross sectional area and

r = Radius of gyration

Then the equation would be

$$P_{cr} = \pi^2. E.A.r^2/(K.L)^2$$

"r" can be taken under and written as below.

$$P_{cr} = \pi^2. E.A/(K.L/r)^2 ------------------------(2)$$

Equations (1) and (2) are identical.

Example: Find the Euler buckling load for the column shown. Both ends are pinned.

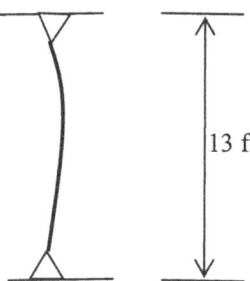

13 ft

E = Young's modulus of steel = 29 x 10^6 lbs/sq. in
I = 60 in^4.

Solution:

Convert all values to lbs and inches.

L = 13 ft = 13 x 12 in = 156 in.
K =1.0 when both ends are pinned. Hence K.L = 156 in.
From equation (1)
$\qquad$ $P_{cr} = \pi^2. E.I/(K.L)^2$
$\qquad$ $P_{cr} = (\pi^2. 29 \times 10^6. 60)/156^2$
$\qquad$ $P_{cr} = 7.06 \times 10^5$ lbs
$\qquad$ $P_{cr} = 706$ kips

Example: Find the Euler buckling load for the column shown. One end is pinned and other end fixed.

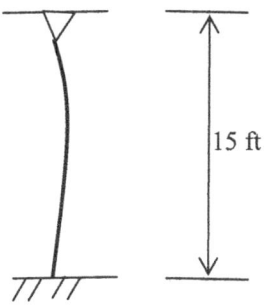

15 ft

E = Young's modulus of steel = 29 x 10^6 lbs/sq. in
r = Radius of gyration of the column = 3 in
Cross sectional area = 20 sq. in

Solution:

Convert all values to lbs. and inches.

L = 15 ft = 15 x 12 in = 180 in.
K =0.8 (Use recommended value)
Hence K.L = 144 in.
From equation (2)
$P_{cr} = \pi^2. E.A/(K.L/r)^2$
$P_{cr} = (\pi^2. 29 \times 10^6. 20)/(144/3)^2$
$P_{cr} = 2.484 \times 10^6$ lbs
$P_{cr} = 2,484$ kips

4.9   Deflection:

Following equations are available to determine the deflections in beams..

Case 1)

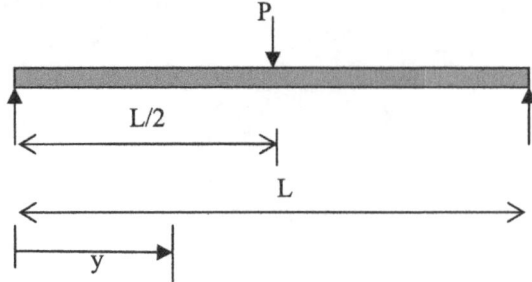

Deflection at the center of the beam (maximum deflection) = $P.L^3/(48E.I)$ $y = L/2$

Deflection at a point "y" from the edge = $P.y (3L^2 - 4.y^2)/(48. EI)$ $y < L/2$

E = Young's modulus;   I = Moment of inertia

Case 2)

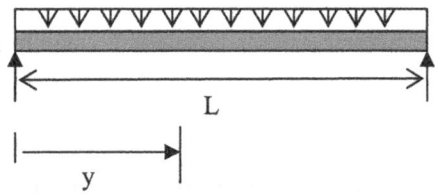

Deflection at the center of the beam (maximum deflection) = $5.w.L^4/(384E.I)$ $y = L/2$

Deflection at a point "y" from the edge = $w.y (L^3 - 2.L.y^2 + y^3)/(24. EI)$ $y < L/2$

Case 3):

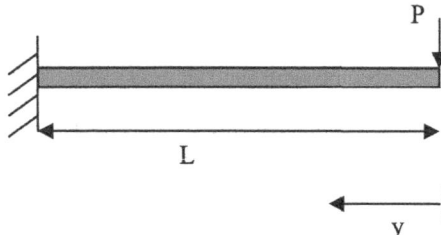

Note that "y" is measured from the free end. (Not from the fixed end).

Deflection at the end of the beam (maximum deflection) = $PL^3/(3E.I)$ $y = 0$

Deflection at a point "y" from the free end = $P (2.L^3 - 3. L^2.y + y^3)/(6. EI)$

Case 4):

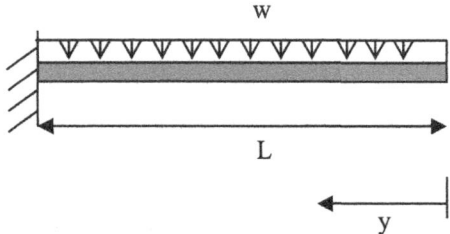

Note that "y" is measured from the free end. (Not from the fixed end).

Deflection at the end of the beam (maximum deflection) = $w.L^4/(8E.I)$; $y = 0$
Deflection at a point "y" from the fixed end = $w.(y^4 - 4.L^3.y + 3.L^4)/(24.EI)$

Practice Problem: A floor slab is supported by W21 x 47 simply supported beams as shown in the figure below. Beams are spaced every 30 ft. Each beam is 50 ft long. The floor slab is 6 inches thick and built of concrete. Live load on the floor is 100 psf. Beams are simply supported.

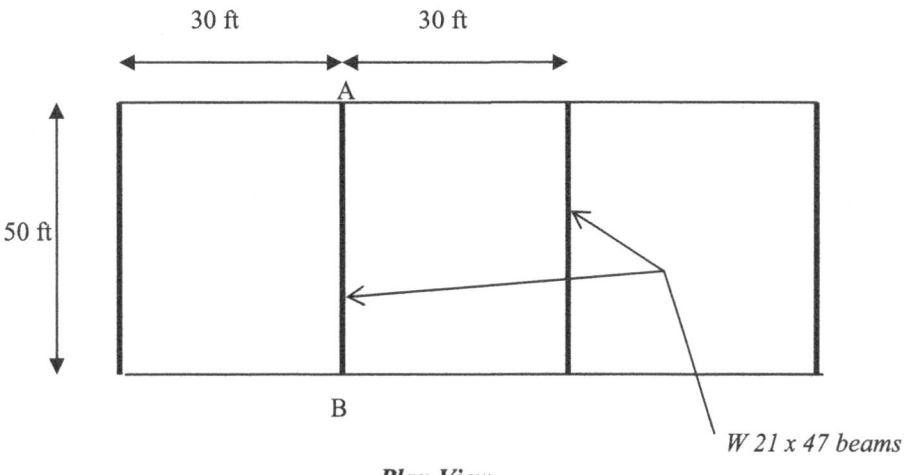

Plan View

Moment of inertia (I) of beams (W21 x 47) = 3,630 in⁴.
Concrete density = 145 pcf
Young's modulus of steel (E) = 29 x 10⁶ lbs/in².

a) What is the maximum deflection of beam AB
b) What is the deflection at a point 10 ft from the end of the beam AB

Solution:

STEP 1:   Find the loading on the beam;
Beam AB, supports an area as shown in the figure below.

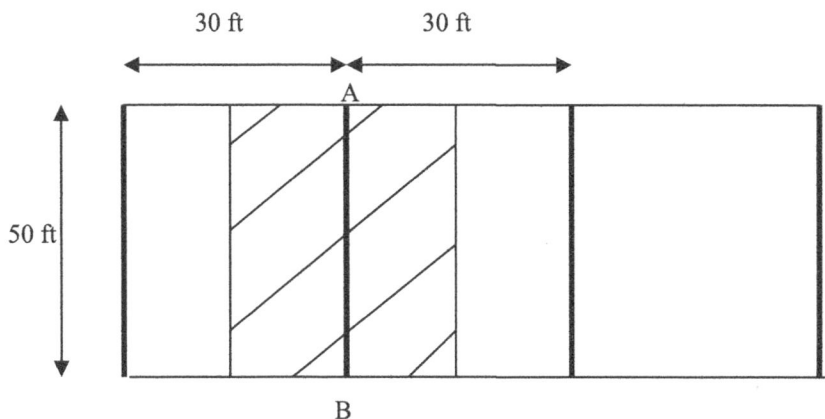

Dead load per ft on beam AB due to concrete slab = 30 x 0.5 x 145 = 2,175 lbs/ft
Above 0.5 is the thickness of the slab and 30 ft is the width of the slab that is supported by the beam.

Dead load due to self weight of the beam = 47 lbs/ft
W21 x 47 beam has a self weight of 47 lbs per ft.
Total dead load = 2,175 + 47 = 2,222 lbs/ft
Live load per ft on beam AB = 30 x 100 = 3,000 lbs/ft
Total load on beam AB per ft = 3,000 + 2,222 = 5,222 lbs/ft.
(Note that load combinations are not considered for deflection computations).

STEP 2: Find the maximum deflection;
Maximum deflection of a simply supported beam is at the center.

Maximum deflection = $5.w.L^4/(384E.I)$
(See above case 2).
Convert all values to inches and lbs.

L = 50 ft = 600 in;
$E = 29 \times 10^6$ lbs/in^2.
I = 3,630 in^4.
w = 5,222 lbs/ft = 435.2 lbs/in

Maximum deflection = $5.w.L^4/(384E.I)$
Maximum deflection = $5 \times 435.2 \times 600^4/(384 \times 29 \times 10^6 \times 3,630)$
= 6.97 inches.
May be too high. Use a larger beam.

b)   What is the deflection at a point 10 ft from the end of the beam AB
Deflection at a point "x" from the edge = $w.x (L^3 - 2.L.x^2 + x^3)/(24. EI)$

w = 435.2 lbs/in;   x = 10 ft = 120 in;   L = 50 ft = 600 in

Deflection at a point 10 ft from the edge = $435.2 \times 120 (600^3 - 2 \times 600 \times 120^2 + 120^3)/(24 \times 29 \times 10^6 \times 3,630)$
= 4.1 inches

5.0   Hydraulics and Hydrology -   7 Questions

5.1 Open-channel flow
5.1.1   Manning's equation:
Canals were constructed by ancient man for agriculture and other uses. The methodologies they used are not available today. Engineers during the eighteenth century noticed that the velocity in an open channel dependent upon the slope and the roughness of the perimeter walls. Higher the slope, higher the velocity. In addition, they noticed smoother slopes resulting high water velocities.

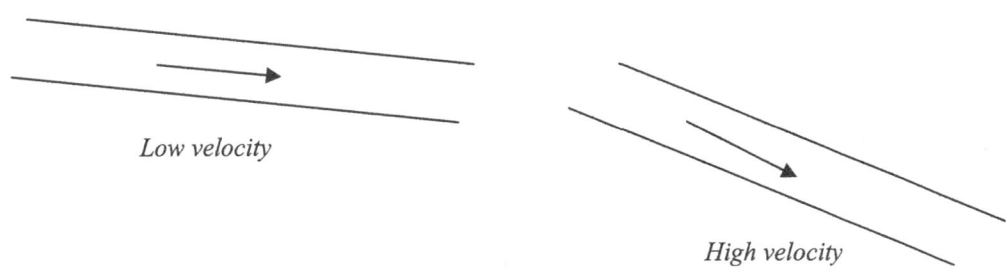

Low velocity

High velocity

Canals with steeper slopes had high water velocities.

It is not the slope…It is square root of the slope:

Major breakthrough was achieved in open channel flow by Dutch Engineer Cornelius Velson in 1749. He noticed that the velocity of water is proportional to square root of the slope ($S^{1/2}$). Prior to this, many believed water velocity was proportional to the slope.

Next Breakthrough: Next breakthrough came few years after. Albert Brahms in 1757 concluded that velocity of water in an open channel is proportional to A/P. (A = Area of the channel, P = Wetted perimeter length of the channel). Pretty soon it was evident that A/P quantity was very useful and it was called hydraulic radius. (R = A/P).

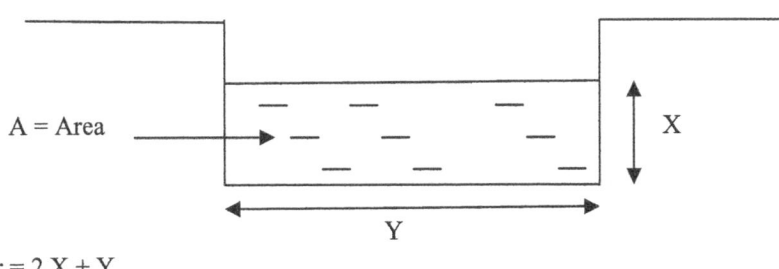

Wetted perimeter = 2.X + Y
Area (A) = X.Y
R = Hydraulic radius = A/P $= \dfrac{X.Y}{[2.X + Y]}$

French Engineer Antoine Chezy in 1765 used Brahms work and Velson's work and came up with an equation. Chezy's equation was adopted by many and still been used.

Chezy's Equation; $v = C \times (RS)^{1/2}$

v = Velocity; C = Chezy coefficient; $R = A/P$; S = Slope

After Chezy, many came up with equations. Some were too complicated and some were easy to use. They all had different levels of accuracy.

Manning in 1,889 came up with the Manning formula, which was a modification of formulas of others. However, Manning formula was simple and easy to use. In addition, it gave reasonably accurate results and became popular.

$$v = \frac{1.49 \times R^{2/3} \, S^{1/2}}{n}$$

v = Velocity, (ft/sec)
R = Hydraulic Radius = A/P = Area/Wetted perimeter
(Area in square feet and wetted perimeter in feet)
S = slope in decimals (Not percentage)

Practice Problem: Water is flowing thru a rectangular channel . The depth of the channel is 5 ft and the width is 10 ft. The slope of the channel is 5%. Manning's coefficient is 0.02. Find the flow rate in the channel.

A) 1,534.5 cu. ft/sec   B) 3,231.7 cu.ft/sec   C) 1,611.2 cu.ft/sec   D) 234.7 cu.ft/sec

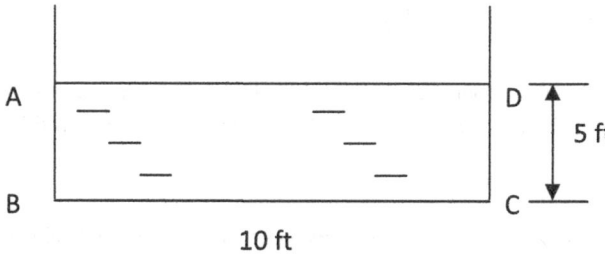

Solution)

STEP 1: Gather all the data given

$$v = \frac{1.49 \times R^{2/3} \, S^{1/2}}{n}$$

v = Velocity,
R = Hydraulic Radius = A/P
A = Area;
P = Wetted perimeter
S = slope
n = Manning's roughness coefficient

R = A/P
A = (5 x 10) = 50
P = Wetted perimeter = 5 + 10 + 5 = 20
R = 50/20 = 2.5
n = 0.02 (given)
S = 5% = 0.05

STEP 2: Input all the known values into Manning's equation:

$$v = \frac{1.49 \times R^{2/3} S^{1/2}}{n}$$

$$v = \frac{1.49 \times (2.5)^{2/3} \times 0.05^{1/2}}{0.02}$$

v = 30.69 ft/sec

Flow = Area x velocity
Flow = 50 x 30.69 = 1,534.5 cu. ft/sec   (Ans A)

Problems can be twisted many ways. In the following problem, flow is given and asked to find the slope.

Practice Problem: Water is flowing thru a rectangular channel at a flow rate of 300 cu. ft/sec. The depth of the channel is 5 ft and the width is 8 ft. What is the slope of the channel if Manning's coefficient is 0.04?

A) 0.18%   B) 1.4%   C) 1.15%   D) 0.19%

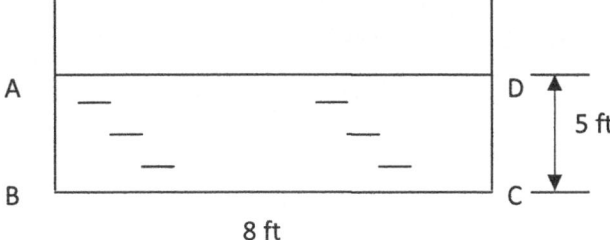

Solution:

STEP 1:

$$v = \frac{1.49 \times R^{2/3} S^{1/2}}{n}$$

v = Velocity,
R = Hydraulic Radius = A/P
A = Area;
P = Wetted perimeter
S = slope
n = Manning's roughness coefficient

v = Velocity = Flow/Area = Q/A = 300/(5 x 8) = 7.5 ft/sec
R = A/P

A = 5 x 8 = 40
P = Wetted perimeter = ABCD = 5 + 8 + 5 = 18 ft
R = 40/18 = 2.22

STEP 2: Input all the known values into Manning's equation:

$$v = \frac{1.49 \times R^{2/3} \, S^{1/2}}{n}$$

$$7.5 = \frac{1.49 \times (2.22)^{2/3} \, S^{1/2}}{0.04}$$

$$S^{1/2} = 0.118$$
$$S = 0.014 \text{ ft/ft}$$

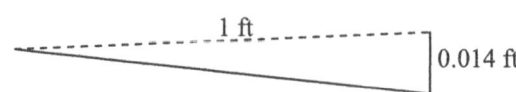

S = 1.4%   (Ans B)

<u>Manning Formula for Trapezoidal Channels</u>:   Most channels are trapezoidal. In addition, most natural channels have approximately trapezoidal shape. Same concepts can be applied for trapezoidal channels as well. Following problem illustrates the application of Manning formula for trapezoidal channels.

Trapezoidal channel:

Practice Problem: Trapezoidal channel is shown in the figure below. The depth of the water is 8 ft. The slope of the channel is given to be 1.4%. Side slopes of the trapezoid is 1V to 2H. Manning roughness coefficient is given to be 0.007. What is the flow (cu. ft/sec) of the channel?

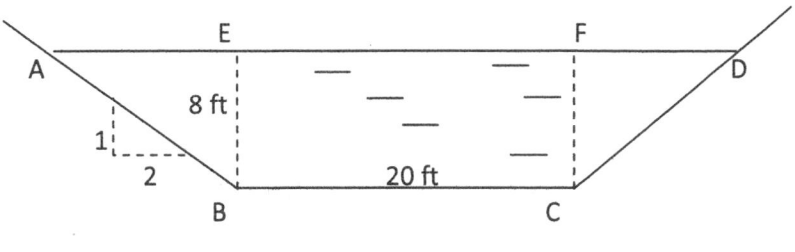

A) 21,659   B) 18,514   C) 21,084   D) 13,480

Solution:

Manning Equation:

v = Velocity,
R = Hydraulic Radius = A/P
A = Area;
P = Wetted perimeter
S = slope
n = Manning coefficient

$$v = \frac{1.49 \times R^{2/3} \, S^{1/2}}{n}$$

STEP 1:   Gather all the given data.

Slope = S = 1.4% = 0.014

Area = (Base + Top)/2 x h

Area (A) = $\frac{(Base + Top)}{2}$ x Height   = $\frac{(AD + BC)}{2}$ x 8

EB = 8 ft.
Hence AE = 16 ft due to 2:1 slope.

Hence AD = 16 + 20 + 16 = 52 ft

Area (A) = $\frac{(AD + BC)}{2}$ x 8 = $\frac{(52 + 20)}{2}$ x 8 = 288 sq. ft

P = Wetted perimeter = AB + BC + CD

$AB^2 = AE^2 + BE^2$
$AB^2 = 16^2 + 8^2 = 320$
AB = 17.9 ft

Wetted perimeter = AB + BC + CD = 17.9 + 20 + 17.9 = 55.8 ft

R (Hydraulic radius) = A/P = 288/55.8 = 5.16

STEP 2: Apply the Manning equation:

$v = \dfrac{1.49 \times R^{2/3} S^{1/2}}{n} = \dfrac{1.49 \times 5.16^{2/3} \times 0.014^{1/2}}{0.007} = 75.21$ ft/sec

Flow Rate = Area x Velocity = 75.21 x 288 = 21,659 cu. ft/sec   (Ans A)

Triangular Channels:

Next we look at a triangular channel.

Practice Problem: Triangular channel is shown in the figure. The flow of the channel is found to be 40,000 gallons/minute. Manning roughness coefficient is 0.05. The depth of flow is 7 ft. What is the slope of the channel?

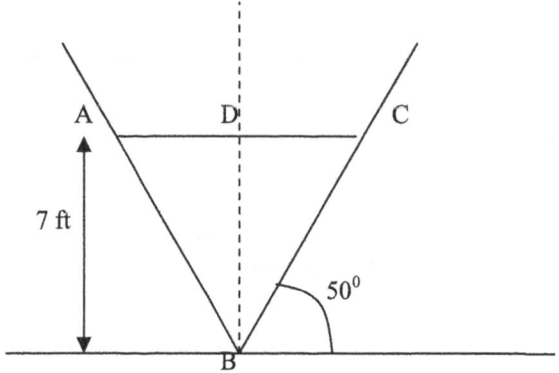

A) 0.18%   B) 1.24%   C) 2.98%   D) 3.98%

Solution:

Manning Equation:

v = Velocity,
R = Hydraulic Radius = A/P
A = Area;
P = Wetted perimeter
S = slope
n = Manning coefficient

$$v = \dfrac{1.49 \times R^{2/3} S^{1/2}}{n}$$

STEP 1:   Find the area

Area = AC x BD/2

DC = BD x tan 40
BD = 7
Hence DC = 7 x tan 40 = 5.87
AC = 2 x DC = 11.74

Area = AC x BD/2 = 11.74 x 7/2 = 41.09

STEP 2:   Find the velocity;

Velocity = Flow/Area
Flow = 40,000 gallons/minute = 666.7 gallons/sec

Convert gallons to cu. ft
1 cu.ft = 7.48gallons
1 gallon = 0.1337 cu. ft
Hence Flow = 666.7 x 0.1337 cu. ft/sec = 89.13 cu. ft/sec
Velocity (v) = Flow/Area = 89.13/41.09 = 2.17 ft/sec

STEP 3:   Find the wetted perimeter and hydraulic radius;

Wetted perimeter = AB + BC
BD = BC. cos 40
7 = BC. cos 40
BC = 7/cos 40
BC = 9.14

Wetted perimeter = AB + BC = 2 x 9.14 = 18.28

Hydraulic radius (R) = A/P = 41.09/18.28 = 2.25

STEP 2:   Apply the Manning equation:

$$v = \frac{1.49 \times R^{2/3} \, S^{1/2}}{n}$$

$$2.17 = \frac{1.49 \times 2.25^{2/3} \, S^{1/2}}{0.05}$$

$S^{1/2}$ = 0.0424
S = 0.0018 = 0.18%   (Ans A)

5.1.2 Critical Flow:

When you throw a stone to the water, waves would form and disperse. If the water is flowing slowly, wave will disperse upstream and downstream where the stone was dropped. If the wave disperse upstream and downstream of the stone dropped location, we would say the flow is subcritical. However, if the water is flowing fast, the wave will not disperse upstream.   In this situation, the flow is called supercritical. The transition velocity between subcritical and supercritical is known as critical velocity.

Wave dispersion in flowing water was studied by Jean Baptiste Belanger in 1830s and his work was carried out further by Joseph Boussinesqu. By mid 1850s it was known that wave velocity in a stream when a stone is thrown would be given by the following equation.

$$c = [g. \, A/T]^{1/2}$$

c = Velocity of wave
A = Area of the stream
T = Free surface width of water

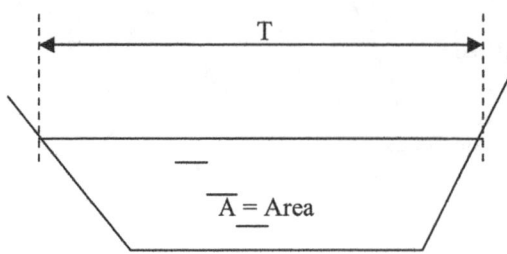

Figure: Free surface width of water shown in the above figure

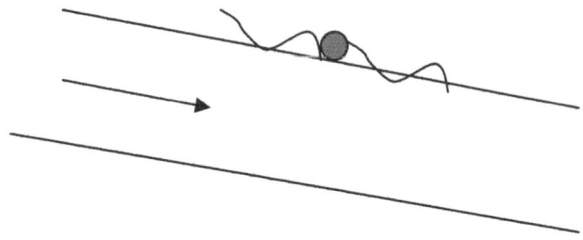

Figure: *Waves will form when a stone is thrown into a flowing channel. If the channel flow velocity is low, waves will spread upstream and downstream. This flow is known as subcritical flow.*

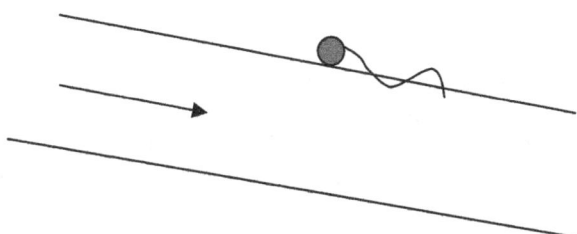

Figure: *At high velocities of flow, the wave will not spread upstream. This flow is known as supercritical flow.*

When the stream flow velocity (v) is equal to wave velocity, critical flow occurs.

v = c   --------→ Critical flow
v > c   --------→ Supercritical flow
v < c   --------→ Subcritical flow

v = Velocity of the channel
c = Velocity of waves

At critical flow;   v = c

Hence; $v = [g. A/T]^{1/2}$

William Froude an English engineer who was working on shape of ships had come up with the quantity $[g. A/T]^{1/2}$. Hence this quantity was called Froude number (Fr).

$$F_R = V/(g. \, A/T)^{1/2}$$

Hence we can see that; when Froude number $F_R = 1$; $v = c$.   (See the equation for wave velocity "c")

$F_R > 1$; $v > c$

$F_R < 1$; $v < c$

When Fr > 1.0 Flow is supercritical
When Fr < 1.0 Flow is subcritical
When Fr = 1.0 Flow is critical

5.1.3 Hydraulic Depth (D):

The quantity A/T is known as hydraulic depth. Note that hydraulic depth is different than the hydraulic radius (A/P).

Hydraulic Depth (D) = A/T = Area/Width of top surface of water
Hydraulic radius (R) = A/P = Area/Wetted perimeter

Practice Problem: A channel is shown below. The velocity of the flow is 7 ft/sec. Channel width is 13 ft and channel depth is 5 ft.

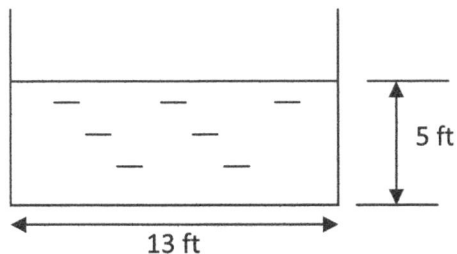

a) Find the Froude number of the channel
b) Is the flow supercritical?

    A)   1.134   B) 0.552   C) 3.665   D) 2.476

Solution:

The Froude number (F_R) is given by the following equation.

$$F_R = V/(g. \, A/T)^{1/2}$$

V = Velocity = 7 ft/sec
A = Area = 13 x 5 = 65 sq. ft
T = width of the top surface.   (In this case T = 13 ft)
g = Gravitational constant (32.2 ft/sec^2 for English units or 9.81 m/sec^2 for metric units).

STEP 1:   Apply the equation for the Froude number;

$(F_R) = V/(g. A/T)^{1/2}$

$A/T = 65/13 = 5$

$(F_R) = V/(g. A/T)^{1/2} = 7/(32.2 \times 5)^{1/2} = 0.552$ (Ans B)

Froude number is less than 1. Hence the flow is subcritical.
Next we look at an irregular channel.

Practice Problem: An irregular shaped channel is shown below. The flow in the channel is 237.8 cu.ft/sec. Channel dimensions are as shown.

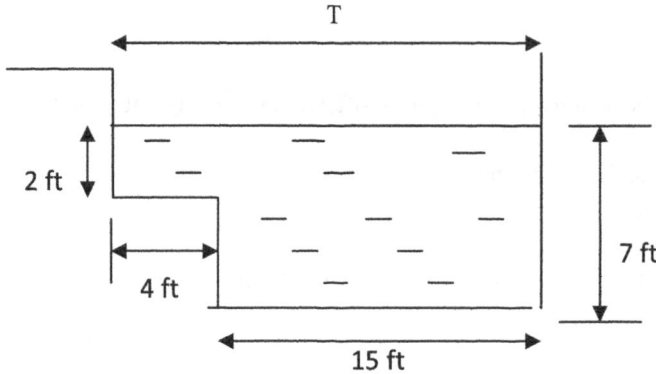

a) Find the Froude number of the channel
b) Is the flow supercritical?

A) 0.13   B) 1.12   C) 0.65   D) 0.15

Solution:

The Froude number (F_R) is given by the following equation.

$$F_R = V/(g. A/T)^{1/2}$$

V = Velocity
A = Area
T = Width of the top surface.
g = Gravitational constant (32.2 ft/sec^2 for English units or 9.81 m/sec^2 for metric units).

STEP 1: Find the area and velocity;
Area (A) = (15 x 7) + (4 x 2) = 113 sq. ft

Velocity = Flow/Area = 237.8/113 = 2.10 ft/sec

STEP 2:   Apply the equation for the Froude number;

$(F_R) = V/(g. A/T)^{1/2}$
T = Surface width = 15 + 4 = 19

A/T = 113/19 = 5.95

$(F_R) = V/(g. A/T)^{1/2} = 2.10/(32.2 \times 5.95)^{1/2} = 0.15$ (Ans D)

Froude number is less than 1. Hence the flow is subcritical.
Next we look at a trapezoidal channel.

Practice Problem)   Critical depth of a trapezoidal channel is found to be 6 ft. Channel dimensions are as shown.
Find the flow in the channel.

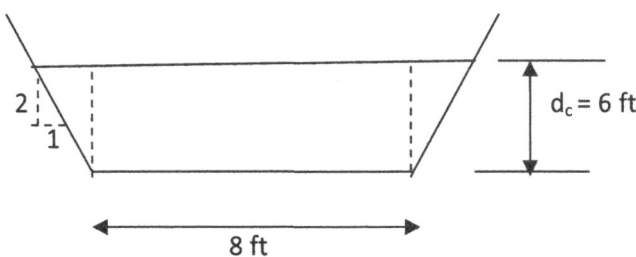

A) 1.34 cu.ft/sec   B) 9.27 cu.ft/sec   C) 812.8 cu.ft/sec   D) 412.7 cu.ft/sec

Solution):   At critical depth, Froude number is equal to 1.0.

$$F_R = V/(g.D)^{1/2}$$

V = Velocity;
D = A/T, where A = Area and T = width of the top surface.
g = Gravitational constant (32.2 ft/sec^2 for English units or 9.81 m/sec^2 for metric units).

STEP 1:   Find the width of the top surface (T);

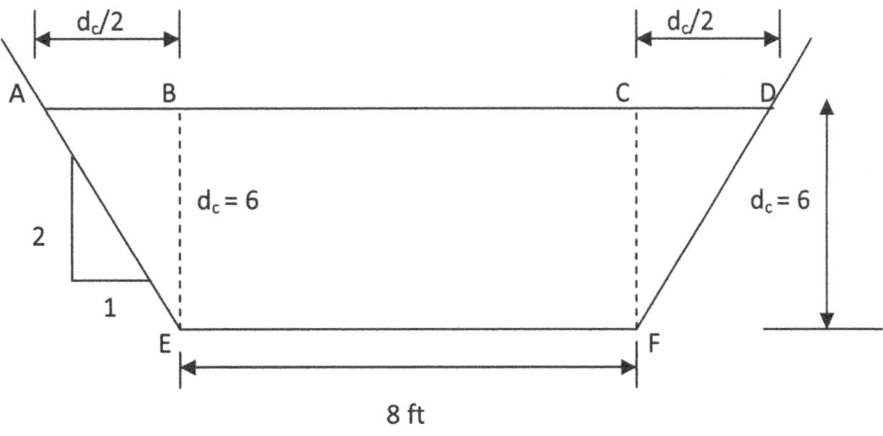

T = Width of the top surface = AD
T = AD = AB + BC + CD
BC = 8 ft
AB = CD = $d_c/2$ (Since the slope is 1 to 2).

T = AB + BC + CD = 3 + 8 + 3 = 14

STEP 2:   Find the Area (A);
Area of a trapezoid = (Base + Top) x ½ x Height

Area = (8 + 14)/2 x 6 = 66 sq. ft

STEP 3:   Find A/T
A/T = Hydraulic depth (D) = 66/14 = 4.71

STEP 4:   Find the velocity;

At critical depth, Froude number = 1.0

$(F_R) = V/(g.D)^{1/2} =$ 1.0 $= \dfrac{V}{[32.2 \times 4.71]^{1/2}}$

V = 12.31 ft/sec
Flow = Area x Velocity = 66 x 12.31 = 812.8 cu. ft/sec   (Ans C)

5.1.4 Hydraulic Jump:
When there is a supercritical flow, at some point it would change to subcritical flow. Hydraulic jump occurs when flow transfers from supercritical flow to subcritical flow.

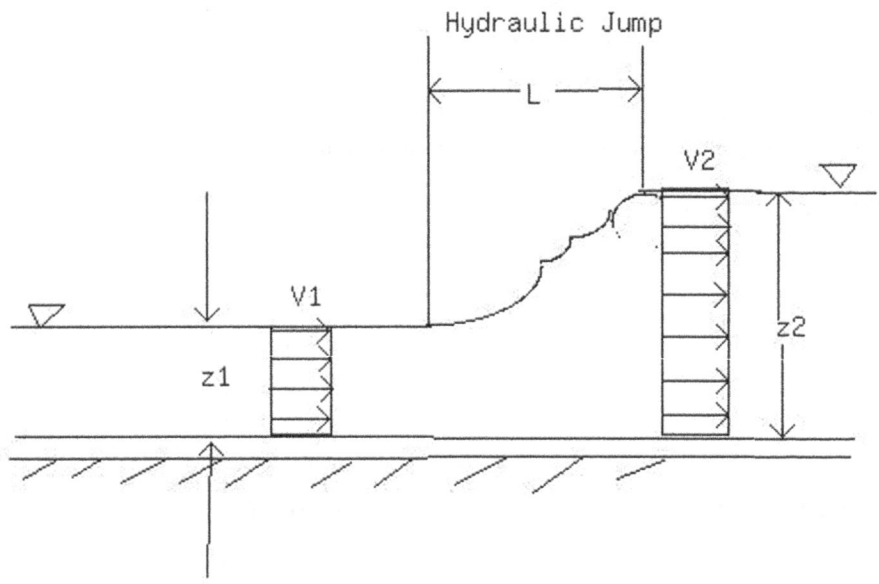

Hydraulic Jump

Equation for Hydraulic Jump;

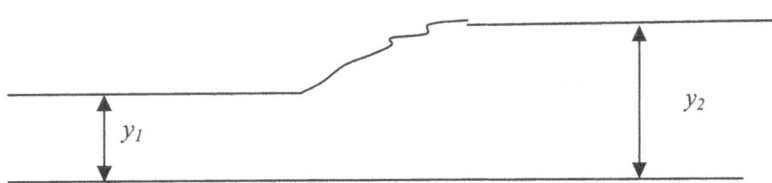

Hydraulic jump can be expressed by following two equations for rectangular channels.

$$y_2/y_1 = 1/2 \times [\, (1 + 8\,Fr_1^2)^{1/2} - 1\,] \text{ -----------(1)}$$

$y_1 =$ Upstream depth of the channel
$y_2 =$ Downstream depth of the channel after the jump. Note that y_2 is always larger than y1.
$Fr_1 =$ Froude number upstream of the channel.
Froude number is defined as follows.

$Fr = v/(gD)^{1/2}$;

$D =$ Hydraulic dsepth = A/T;
$A =$ Area;
$T =$ Length of the top surface

Practice Problem)   Hydraulic jump is shown in the figure below. Upstream depth of the channel is 1.5 ft and velocity of the channel is 13.0 ft/sec. What is the depth after the hydraulic jump?

A) 2.53 ft   B) 2.19 ft   C) 5.92 ft   D) 3.61 ft

Solution)

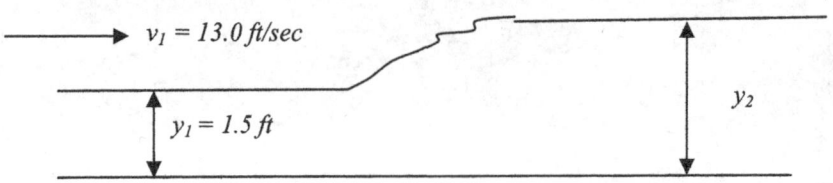

Hydraulic jump equation for a rectangular channel is shown below. Note that this equation is valid only for rectangular channels.

$$y_2/y_1 = 1/2 \ x \ [\ (\ 1 + 8 \ Fr_1^2)^{1/2} \ - \ 1 \ ] \ \text{-----------}(1)$$

y_2 = Depth after the jump;
y_1 = Depth before the jump
v_1 = Velocity of the channel before the jump

Fr_1 =   Froude number upstream of the channel.

Froude number is defined as follows.

$Fr = v/(gD)^{1/2}$;

A = Area = Width of the channel x y_1
T = Length of the top surface = Width of the channel
D = Hydraulic dsepth = A/T
D = A/T = [Width of the channel x $y_{1]}$/Width of the channel $= y_1 = 1.5$ ft

Find the Froude number upstream of the channel;

   $Fr = v/(gD)^{1/2}$;
   $F_{r1} = 13/(32.2 \ x1.5)^{1/2}$;

y_1= 1.5 ft; v_1 = 13.0 ft

F_{r1} = 1.87

Using above equation (1) for hydraulic jump;

$y_2/y_1 = 1/2 \ x \ [\ (\ 1 + 8 \ Fr_1^2)^{1/2} \ - \ 1 \ ]$ ----------------(1)
$y_2/1.5 = 1/2 \ x \ [\ (\ 1 + 8 \ x \ 1.87^2)^{1/2} \ - \ 1 \ ]$ -----------(1)
y_2 = 2.19 ft   Ans B

Practice Problem: Water is discharged from a sluice gate at a rate of 20 cu. ft/sec. The width of the channel is 5 ft. The depth of water where hydraulic jump occurs is found to be 1.5ft. Find the following;

a) Velocity before the jump
b) Velocity after the jump
c) Depth after the jump

Solution:

STEP 1:   Find the velocity before the jump;
Flow = Area x Velocity

Flow is given to be 20 cu.ft/sec.
Area = Depth x Width = 1.5 x 5 =7.5 sq. ft
Velocity before the jump (v_1) = 20/7.5 = 2.67 ft/sec

STEP 2:   Find the velocity after the jump;
Flow = Area x Velocity
Flow is given to be 20 cu.ft/sec.
Assume depth after the jump = y_2.
Area = Depth after the jump x Width = y_2 x 5
Velocity after the jump (v_2) = 20/Area = 20/(5. y_2) ft/sec
v_2 =20/(5. y_2) ft/sec ----------------------------(1)

Unfortunately we cannot find v_2 since we do not know y_2.

STEP 3:   Find the depth after the jump (y_2);

$y_2/y_1 = 1/2 \ x \ [\ (\ 1 + 8 \ Fr_1^2)^{1/2} \ - \ 1 \ ]$

y_1 = 1.5 ft;
$F_{r1} = v_1/(g. D_1)^{1/2}$;

D_1 = A/T = (1.5 x 5)/5 = 1.5
$F_{r1} = v_1/(g. D_1)^{1/2}$;
v_1 = 2.67 ft/sec   (found in step 1)
$F_{r1} = v_1/(g. D_1)^{1/2}$;

Hence F_{r1} = 2.67/(32.2 x 1.5)$^{1/2}$ = 0.38

Apply known values in hydraulic jump equation;

$y_2/y_1 = 1/2 \ x \ [\ (\ 1 + 8 \ Fr_1^2)^{1/2} \ - \ 1 \ ]$
$y_2/1.5 = 1/2 \ x \ [\ (\ 1 + 8 \ x 0.38^2)^{1/2} \ - \ 1 \ ]$
y_2 = 0.35

5.1.5 Spillway Capacity :

When water goes over earth dams, erosion would set in. Within hours earth dams would fail due to erosion. Hence water should not be allowed to go over earth dams. During heavy rain events, water has to be removed from a special structure known as a spillway.

Earth dam and a spillway: *Spillway is seen to the left. Water is allowed to go over the spillway, which is built using concrete. The dam is made of earth. Under no circumstance water should go over the earth dam. If water goes over the earth dam, it would fail within hours.*

Today most dams are constructed using concrete. Some water going over concrete may not be a big issue. Yet it is necessary to have a specially designed spillway to remove excess water from a reservoir in a daily basis. It is important to know the flow capacity of a spillway. If the water is not removed fast enough, rising water will go over the dam and flood the area. Spillway capacity has to be larger than the highest inflow of water to the reservoir.

Flow over spillway is given by the following equation;

$$Q = C. \, L. \, H^{3/2}$$

C = Spillway constant
L = Length of the spillway
H = Height of water above spillway

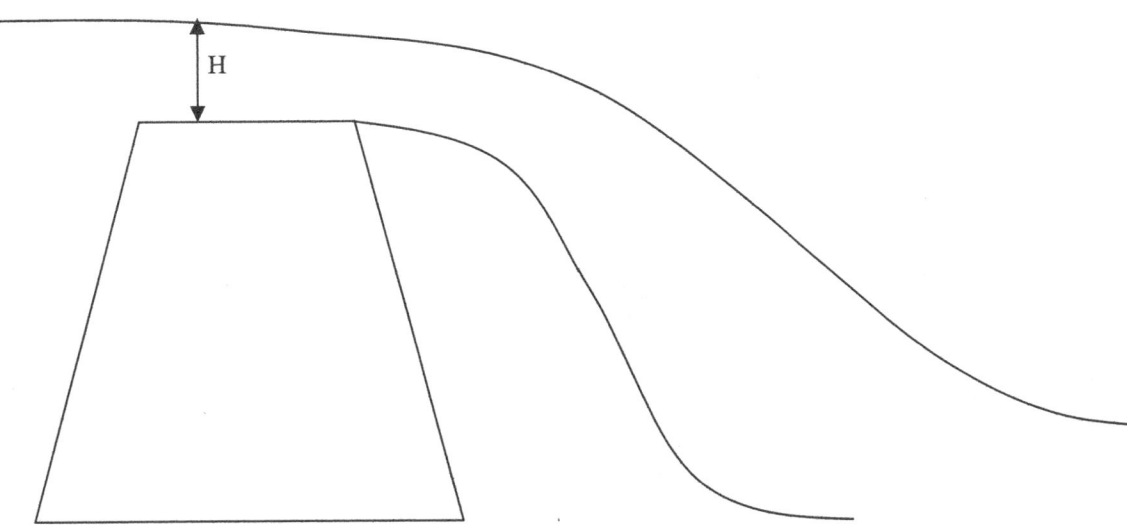

Practice Problem: Spillway is 100 ft long and the water level above the spillway is 2.5 ft. Spillway coefficient is 3.2. Find the discharge from the spillway.

Solution:

$Q = C.\ L.\ H^{3/2}$
$Q = 3.2 \times 100 \times 2.5^{3/2} = 1{,}264.9$ cu.ft/sec

5.1.6 Flow measurement – Open Channel:

There are many instruments available in the market to measure the velocity of water at a given point. Use the velocity gauge and obtain velocities of water at number of locations. Find the average velocity and multiply by the area to obtain the flow.

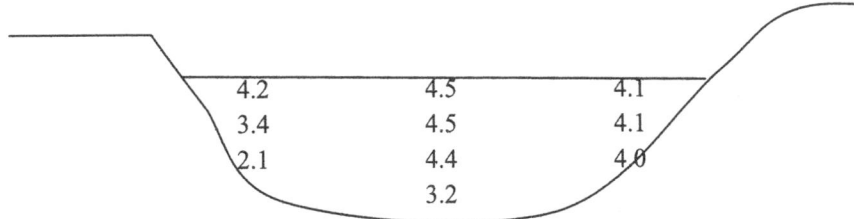

In the above figure velocities of water at 10 locations were taken. Assume the area of the channel is 20 sq. ft.
Hence the flow = Area x average velocity
Flow = 20 x 3.85 = 77. cu.ft/sec

Flow Measurement Using Weirs;

Flow measurements can be done in a channel by placing a weir. All the water in the channel will pass thru the weir.

Above figures shows weirs placed on channels. One can predict the flow in the channel by measuring the height of flow in the weir.

Weir Formula: The equation for V – notch weir or triangular weir is given by the following formula.

$$Q = 8/15 \times C_d \times (2.g)^{1/2} \times \tan(\theta/2) \times h^{5/2}$$

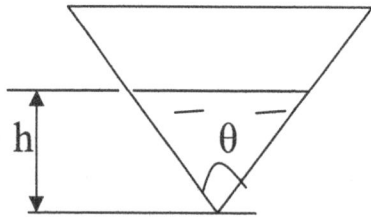

Q = Flow in cu.ft/sec
C_d = Weir coefficient
θ = Weir angle
h = Height of water in the weir.

Practice Problem: Find the flow in a weir if the height of water in the weir is 4 ft and the weir angle is 55 degrees. Weir coefficient is 0.98.

Solution:

$Q = 8/15 \times C_d \times (2.g)^{1/2} \times \tan(\theta/2) \times h^{5/2}$
$Q = 8/15 \times 0.98 \times (2 \times 32.2)^{1/2} \times \tan(55/2) \times 4^{5/2}$
$Q = 69.87$ cu.ft/sec

Water resources and environmental section accounts for 20% of questions in the morning session. You are expected to know pipe flow, open channel flow, hydrology and environmental and wastewater engineering.

5.2.0 <u>Hydrology</u>:

Hydrology is the study of movement of water on earth. In Civil Engineering we are interested in movement of water mostly on surface of earth. During rain events water will travel on surface and finally enter into storm sewers. Storm sewers need to be sized to transport the flow that could occur during rain events.

5.2.1   Storm characterization:

Storm is a rain event. Some rains can last for hours while others could last only few minutes. Researchers have found that high intensity rainfalls do not last long. Hence it is generally true that higher the intensity, smaller the duration. Very rarely one would see a very high intensity rainfall lasting for a longer period of time. Such events may happen every 100 years or so.

5.2.2   Rainfall Intensity, Duration, and Frequency (IDF) Curves:

Let us assume that a researcher measures rainfall intensity and duration for one year. Then he would record the maximum rainfall for a given duration. The researcher would get a table as shown below.

Duration (hrs)	1.0	2.0	3.0	4.0
Rainfall Intensity (in/hr)	4.3	2.2	1.8	1.5

Researcher can then draw a graph between intensity and duration.

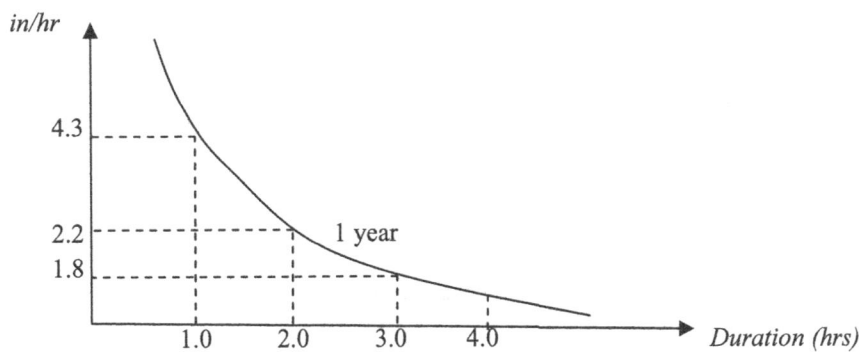

Let us look how to draw the rainfall intensity graph;

STEP 1:   Record all the rainfall data within that year. Record the durations and intensities.
STEP 2:   Find the maximum rainfall intensity for duration 1 hr, 2 hr etc.
STEP 3:   Draw a graph between duration and the maximum intensity for that duration.

As you could see higher the intensity, smaller the duration. The observations were done only for a time period of 1 year. Hence the graph will be known as 1 year rainfall graph.

Next conduct the survey for 5 years. Like in the previous case, record all rainfall durations and their intensities. Then follow the same steps as before. Then you would end up with a graph as shown below.

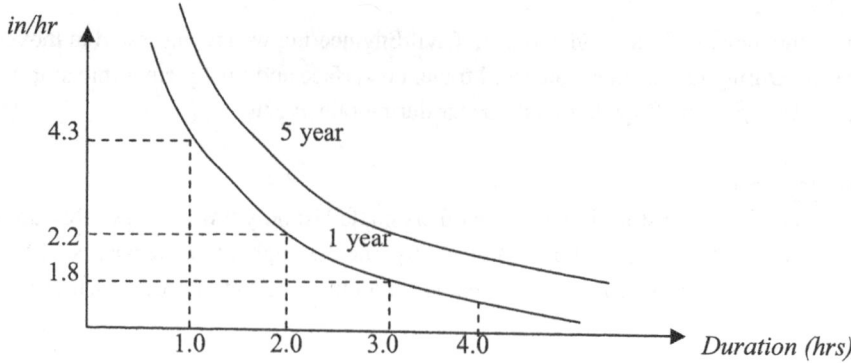

The 5 year graph would be higher than the 1 year graph. For an instance, if you take 1 hr duration rainfalls, one would get a larger maximum if the observations were conducted for five years instead of one year. Similarly one can draw graphs for 10, 20 or 100 years. These curves are known as IDF (Intensity, Duration, Frequency Curves).

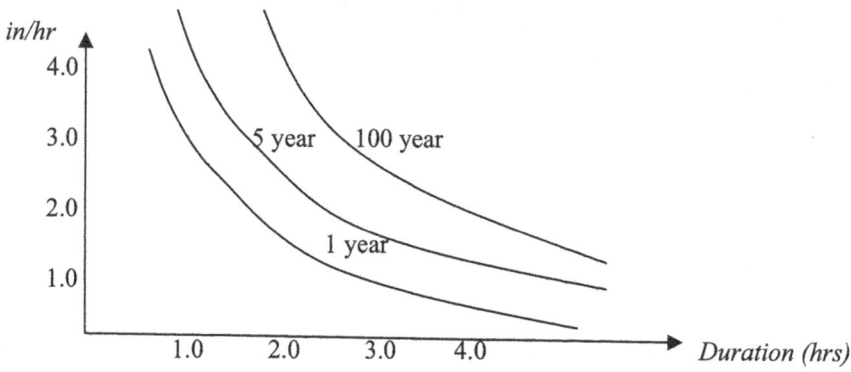

IDF (Intensity, Duration, Frequency Curve)

Practice Problem: Use the above graph to answer following problems.

a) What is the intensity of a 1 year frequency 3 hour duration rainfall
b) What is the intensity of a 5 year frequency 3 hour duration rainfall
c) What is the intensity of a 100 year frequency 3 hour duration rainfall

Solution:
a) 1.0 in/hr
b) 1.5 in/hr
c) 2.7 in/hr

Practice Problem: IDF curve for San Diego area was developed for a period of five years. Researchers recorded 2 hour and 3 hour duration rainfalls. The intensities of these rainfalls are shown below.

Duration (hrs)	Intensity (in/hr)	Duration (hrs)	Intensity (in/hr)
2 hrs	3.2 in/hr	3 hrs	2.1 in/hr
2 hrs	1.9 in/hr	3 hrs	2.3 in/hr
2 hrs	4.1 in/hr	3 hrs	3.5 in/hr
2 hrs	2.2 in/hr	3 hrs	1.7 in/hr

a)   What values should be used for the IDF curve?

    A)   3.2 (2 hr duration), 2.1 (3 hr duration)

    B)   3.2 (2 hr duration), 3.5 (3 hr duration)

    C)   4.1 (2 hr duration), 3.5 (3 hr duration)

    D)   4.1 (2 hr duration), 1.7 (3 hr duration)

b)   As per above data, rainfall intensities of 3 hr duration storms were generally smaller than 2 hr duration storms.

    A)   Something wrong with data recording,

    B)   Data seem to be correct. This is expected

    C)   Intensities of 3 hr storms are generally larger than intensities of 2 hr storms for a given period.

    D)   Depends on the location.

c)   Hydrologists monitored rainfall data for the same area 10 years. Data for 10 year period is given below.

Duration (hrs)	Intensity (in/hr)	Duration (hrs)	Intensity (in/hr)
2 hrs	3.2 in/hr	3 hrs	2.1 in/hr
2 hrs	1.9 in/hr	3 hrs	2.3 in/hr
2 hrs	4.1 in/hr	3 hrs	3.5 in/hr
2 hrs	2.3 in/hr	3 hrs	1.7 in/hr
2 hrs	4.8 in/hr	3 hrs	1.9 in/hr
2 hrs	1.8 in/hr	3 hrs	1.5 in/hr
2 hrs	3.7 in/hr	3 hrs	4.2 in/hr
2 hrs	2.9 in/hr	3 hrs	2.7 in/hr

What values should be used to be plotted in the IDF curve for the 10 year period?

d)   Data suggests that 10 year maximum intensity for 2 hr duration is greater than the 5 year period for the same duration rainfall. Explain the reason for this observation.

Solutions:

a)   Maximum intensities for a given duration should be used in the IDF curve. Hence use 4.1 and 3.5 (Ans C)

b)   Intensities of 3 hr storms are generally smaller than intensities of 2 hr storms for a given period. (Ans B)

c)   Maximum intensities for 10 year period is

    2 hr duration rainfall = 4.8 in/hr

    3 hr duration rainfall = 4.2 in/hr

d)   In this case we are interested in rainfalls that have a duration of 2 hrs approximately. If you record data for one week, you probably will not get a rainfall of 2 hr duration. Hence your intensity for 2 hr rainfall for a period of one week is 0.0. If you record data for one month you may get a single 2 hr rainfall. Let us say you got a rainfall with an intensity of 1.5 in/hr and a duration of 2 hrs. Hence design rainfall for one month period for 2 hr duration is 1.5 in/hr. Now let us say that you record data for one whole year. In that case, you probably would get bunch of rainfalls that has a duration of 2 hrs. Let us say you got four rainfalls with duration 2 hrs with intensities 2.3, 3.7, 1.2 and 1.9. Your design rainfall intensity for one year period is the maximum of the four values. That is 3.7 in/hr. As you could see if you record data for a longer period of time, the maximum intensity for that time period would go up. I have summarized what we discussed here.

Period = 1 week,	No rain. Design intensity = 0.0
Period = 1 month,	One rainfall event with 2 hr duration. Design intensity = 1.5 in/hr
Period = 1 year,	Four rainfall events with 2 hr duration. Design intensity = 3.7 in/hr

Longer the period of consideration, higher the maximum intensity or the design intensity.

5.2.3. Time of Concentration:

Let us assume that a person is dead. Emails would be sent out to all his relatives. Relatives who are living in the same block would arrive first. Then relatives living in the same town would arrive. Next, the relatives living in the same state would arrive. Next relatives from other states would arrive. Finally, relatives residing in overseas would arrive. Let us throw in some numbers and try to draw a graph.

Same day as the death of the person	→ Number of arrivals = 5	(People living in the same block)	Total = 5
One day after the death	→ Number of arrivals = 6	(People living in the same town)	Total = 11
Two days after the death	→ Number of arrivals = 4	(People living in the same state)	Total = 15
Three days after the death	→ Number of arrivals = 5	(People living in other states)	Total = 20
Four days after the death	→ Number of arrivals = 3	(People living in other countries)	Total = 23

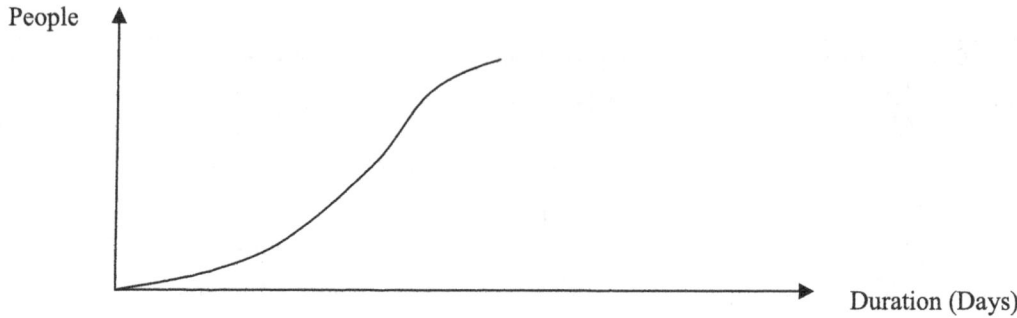

As you could see it would take few days for everyone to arrive. Hence funeral has to be scheduled few days after the death so that all can attain the funeral.

After the funeral, people would start to leave. Hence, our curve would look like as shown below.

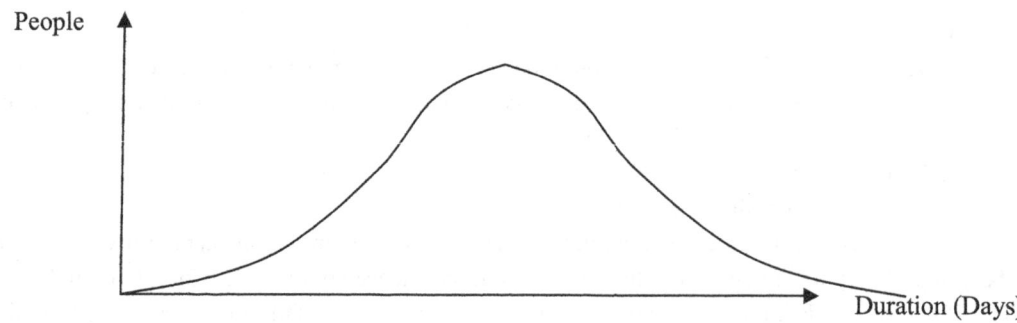

Now we can discuss time of concentration in a catchment basin.

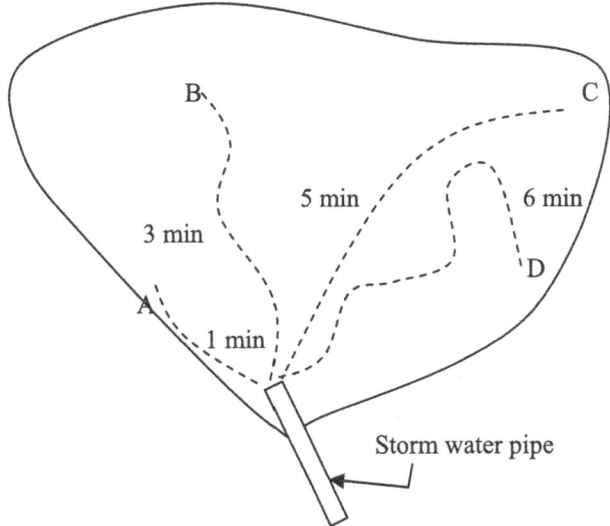

Above figure shows a catchment basin and a storm water pipe. Rainwater falling onto the catchment basin would flow to the storm water pipe. Four points are shown in the figure. Once the rain starts, water from point A would arrive first. Then water from point B and C would arrive. Finally, water from point D would arrive.

1 minute after rainfall =   Water from point A would arrive at the storm pipe
3 minutes after rainfall =   Water from points A and B would arrive at the storm pipe
5 minutes after rainfall =   Water from points A, B and C would arrive at the storm pipe
6 minutes after rainfall =   Water from points A, B, C and D would arrive at the storm pipe

Full flow in the storm pipe would happen 6 minutes after start of the rainfall. Hence, time of concentration for this catchment basin is 6 minutes.
For full flow to occur, rainfall should last 6 minutes or more.

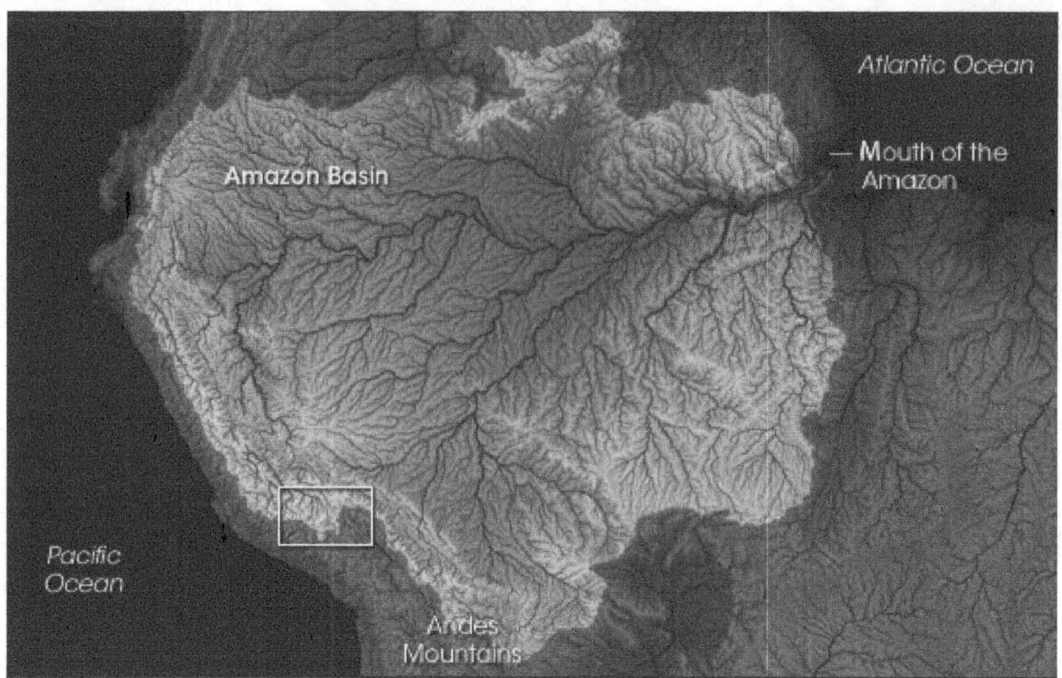

Above figure shows the Amazon catchment basin. Typical catchment basins that we would consider to design culverts and storm pipes are much smaller in size.

Drainage area and watershed area are other names for the catchment basin.

Definition of Time of Concentration: Time of concentration is the time taken for water to reach a culvert or storm water pipe from the most hydrologically distant point.
It is important to remember that most distant point may not be the most hydrologically distant point. Hydrological distance depends on the route that water would take and also the slope. Higher the slope, higher the speed of flow.

5.2.4 Runoff Analysis - Rational Method:

Catchment Basin:
Whole idea is to find the runoff flow from a catchment basin. During a rainfall event, water would flow thru the surface and enter small tributaries. Small tributaries feed larger tributaries and finally enter a storm pipe or a canal.

Runoff flow can be found using the rational formula.

$$\text{Rational Formula}$$
$$Q = C.I.A$$

Q = Design flow (Acre. in/hr)
C = Runoff coefficient (No units)
I = Design rainfall intensity (in/hr)
A = Area of the drainage basin (Acres)

Q (Design Flow):   This is the flow that we would like to know. If we know the flow, we can size up storm water pipes that would carry the flow.

C (Runoff Coefficient):   Runoff coefficient is dependent on the surface characteristics. Asphalt surface would have a larger runoff coefficient compared to a woodland. Higher the "C" value, higher the runoff.

I (Design Rainfall Intensity):   Design rainfall intensity should be obtained from IDF (Intensity, duration frequency) graphs.

A (Area):   A is the area of the drainage basin.

Let us look at an example.

Practice Problem: Drainage basin has an area of 30 acres. Design intensity of the rainfall is 2.2 in/hr. The area is mostly covered with vegetation. Runoff coefficient of the area is 0.35. Find the design flow in cu. ft/sec.

A) 20.13 cu. ft/sec   B) 23.29 cu. ft/sec   C) 19.01 cu. ft/sec   D) 28.8 cu. ft/sec

Solution:

STEP 1:   Apply the rational formula;

Q = C. I. A

C = 0.35;   I = 2.2 in/hr;   A = 30 acres
Q = 0.35 x 2.2 x 30 = 23.1 acre. in/hr

STEP 2:   Convert acre. in/hr to cu.ft/sec.

1 acre = 43,560 sq, ft

23.1 acre. in/hr = 23.1 x 43,560/12 cu. ft/hr = 83,853 cu. ft/hr
= 83,853/3,600 cu. ft/sec
= 23.29 cu. ft/sec   (Ans B)
Note: Due to a mathematical coincidence, value for acre in/hr is almost same as cu. ft/sec.

Practice Problem: Drainage basin has an area of 15 acres. Engineers would like to design a storm water pipe that would function for a 100 year flood. The watershed area has a runoff coefficient of 0.32. Time of concentration for the watershed area is 30 minutes. Find the design flow in cu. ft/sec. (IDF curve for the area is given below).

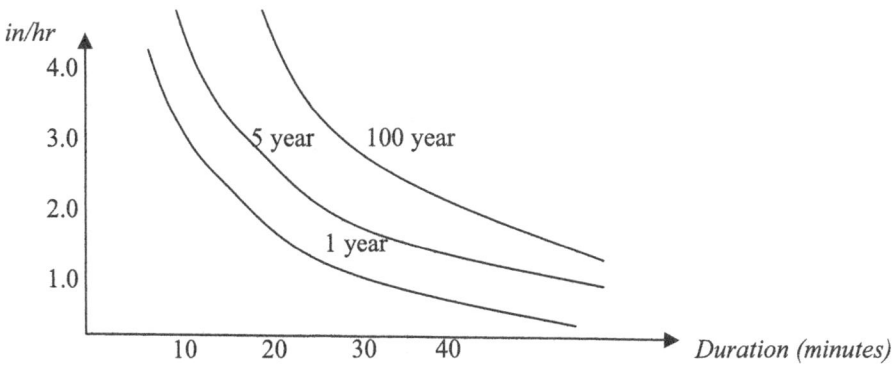

IDF (Intensity, Duration, Frequency Curve)

A) 10.18 cu. ft/sec   B) 18.29 cu. ft/sec   C) 13.55 cu. ft/sec   D) 224.8 cu. ft/sec

Solution:

STEP 1:   Collect all necessary information for the rational formula;

Q = C. I. A

C = 0.32 (given)
To find "I", draw a line thru 30 minutes.   Rainfall intensity for 100 year flood with a duration of 30 minutes is approximately 2.8 in/hr. (I = 2.8 in/hr)

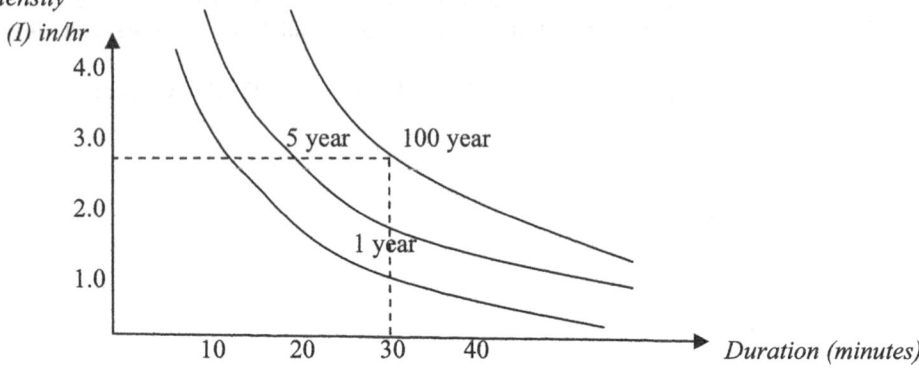

IDF (Intensity, Duration, Frequency Curve)

C = 0.32;   I = 2.8 in/hr;   A = 15 acres
Q = 0.32 x 2.8 x 15 = 13.44 acre. in/hr

STEP 2:   Convert acre. in/hr to cu.ft/sec.

13.44 acre. in/hr = 13.44 x 43,560/(12 x 3,600) cu. ft/sec = 13.55 cu. ft/sec   (Ans C)

Average Runoff Coefficient: In most cases drainage basins are not homogeneous. Portion of the drainage basin could be grassland. Another part could be wooded. Third area could be urban landscape. In such situations average runoff coefficient need to be used. The average runoff coefficient should be found with regard to the area.

Let us assume a drainage basin with three different surface conditions.

Surface condition 1 (Wooded area):   Area = A1;   Runoff coefficient = C1
Surface condition 2 (Parking lots):   Area = A2;   Runoff coefficient = C2
Surface condition 3 (grass areas):   Area = A3;   Runoff coefficient = C3

$$C_{average} = \frac{(C1. A1) + (C2. A2) + (C3. A3)}{A1 + A2 + A3}$$

Practice Problem: Drainage area is shown in the figure below.

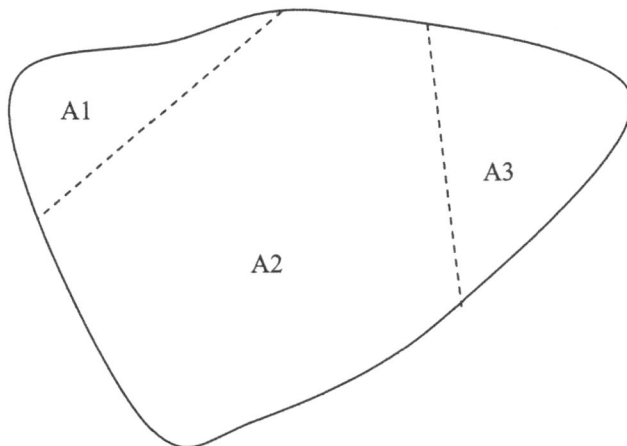

A1 (agricultural land).   A1 = 12 acres; C1 = 0.2
A2 (Urban development).   A2 = 38 acres; C2 = 0.5
A1 (Playgrounds).   A3 = 20 acres; C3 = 0.3

Find the average runoff coefficient.
A) 0.39   B) 0.67   C) 0.28   D) 0.51
Solution:

Average runoff coefficient is given by;

$$C_{average} = \frac{(C1. A1) + (C2. A2) + (C3. A3)}{A1 + A2 + A3}$$

$$C_{average} = \frac{(0.2 \times 12) + (0.5 \times 38) + (0.3 \times 20)}{12 + 38 + 20} = 27.4/70 = 0.39 \qquad \text{(Ans A)}$$

Practice Problem: 15% of a drainage area consists of woodlands. 60% of the drainage basin is mostly residential. Other 25% is wetlands.

Runoff coefficients;
Woodlands - 0.54
Residential - 0.82
Wetlands   - 0.10
What is average runoff coefficient of the drainage basin?
A) 0.598   B) 0.871   C) 0.561   D) 0.312

Solution:

In this problem, areas are not given. Hence instead of areas, you need to use percentages or decimals.

$$C_{average} = \frac{(C1. A1\%) + (C2. A2\%) + (C3. A3\%)}{A1\% + A2\% + A3\%}$$

$$C_{average} = \frac{(0.54 \times 0.15) + (0.82 \times 0.60) + (0.10 \times 0.25)}{0.15 + 0.60 + 0.25} = 0.598 \qquad \text{(Ans A)}$$

Time of Concentration and Rainfall Intensity Relationships:

Actually there is no relationship between time of concentration of a watershed area and rainfall intensity. Time of concentration is a function of the watershed area. Each watershed area has its own time of concentration. However, during our computations we equate the time of concentration to the duration of the rainfall.

Let us say we are interested in finding the maximum flow in watershed area Q. Time of concentration for this watershed area is 15 minutes. Let us assume there was a rainfall that lasted only for 5 minutes. Then do we get the maximum flow? The answer is NO. To get the maximum flow, we need to have a rainfall that would last 15 minutes or more. If the rainfall is less than the time of concentration, water from all points will not be able to make to the storm water pipe or culvert we are trying to design.

What about rainfalls that are longer than time of concentration? In this case, water from all points in the drainage basin will gather to the storm water pipe. However, there is a caveat. Longer duration rainfalls would generally have low intensity. Hence to get the maximum flow, we should consider rainfalls that have the same duration as the time of concentration of the catchment basin.

Let us summarize our findings;

- If the duration of the rainfall is less than the time of concentration, water from all points in the drainage area will not contribute to the flow.

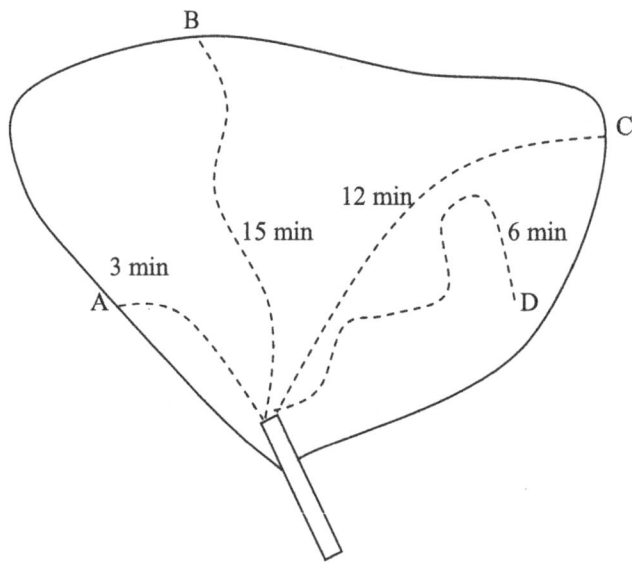

The time of concentration of the above drainage basin is 15 minutes. That means water from point B would take 15 minutes to get to the storm water pipe.

What would happen if the duration of the rainfall is only 3 minutes? In this case, water from point A would reach the storm water pipe. Water from points B, C and D will not contribute to the flow in the pipe. We should not design the size of the pipe based on a rainfall of 3 minute duration.

- What about a rainfall with a duration of 15 minutes?
In this case, water from all the points would contribute to the flow in the pipe. Hence we would get the maximum flow.

- What about a rainfall with a duration greater than 15 minutes?
In this case, water from all the points would contribute to the flow in the pipe. However, as we learnt before, intensity of longer duration rainfalls are smaller. Hence to get the maximum flow we should consider rainfalls with duration that is equal to the time of concentration of the drainage basin of concern.

May be some of you who think little ahead might ask, what about a rainfall that has a duration less than 15 minutes but has a higher intensity? Let us assume a rainfall with a duration of 10 minutes for this drainage basin. In this case, not all the points in the drainage basin would contribute to the flow. However, intensity of a 10 minute rainfall is *higher* than a 15 minute rainfall. Is it theoretically possible that a rainfall with a duration less than the time of concentration could produce the maximum flow? The answer is yes. Such possibility exists. However, for design purposes we always use rainfalls with duration equal to time of concentration of the watershed area.

Relationships between rainfall duration and rainfall intensity:

Many researchers have come up with many equations. The equations are dependent on the regional climate.

Some of the equations I have seen in research papers;

I = Intensity (in/hr);   d = Duration (min)

$$I = 100/(d + 10)$$
$$I = 200/(d + 20)^{1/2}$$
$$I = 75/(d + 12)^{0.85}$$

Since we choose duration to be equal to time of concentration (TC), I would use TC instead of the duration.

Practice Problem: Watershed area A has a time of concentration of 20 minutes and watershed area B has a time of concentration of 10 minutes. Following relationship is given for this region.

$$I = 75/(TC + 12)^{0.85}$$

I = Intensity (in/hr);   d = Duration (min)

Find the intensity of rainfall in watershed areas A and B.

Solution:

Watershed area (A): TC = 20 minutes

$$I = 75/(20 + 12)^{0.85} = 3.94 \text{ in/hr}$$

Watershed area (B): TC = 10 minutes

$$I = 75/(10 + 12)^{0.85} = 5.42 \text{ in/hr}$$

Note that when TC is smaller intensity (I) is greater.

Time of Concentration vs. Slope: Slope of water path affect the speed of flow. High speeds would produce smaller time of concentration values.

Let us look at an example.

Practice Problem)   Following information is provided for a drainage basin.

Area	Runoff coefficient	Slope	Distance to hydrologically most distant point
12 acres	0.65	9%	1.6 miles

Slope	Velocity
5%	2 ft/sec
10%	11 ft/sec
15%	16 ft/sec
20%	20 ft/sec

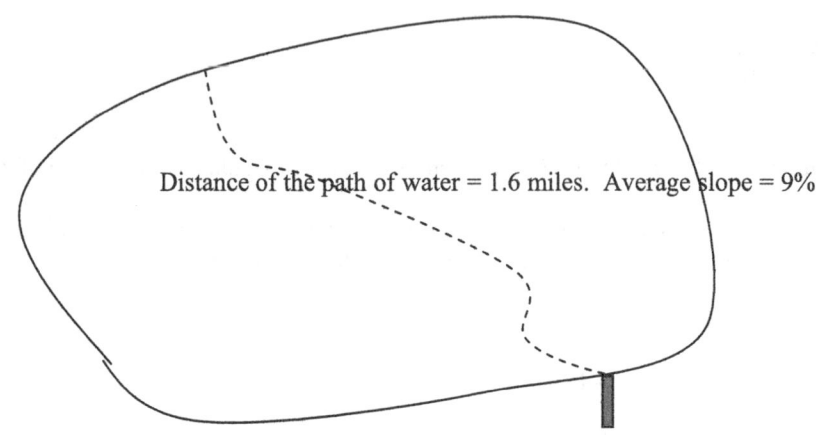

Distance of the path of water = 1.6 miles. Average slope = 9%

Rainfall data for the region as shown below are for 10 year, 20 year and 100 year rainfalls.

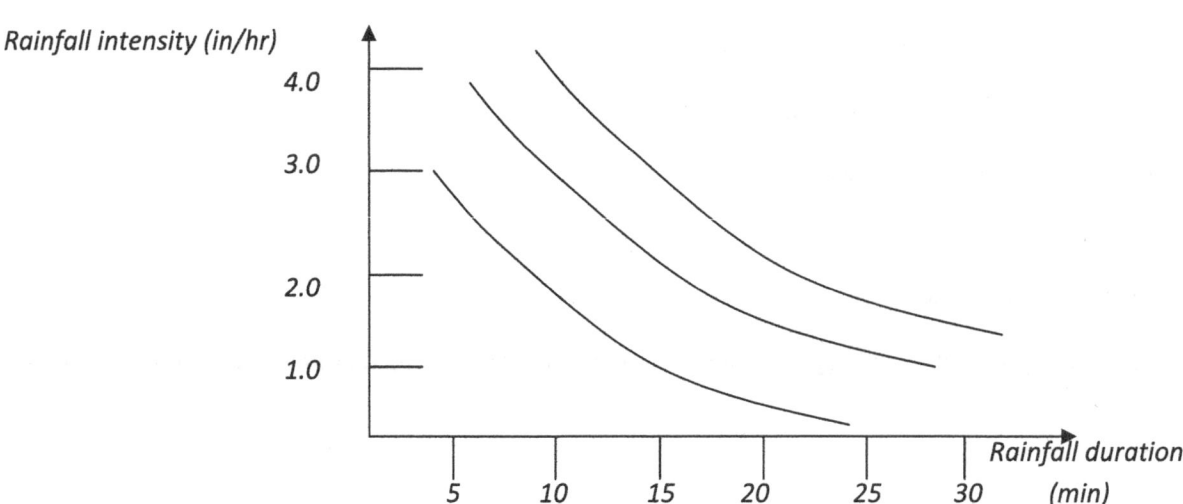

Find the maximum flow for 20 year rainfall in cu.ft/sec.

A)   17.3 cu.ft/sec   B) 71.6 cu.ft/sec   C) 141.3 cu.ft/sec   D) 110.6 cu.ft/sec

Solution: STEP 1:   Find the time of concentration;

Velocity = Distance/Time
Time = Distance/Velocity

Distance of water path is given to be 1.6 miles. (1.6 miles = 8,448 ft)

Find the velocity of water.
The slope is given to be 9%.

From the table given;
Slope = 5%,   Velocity = 2 ft/sec
Slope = 10%,   Velocity = 11 ft/sec

We should be able to find the velocity for 9% slope thru interpolation.

Slope = 5%,   Velocity = 2 ft/sec
Slope = 9%,   Velocity = V ft/sec
Slope = 10%,   Velocity = 11 ft/sec

$$\frac{10 - 5}{9 - 5} = \frac{11 - 2}{V - 2}$$

5/4 = 9/(V - 2)
5.V - 10 = 36
V = 46/5 = 9.2 ft/sec

Time = Distance/Velocity
Time = 8,448/9.2 = 918.3 sec = 15.3 min

STEP 2:   Find the rainfall Intensity "I":

"I" can be obtained by using the graph for 20 year rainfall. Time of concentration is found to be 15.3 minutes. Periods are not marked in the graphs. Lowest graph is for 10 year flood. Middle graph is for 20 year flood. Top graph is for 100 year flood. As we discussed earlier, longer the period, higher the intensity.

Use the middle graph since the problem state to use 20 year period for design.

I = 2.2 in/ hr (approximately). [Use rainfall duration = 15.3 and the middle graph]

STEP 3:   Apply the rational formula;

Q = C.I.A

C = 0.65 (Given);
I = 2.2 in/hr
A = 12 Acres; (Given)

Q = 0.65 x 2.2 x 12 = 17.2 Acre in/hr
1 Acre = 43,560 Sq. ft
Q = 17.2 x 43,560/(12 x 3600) = 17.3 cu.ft/sec   (Ans A)

5.2.5   Hydrographs;

Hydrographs are another method to design storm water pipes and culverts. Maximum design flow can be deduced using hydrographs.

What is a hydrograph?   Hydrograph is a graph between time and flow.

Let us look at an example. Let us assume there are two drainage basins. Let us call them drainage basin 1 and drainage basin 2. Drainage basin 1 consists of mainly parking lots and buildings. Drainage basin 2 consists of mainly wooded areas. Needless to say that there would be more flow in drainage basin 1.

Drainage basin 1: (Mostly parking lots)   Drainage basin 2: (Mostly wooded area)

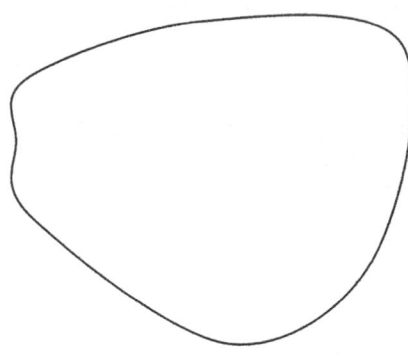

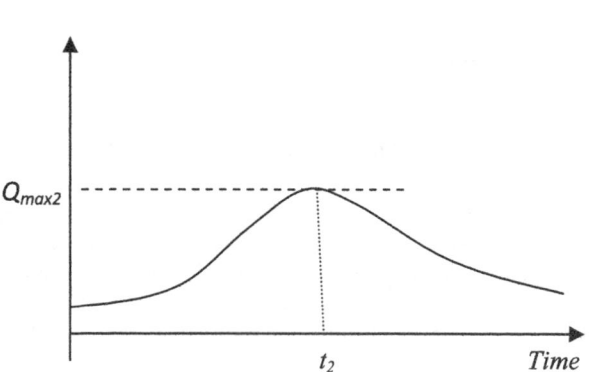

In above graphs, X - axis represents the time and Y-axis represents the flow.

Let us assume 3 inch rainfall fell on both basins. Maximum flow will be achieved soon in basin 1. Water flows fast in parking lots. On the other hand, water flows much slowly in wooded areas.

Hence we can say that for same rainfall event $t_1 < t_2$.

In wooded areas there is more infiltration to the ground. Hence Q_{max2} will be less than Q_{max1}.

$$Q_{max2} < Q_{max1}.$$

Above graphs between time and flow are known as hydrographs. Above graphs were drawn for 3 inch rainfall events. Now let us try to obtain a hydrographs for 1 inch rainfall.

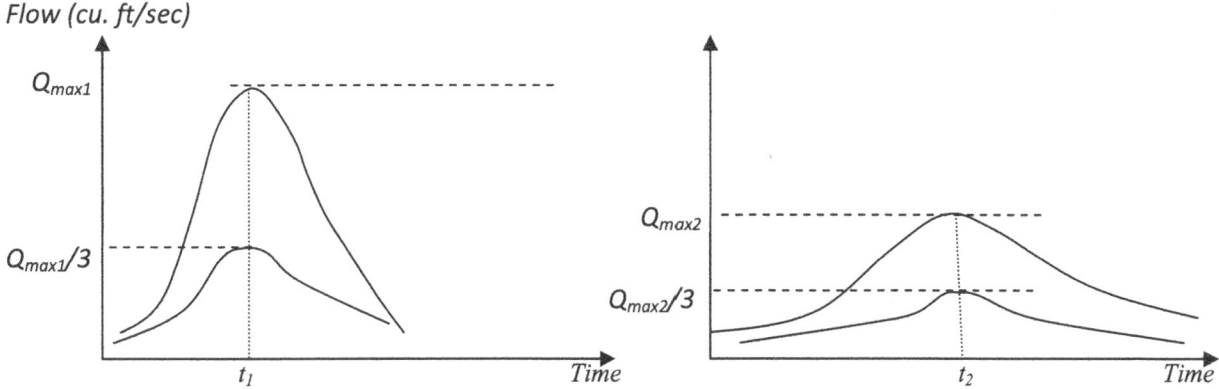

Hydrograph for 1 inch rainfall is obtained by dividing all Y axis values by 3.0.

Basin 1:

Maximum flow for a 3 inch rainfall = Q_{max1}

Maximum flow for a 1 inch rainfall = $Q_{max1}/3$

Basin 2:

Maximum flow for a 3 inch rainfall = Q_{max2}

Maximum flow for a 1 inch rainfall = $Q_{max2}/3$

If we know the hydrograph for any rainfall, we can obtain the hydrograph for 1 inch rainfall.

Practice Problem: Hydrograph for a 2.5 inch rainfall is shown below.

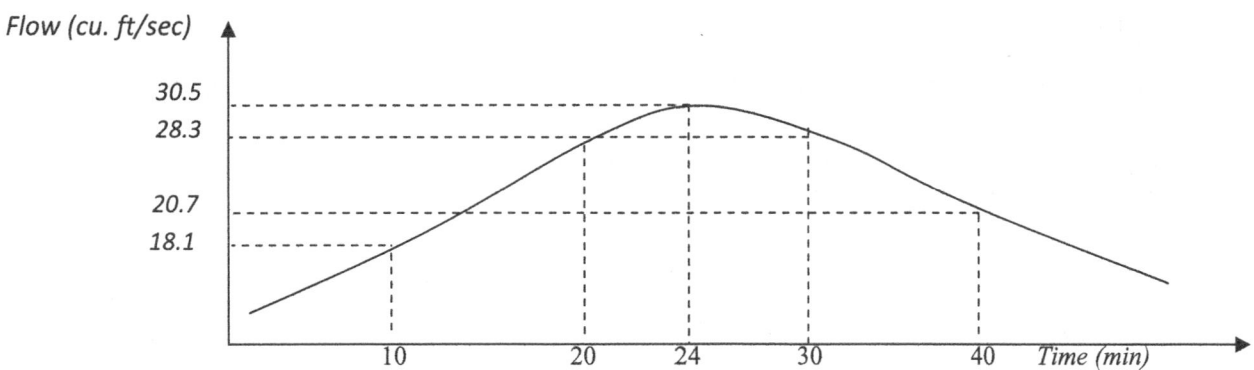

Table for the above graph is shown below. Left column represents time elapsed after the rain started. As soon as the rain starts, flow in streams and storm pipes would start to pick up. As the rain continues, flow gradually would increase to a maximum. Then after the end of the rain, flow would gradually die down.

Elapsed time after rainfall started (min)	Flow (cu. ft/sec)
10	18.1
20	28.3
24 (Maximum flow)	30.5
30	28.3
40	20.7

Develop the table for a 1 inch rainfall and draw the hydrograph.

Solution: Hydrograph and table for 2.5 inch rainfall is given above. To obtain the hydrograph for 1 inch rainfall, divide all numbers by 2.5.

Elapsed time after rainfall started (min)	Flow (cu. ft/sec) for 2.5 inch rainfall	Flow (cu. ft/sec) for 1.0 inch rainfall
10	18.1	7.24
20	28.3	11.32
24 (Maximum flow)	30.5	12.2
30	28.3	11.32
40	20.7	8.28

Hydrograph for 1inch rainfall is known as unit hydrograph.

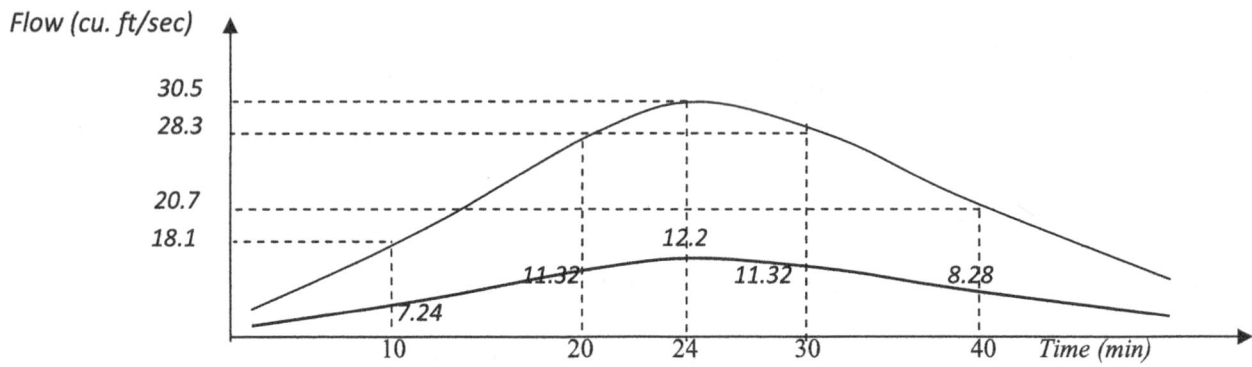

Hydrograph for 1 inch rainfall is known as unit hydrograph.

Practice Problem: Draw the hydrograph for 5.5 inch rainfall for the catchment basin in above problem.

Solution: We obtained the hydrograph for 1 inch rainfall. Multiply all Y-axis values by 5.5 to obtain the hydrograph for a 5.5 inch rainfall.

Elapsed time after rainfall started (min)	Flow (cu. ft/sec) for 2.5 inch rainfall	Flow (cu. ft/sec) for 1.0 inch rainfall	Flow (cu. ft/sec) for 5.5 inch rainfall
10	18.1	7.24	39.82
20	28.3	11.32	62.26
24 (Maximum flow)	30.5	12.2	67.1
30	28.3	11.32	62.26
40	20.7	8.28	45.54

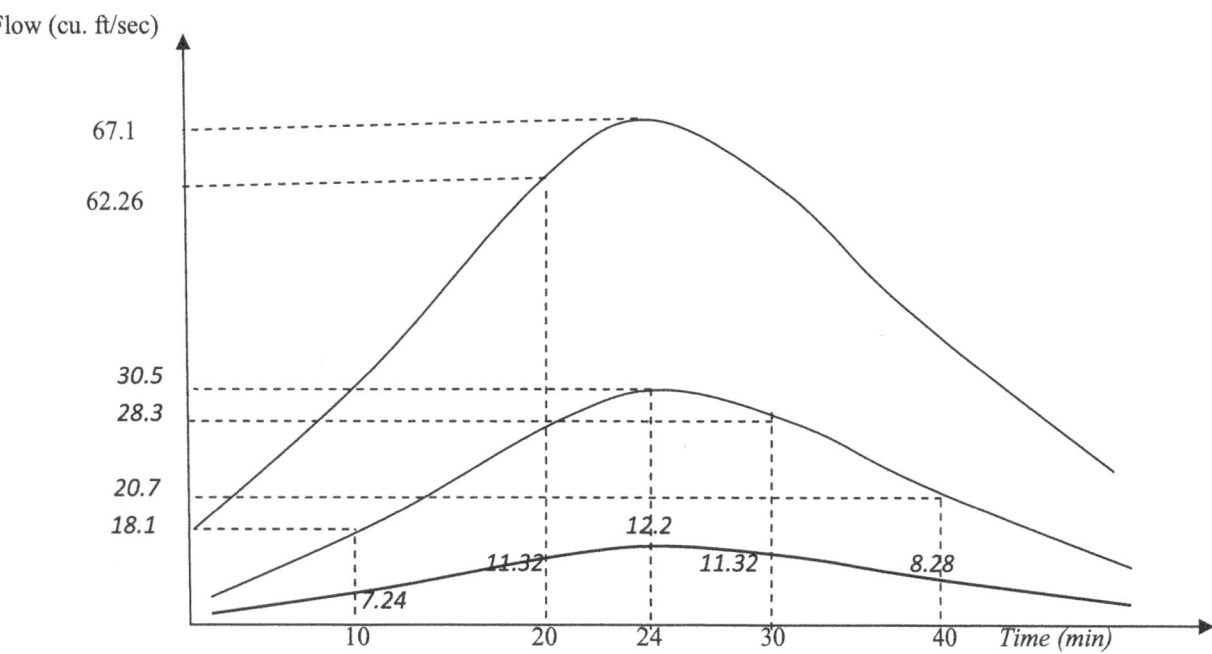

For the most part, we are interested in maximum flow.

Practice Problem: Hydrologists developed a hydrograph for a catchment basin for a rainfall of 6.5 inches. (See below). What is the expected maximum flow for a rainfall of 3.8 inches?

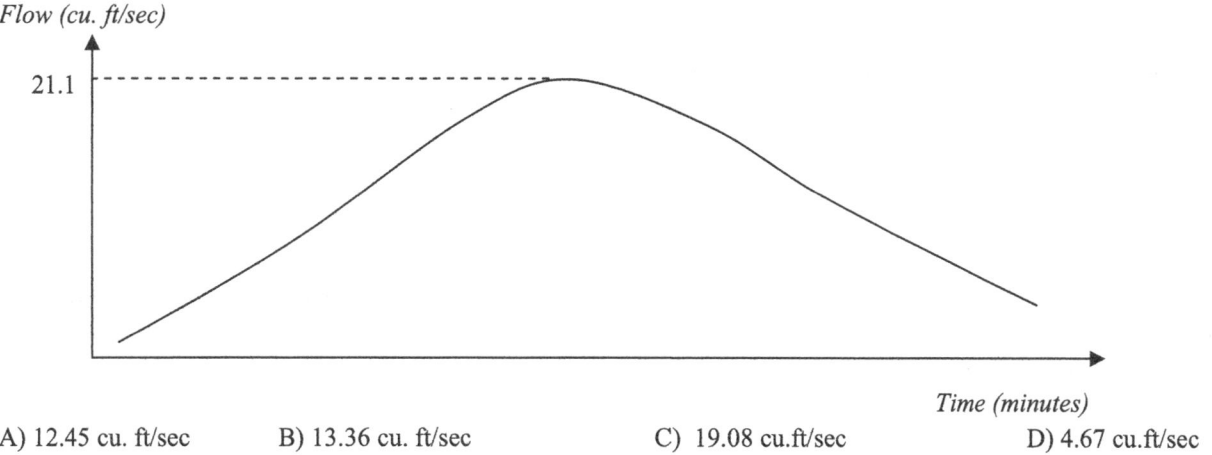

A) 12.45 cu. ft/sec   B) 13.36 cu. ft/sec   C) 19.08 cu.ft/sec   D) 4.67 cu.ft/sec

Solution:

As per the hydrograph given above,

The maximum flow for a rainfall of 6.5 inches = 21.1 cu.ft/sec

The maximum flow for a rainfall of 1.0 inches (unit hydrograph) = 21.1/6 cu.ft/sec

The maximum flow for a rainfall of 3.8 inches (unit hydrograph) = 21.1/6 x 3.8 = 13.36 cu.ft/sec   (Ans B)

5.2.6 Area of a Hydrograph:

What parameter is represented by the area of a hydrograph? Let us look at a curve between speed and time. Let us say a researcher measured the speed of a car every minute and drew a graph as shown.

Speed (ft/min)

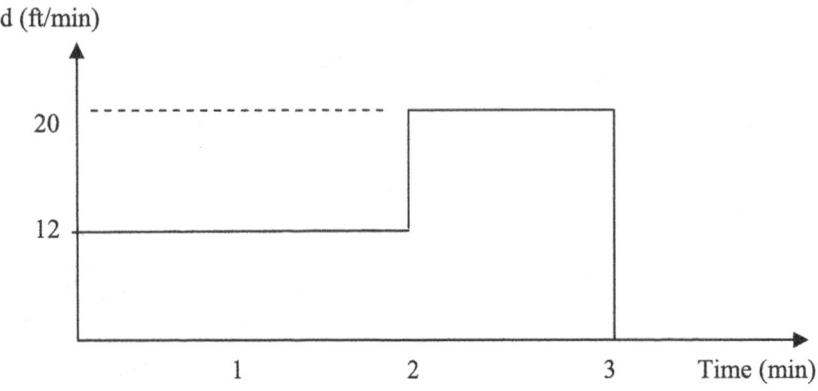

The car travels at a speed of 12 ft/min for 2 minutes.
Distance travelled during first 2 minutes = 2 x 12 = 24 ft

Then car travels at a speed of 20 ft/min for a duration of 1 minute.
Distance travelled during 1 minute = 1 x 20 = 20 ft
Total distance travelled = 24 + 20 = 44 ft.

Now what is the area of the graph above?
Area = (2 x 12) + (1 x 20) = 44

Hence you can see that area of a speed vs. time graph gives the distance.

Similarly, area of a
- Flow vs. time graph gives the total volume of water.
- Power vs. time graph gives total work done

Flow and Total Runoff: Flow is important to size up storm water pipes and culvert sizes. In addition, height of a bridge above a river is dependent on the maximum flow of water. Total volume of water is important to design detention basins and wastewater plants.   Area of a hydrograph gives us the total runoff volume of the drainage basin.

Flow (cu. ft/sec)

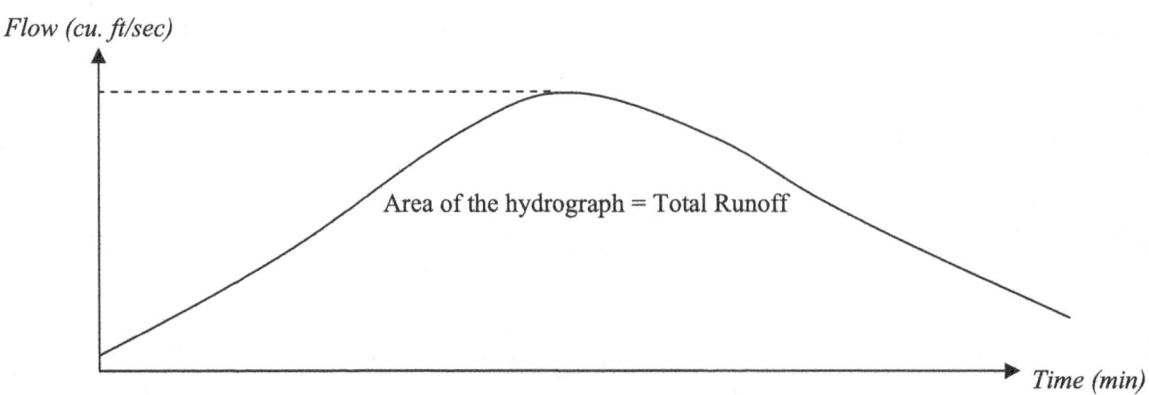

In addition, there is another way of getting the total runoff. Let us consider a football field of 100 ft x 100 ft. Assume 3 inches of rainfall dropped on the football field and 1 inch was infiltrated. What is the total runoff?

Net precipitation = Gross precipitation - Infiltration

Total runoff = Area of the football field x net precipitation = (100 x 100) x 2/12 cu. ft = 1,666 cu. ft.
(Note that this is a volume not a flow).

Net Precipitation = Gross Precipitation - Infiltration

Runoff = Area of the Hydrograph

Runoff = Net Precipitation x Area of the drainage basin

Practice Problem: Hydrograph of a stream is shown below.

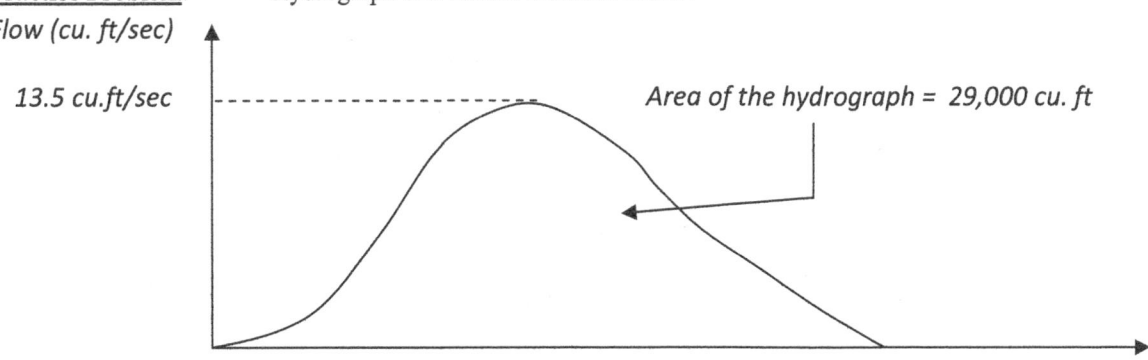

Flow (cu. ft/sec)

13.5 cu.ft/sec   *Area of the hydrograph = 29,000 cu. ft*

Area of the drainage basin is 5 acres. What is the net precipitation?
A) 2.3 in.   B) 1.6 in.   C) 3.4 in   D) 6.1 in.

Solution:
Runoff = Area of the hydrograph = 29,000 cu. ft
Runoff = Net Precipitation x Area of the drainage basin
29,000 cu. ft = Net Precipitation x 5 acres
29,000 cu. ft = Net Precipitation x 5 x 43,560 sq. ft
Net Precipitation = 29,000/(5 x 43,560) = 0.133 ft = 1.6 inches   (Ans B)
Note that area of the drainage basin is different from the area of the hydrograph.

Practice Problem: What is the maximum flow of a unit hydrograph for the catchment basin in previous problem?
A) 8.44 cu.ft/sec   B) 12.67 cu.ft/sec   C) 9.09 cu.ft/sec   D) 4.10 cu.ft/sec

Solution: Note that unit hydrograph is developed for a net precipitation of 1 inch. Catchment basin in the previous problem had a maximum flow of 13.5 cu.ft/sec for a net precipitation of 1.6 inches.

Hence maximum flow of a unit hydrograph = 13.5/1.6 cu.ft/sec = 8.44 cu.ft/sec.

Practice Problem: What is the maximum flow of a hydrograph for the catchment basin in the previous problem for a net precipitation of 5.1 inches?

A) 43.04 cu.ft/sec   B) 51.17 cu.ft/sec   C) 29.19 cu.ft/sec   D) 41.10 cu.ft/sec

Solution:

Maximum flow of a unit hydrograph = 13.5/1.6 cu.ft/sec = 8.44 cu.ft/sec.

Maximum flow of a hydrograph for a precipitation of 5.1 in = 8.44 x 5.1 = 43.04 cu. ft/sec   (Ans A)

Practice Problem: Hydrograph for a drain pipe in a catchment basin is shown below. Find the maximum flow in the drain pipe (cu.ft/sec) if the net precipitation is 3.5 inches. Area of the drainage basin is 4 acres.

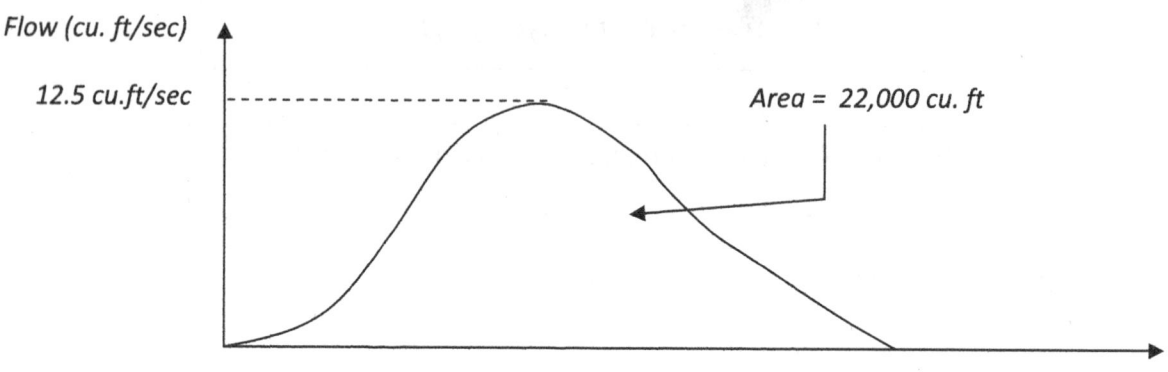

A) 12.6 cu.ft/sec   B) 3.6 cu.ft/sec   C) 28.9 cu.ft/sec   D) 23.7 cu.ft/sec

Solution: Following equations are necessary to solve hydrograph problems.

$$\boxed{\text{Runoff} = \text{Area of the Hydrograph}}$$

$$\boxed{\text{Runoff} = \text{Net Precipitation x Area of the drainage basin}}$$

STEP 1:   Find the Runoff:

Runoff volume is given by the area of the hydrograph.
Runoff = 22,000 cu.ft.

STEP 2:   Find the net precipitation;

Runoff = Net precipitation x Area of the drainage basin

Area of the drainage basin = 4 acres;

1 acre = 43,560 sq. ft
Runoff = Net precipitation x Area of the drainage basin
22,000 = Net precipitation x (4 x 43560)
Net precipitation = 22,000/(4 x 43,560) = 0.1263 ft

= 1.515 inches

STEP 3: Find the maximum flow due to 1 inch rainfall;

According to the hydrograph given, maximum flow of 12.5 cu.ft/sec is achieved due to a net precipitation of 1.515 inches.

Maximum flow due to 1.0 inch = 12.5/1.515 = 8.25 cu.ft/sec

STEP 4: Find the maximum flow due to 3.5 inch rainfall;

Maximum flow due to 1 inch rainfall = 8.25 cu.ft/sec
Maximum flow due to 3.5 inch rainfall = 8.25 x 3.5 cu.ft/sec = 28.9 cu.ft/sec   (Ans C)

Practice Problem: Drainage basin has an area of 12.8 acres. A hydrograph was developed for a net precipitation of 4.6 inches. What is the area of the hydrograph?
A) 98,098 cu. ft   B) 193,510 cu. ft   C) 291,256 cu. ft   D) 213,734 cu. ft

Solution:
STEP 1:   Find the runoff;

Runoff = Net Precipitation x Area of the drainage basin
Runoff = 4.6 in x 12.8 acres = (4.6/12) ft x (12.8 x 43,560) sq. ft = 213,734.4 cu. ft
In addition, we know that area of the hydrograph = Runoff.
Hence area of the hydrograph = 213,734.4 cu. ft   (Ans D)

Non Bell Shaped Hydrographs: Typically hydrographs are bell shaped. However, in some instances hydrographs are approximated to trapezoids or various other shapes.

Practice Problem: Approximate hydrograph for a stream is shown below.

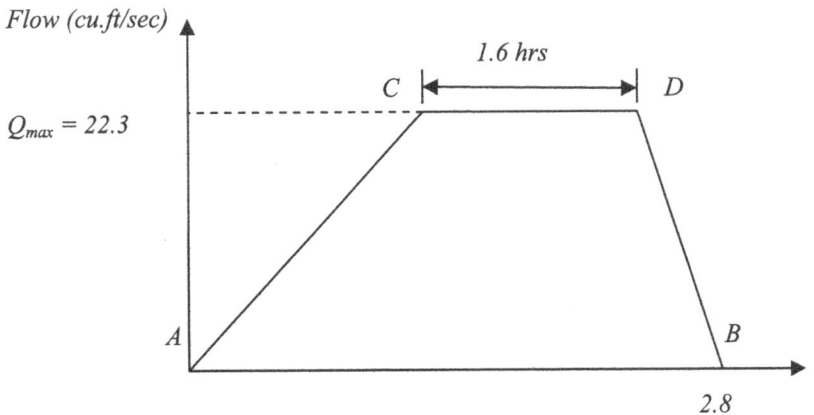

The net precipitation was measured to be 1.3 inches and the maximum flow was 22.3 cu. ft/sec. What is the area of the drainage basin?
A) 37.43 acres   B) 27.41 acres   C) 174.09 acres   D) 7.93 acres

Solution):   Area of the hydrograph provides the total runoff or the flood volume.

> Area of the hydrograph = Runoff

Runoff is given by the following equation.

> Runoff = Net precipitation x Area of the drainage basin

STEP 1: Find the area of the hydrograph;

Area of the hydrograph using geometry   = (Top + Base)/2 x Height
     = (CD + AB)/2 x Q

Time is given in hours. This needs to be converted to seconds since the flow is given in cu. ft/sec.

CD = 1.6 hrs = 1.6 x 3,600 sec = 5,760 sec

AB = 2.8 hours = 2.8 x 3600 seconds = 10,080 sec

Area of the hydrograph = Runoff = (CD + AB)/2 x Q_{max}
     = (5,760 + 10,080)/2 x 22.3 cu. ft

         Runoff = 176,616 cu. ft

STEP 2:   Runoff = Net precipitation x Area of the drainage basin

Runoff = 176,616 cu. ft   (Found in previous step)

Net precipitation = 1.3 inches;   (given)

Area of the drainage basin = A acres

176,616 cu. ft = 1.3 x A acre.in

Convert all units to ft.

176,616 cu. ft = (1.3/12) x A x 43,560 cu.ft

A (Area of the drainage basin) = 37.43 acres   (Ans A)

Practice Problem: Hydrograph for a catchment basin is approximated to a triangle as shown below. Area of the catchment basin is 20 acres. Net precipitation on the catchment basin is 2.9 inches. Find the maximum flow.

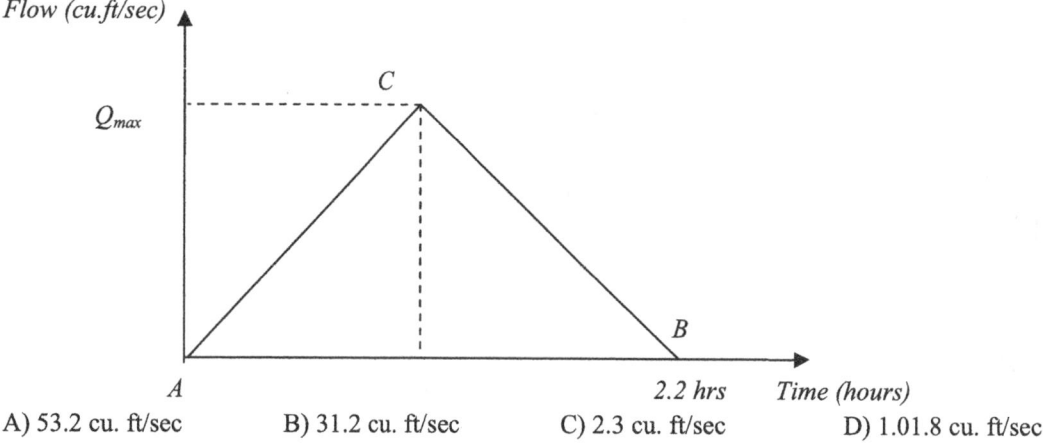

A) 53.2 cu. ft/sec   B) 31.2 cu. ft/sec   C) 2.3 cu. ft/sec   D) 1.01.8 cu. ft/sec

Solution:

STEP 1:   Find the runoff;

Runoff can be found in two different ways.

Runoff = Area of the hydrograph ------------------------------(1)

Also

Runoff = Area of the catchment basin x Net Precipitation-----(2)

Runoff = Area of the hydrograph = Area of the triangle ABC = 2.2 x Q_{max}/2

2.2 is hours. This needs to be converted to seconds since Q is in cu. ft/sec.

Runoff = Area of the hydrograph = Area of the triangle ABC = (2.2 x 3,600) x Q_{max}/2 -------(1)

Area of the catchment basin = 20 acres = (20 x 43,560) sq. ft

Net precipitation = 2.9 in = 2.9/12 = 0.242 ft

From (2)

Runoff = Area of the catchment basin x Net Precipitation-----(2)

Runoff = (20 x 43,560) x 0.242 cu. ft = 210.830.4 cu. ft

From (1)

(2.2 x 3,600) x Q_{max}/2 = 210,830.4

Q_{max} = 53.2 cu. ft/sec   (Ans A)

Hydrograph Problems:

(Complete solutions provided in Four Sample Exams for the Civil PE Exam, Second Edition)

Problem 1)   Find the maximum flow in the drain pipe (cu.ft/sec) for a unit hydrograph. Area of the drainage basin is 4 acres.

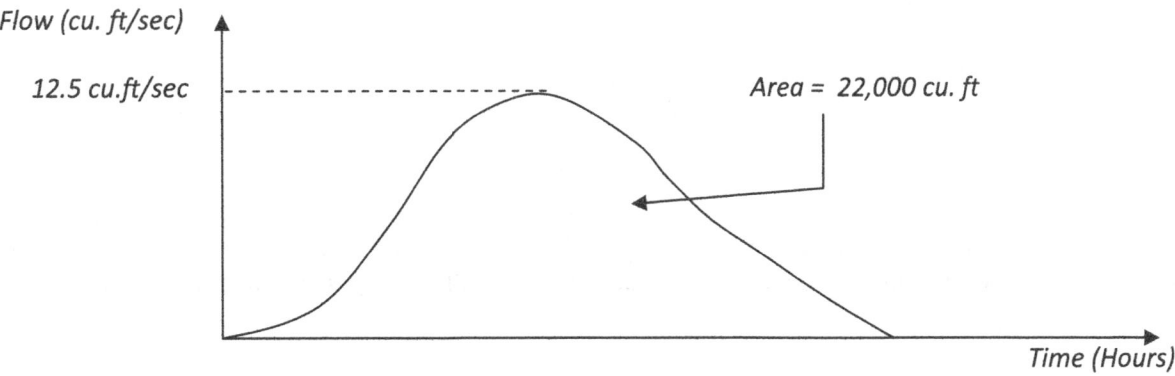

A) 7.86 cu.ft/sec   B) 8.27 cu.ft/sec;   C) 2.78 cu.ft/sec;   D) 7.97 cu.ft/sec

Ans B

Problem: Approximate hydrograph for a stream is shown below.

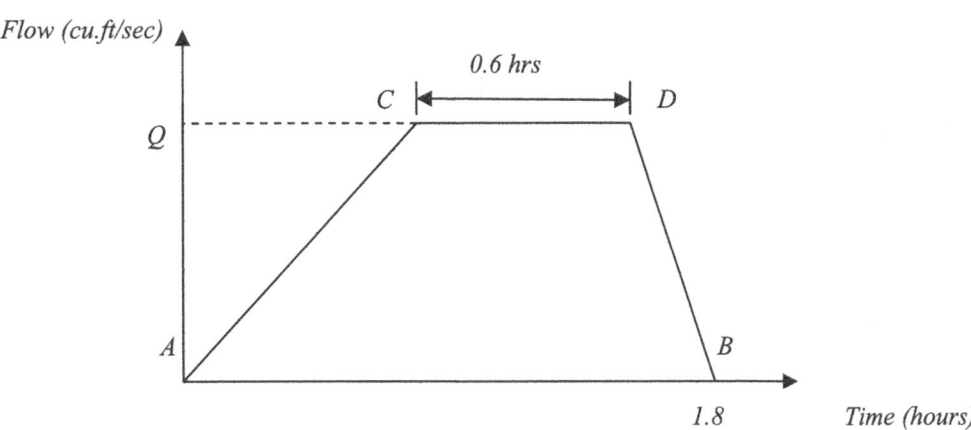

Area of the drainage basin was found to be 15 acres and the net precipitation was 1.3 inches.
Find the maximum flow (Q) in cu.ft/sec.

A) 16.4 cu.ft/sec   B) 22.9 cu.ft/sec   C) 76.7 cu.ft/sec   D) 42.2 cu.ft/sec

Ans A

Problem 3):   Find the runoff for a unit hydrograph for the data given in the previous problem.

   A) 54,450 cu.ft   B) 64,950 cu.ft   C) 12,512 cu.ft   D) 23,887 cu.ft
Ans A

5.2.7   SCS Curve Number Method:

SCS curve number method was developed by USDA, Soil Conservation Service (SCS). This method was developed to predict the runoff from an area.

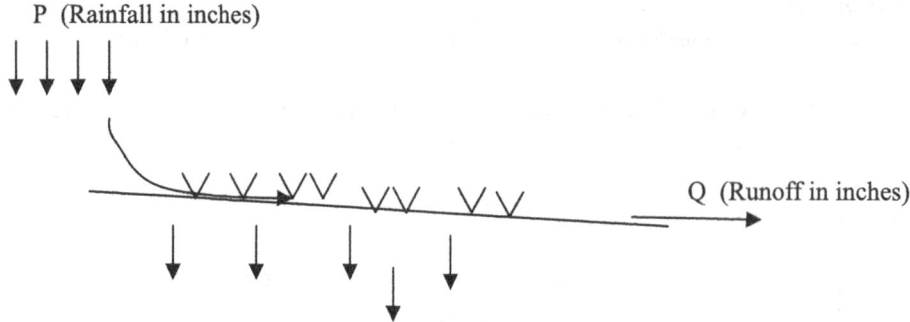

Above figure shows, P inches of rainfall falling into an area. This rainwater would flow down the area. During the process some water would be infiltrated. Some evaporation also happens. There may be other losses such as water been collected by people, water getting into underground tunnels etc. Generally main loss of water is due to infiltration to the ground.

Runoff = Rainfall - Losses

Infiltration in Sandy Soils: More water infiltrate into sandy soils than to clay soils. Sandy soils typically have a higher permeability.

$$Q = (P - 0.2\ S)^2/(P + 0.8\ S)$$

$$S = 1{,}000/CN - 10$$

P = Rainfall   (Measured in inches)
Q = Runoff (Measured in inches)
S = Maximum potential retention (Measured in inches)
CN = Curve Number

Hydrologic Soil Groups: USDA SCS divides the soil into four categories.

Group A— (More than 90% sand and gravel) (High Infiltration, Low Runoff):
Soils in this group have low runoff potential and high infiltration.   Group A soils typically have less
than 10 percent clay and more than 90 percent sand or gravel and have gravel or sand textures.
Group B— (50% to 90% sand and loamy sand) (Moderately high infiltration, moderately low runoff)
Soils in this group have moderately low runoff potential. Group B soils typically have 50 percent to 90 percent sand and
loamy sand.
Group C—(Less than 50% sand) (Low infiltration, Moderately high runoff)
Soils in this group have moderately high runoff potential. Soils typically have between less than 50 percent sand.
Group D— (Less than 50% sand and greater than 40% clay), (Very low infiltration, High runoff):

Group D soils have very low infiltration and high runoff potential. Group D soils typically have greater than 40 percent
clay and less than 50 percent sand.

Problem: Runoff in group D soil is higher than the runoff in group B soil.
Answer: Correct

Problem: Infiltration in group D soil is higher than the infiltration in group B soil.
Answer: Incorrect

Hydrologic Condition:

Cover description			Curve numbers for hydrologic soil group—			
Cover type	Treatment*	Hydrologic condition**	A	B	C	D
Fallow	Bare Soil	—	77	86	91	94
	Crop residue cover (CR)	Poor	76	85	90	93
		Good	74	83	88	90
Row crops	Straight row (SR)	Poor	72	81	88	91
		Good	67	78	85	89
	SR + CR	Poor	71	80	87	90
		Good	64	75	82	85
	Contoured (C)	Poor	70	79	84	88
		Good	65	75	82	86
	C + CR	Poor	69	78	83	87
		Good	64	74	81	85
	Contoured & terraced (C&T)	Poor	66	74	80	82
		Good	62	71	78	81
	C&T + CR	Poor	65	73	79	81
		Good	61	70	77	80
Small grain	SR	Poor	65	76	84	88
		Good	63	75	83	87
	SR + CR	Poor	64	75	83	86
		Good	60	72	80	84
	C	Poor	63	74	82	85
		Good	61	73	81	84
	C + CR	Poor	62	73	81	84
		Good	60	72	80	83
	C&T	Poor	61	72	79	82
		Good	59	70	78	81
	C&T + CR	Poor	60	71	78	81
		Good	58	69	77	80
Close-seeded or broadcast legumes or rotation meadow	SR	Poor	66	77	85	89
		Good	58	72	81	85
	C	Poor	64	75	83	85
		Good	55	69	78	83
	C&T	Poor	63	73	80	83
		Good	51	67	76	80

Condition: Use the above table by USDA to use SCS curve number method.

Condition could be good or poor. When there is more than 75% grass cover, the condition is considered to be good. When grass cover is between 50 to 75%, the condition is considered to be fair. When grass cover is less than 50%, the condition is poor.

Poor condition (grass cover <50%)
Fair condition (grass cover 50 to 75%)
Good condition (grass cover > 75%)

Curve Number: Once you know the soil type and soil condition, you would be able to find the curve number using tables provided by SCS. Once you have the curve number runoff can be found.

Practice Problem: Certain land is covered with row crops. Crops are planted in straight rows. Soil type in the land is type B. Hydrologic condition is considered to be good. If the rainfall in the area is 3 inches what is the runoff?

Solution:
STEP 1:   Find the curve number using SCS table.
For row crops, straight rows, soil condition good, soil type B   ---->   Curve number = 78

STEP 2:   Find S; (Maximum potential retention (Measured in inches)
$S = 1,000/CN - 10$
$S = 1,000/78 - 10 = 2.82$ in.

STEP 3:   Find Q (Runoff)
$Q = (P - 0.2 S)^2/(P + 0.8 S)$
$Q = (3 - 0.2 \times 2.82)^2/(3 + 0.8 \times 2.82) = 1.13$ in.
As you could see in this case, rainfall was 3 inches but the runoff was only 1.13 inches.

Practice Problem: Watershed area has an area of 8 acres. The land is covered with small grain crops. Small grain crops are contoured and terraced. Soil in the land is less than 50% sand. Hydrologic condition is considered to be poor. If the rainfall in the area is 3 inches what is the runoff in cubic feet?

Solution:
STEP 1:   Find the curve number using SCS table.
For small grain crops, contoured and terraced (C and T), soil condition poor, soil type C   ---->   Curve number = 79

STEP 2:   Find S; (Maximum potential retention (Measured in inches)

$S = 1,000/CN - 10$
$S = 1,000/79 - 10 = 2.66$ in.

STEP 3:   Find Q (Runoff)

$Q = (P - 0.2 S)^2/(P + 0.8 S)$
$Q = (3 - 0.2 \times 2.66)^2/(3 + 0.8 \times 2.66) = 1.19$ in.

STEP 4:   Find the volume of water;
1.19 inches per area of 8 acres.

Volume = 8 x 1.19/12 Acre. ft

Acre = 43,560 sq. ft

Volume = (8 x 43,560) x 1.19/12 cubic. feet = 34,557.6 cu. ft

5.3 Wastewater Treatment:

Clean water comes to our houses. We use the water to clean dishes, flush toilets, bathing and laundry. Remaining water is known as wastewater. This water coming out from our houses cannot be discharged into rivers and oceans without treatment.

Wastewater generated from houses is known as municipality wastewater. Other than municipality wastewater there are industrial wastewater. In addition, there is agricultural wastewater.

What is wastewater?

Kitchen sinks - Food residue, dish washing liquids

Bathroom sinks - Shaving cream, toothpaste, soap, hair spray, spit

Toilets - Feces, water

In some cities municipality wastewater is combined with storm water. These systems are known as combined sewer systems. Main advantage of combined sewer is that only one large pipe network has to be built. On the other hand, large quantity of water has to be treated unnecessarily. Many cities have started to build two separate systems for sanitary sewer and storm water.

5.3.1 Collection Systems

Earliest known sewer system, 3,000 B.C – Harappa

Earliest known sewer collection system was found in Harappa, India. This system was built using bricks. Later civilizations such as Roman, Greek used stones or clay bricks to build sewer lines. Today wastewater is carried from houses to main sewer lines using pipes. These pipes are known as laterals. Laterals are connected to main sewer lines. Main sewer lines take wastewater to lift stations of pump stations. Pump stations lift or pump the wastewater to a higher elevation. This is important since wastewater flows on gravity.

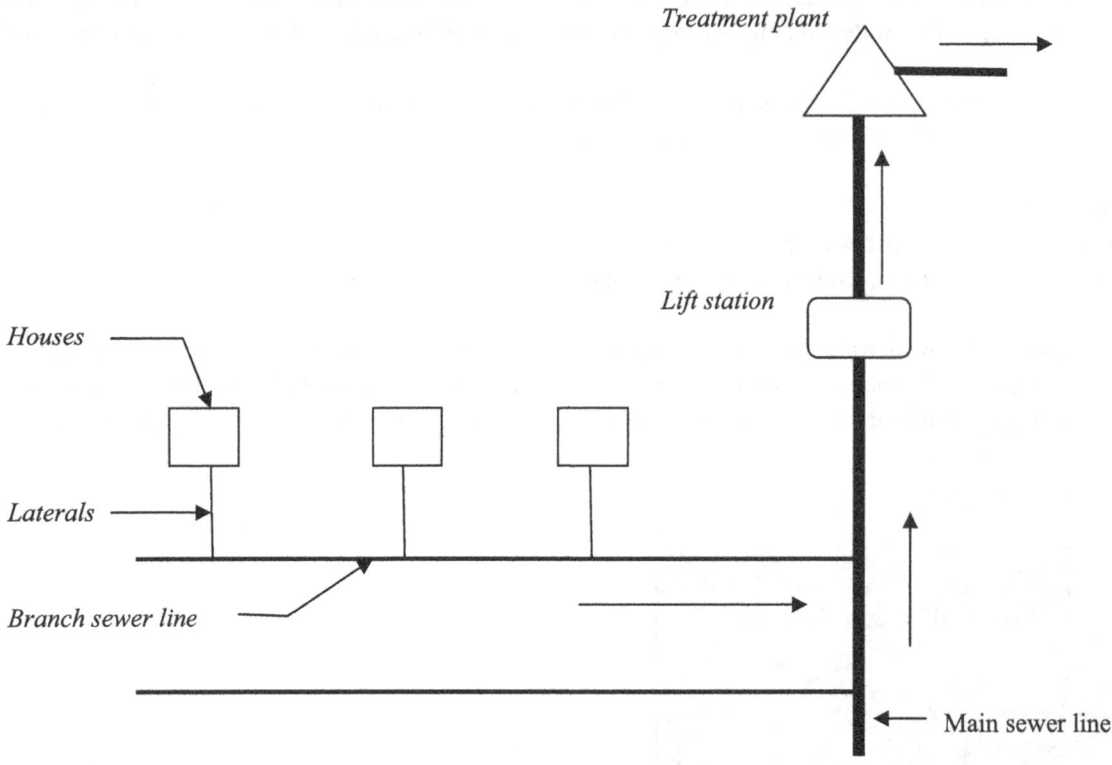

Wastewater Collection System

A wastewater collections system is shown above. Wastewater in residential houses go to sewer laterals. Sewer laterals feed branch sewer lines. Branch sewer lines feed main sewer lines. Main sewer lines are connected to lift stations or pump stations. At lift stations, sewer is lifted to a higher elevation. If not for lift stations, sewer trenches would be very deep. This point is illustrated below.

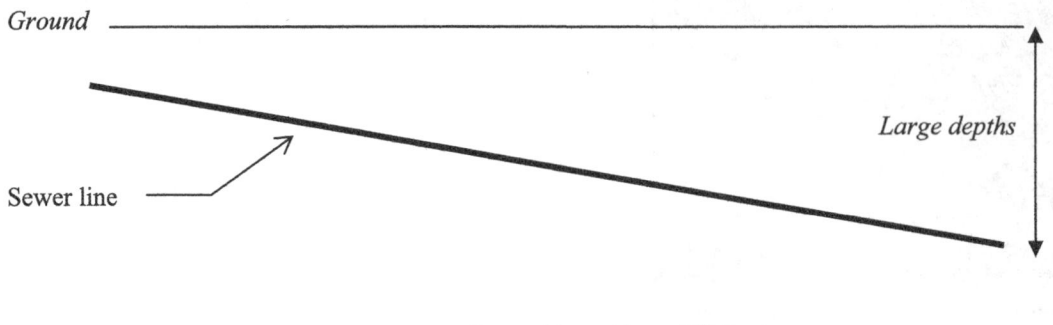

Sewer Line without Lift Stations

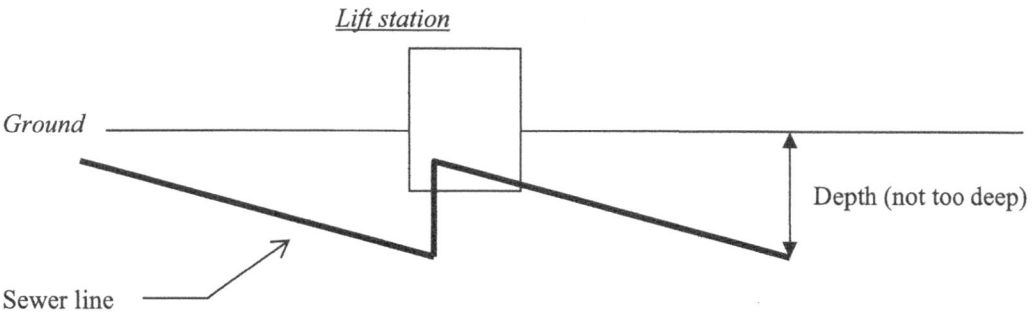

Lift station

Ground

Sewer line

Depth (not too deep)

Sewer Line with Lift Stations

Since sewer is lifted to higher depths at lift stations, depth of sewer trenches will be reduced. Deep sewer trenches are very costly to construct. Excavation cost, shoring cost and manhole construction cost would be much higher for deeper trenches. Also deep excavations are unsafe for plumbers to work. It is not desirable to place sewer lines too deep since the possibility of contaminating groundwater. Due to these reasons, sewer is lifted to a higher elevation at lift stations. (Also known as pump stations).

5.3.2 Wastewater Treatment:
Wastewater cannot be discharged into waterways without treatment. Many different methods are used to treat wastewater.

Gravity: Take a glass of unclean water. Just leave it on a table for few hours. After few hours, sludge would form at the bottom and scum would form on top.

Wastewater bottle is shown above. After few hours later sludge layer would form at the bottom and a scum layer would form on top.

This concept is used in clarifier tanks. Clarifiers are also known as primary sedimentation tanks. Wastewater is sent to a clarifier at a slow flow rate. Slow flow rate would allow sludge and scum to form. Scum is removed using skimmers and sludge is removed using sludge removal system.

Clarifiers (Primary Sedimentation Tanks):

Clarifier.

Below figure shows the mechanism of a clarifier.

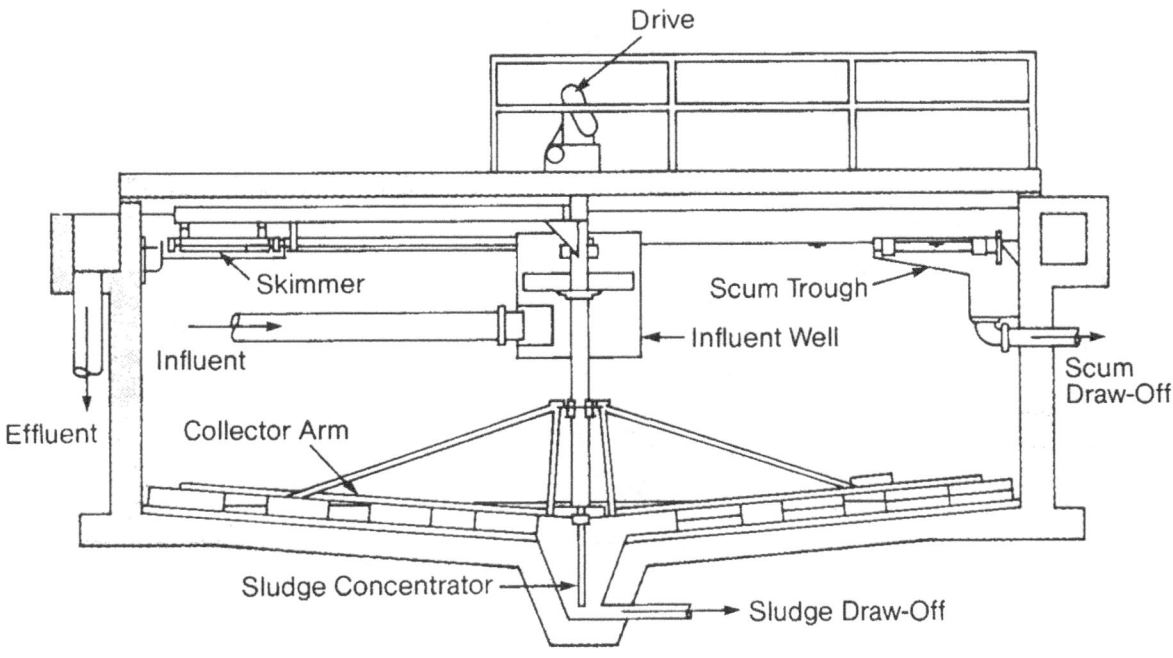

Take a good look at the clarifier shown above. Wastewater comes to the tank from the influent pipe. Sludge is deposited at the bottom of the tank. Sludge is removed using suction pumps. Scum is formed on the surface. Scum is drawn off using scum draw-off mechanism. Effluent is sent to the next tank.

Biological Treatment:

Aeration tanks: You could see that clarifiers simply let gravity do the work. On the other hand we know that we live in a world of trillions of Bacteria. Bacteria are living organisms and they need food. Wastewater contains plenty of organic matter. Think of food residue, feces and tree leaves. All these contain organic matter that can be used up by Bacteria. If we can introduce Bacteria to wastewater, Bacteria would consume organic matter and produce CO_2 and H_2O. Once Bacteria has done their job of consuming organic matter, they would be killed off using ultra violate rays.

Organic matter ⟶ Bacteria ⟶ CO_2 and H_2O

How to Multiply Bacteria in Wastewater?   Bacteria need three things. Water, food (Organic matter) and Oxygen. In that sense Bacteria is no different from humans. Wastewater has plenty of water and organic matter. What is lacking is Oxygen. Oxygen is introduced to wastewater thru aeration tanks.

Aeration tank is shown on left. Photo on right shows cups at the bottom of the tank providing air jets. Bacteria needs Oxygen to thrive. Oxygen is introduced to wastewater so that Bacteria would multiply.

Trickling Filters: Trickling filters are another good way to introduce Oxygen and new cultures of Bacteria. In the case of trickling filters, water is trickled thru a gravel media. Gravel is could be natural gravel or artificial gravel. Various Bacteria cultures are introduced to the gravel media.

Trickling Filter: Wastewater is sent thru a bed of gravel as shown. The gravel bed has various bacteria cultures who consume organic matter. Also during the trickle down process, Oxygen is introduced to wastewater. In some instances anaerobic digestion also occur in trickling filters.

Bar Screens: Bar screens are used to remove large objects such as tree branches, coke cans, bottles etc.

Bar screen to capture coke cans, tree branches and other large objects

Grit Chamber: Grit chambers are used to remove grit or sandy particles. Wastewater is sent to a grit chamber. Grit or sandy particles settle at the bottom of the tank.

Grit Chamber: _Grit or heavy sandy particles are settled at the bottom of the tank._

Wastewater Treatment Plant: Typical wastewater treatment plant will have following elements.

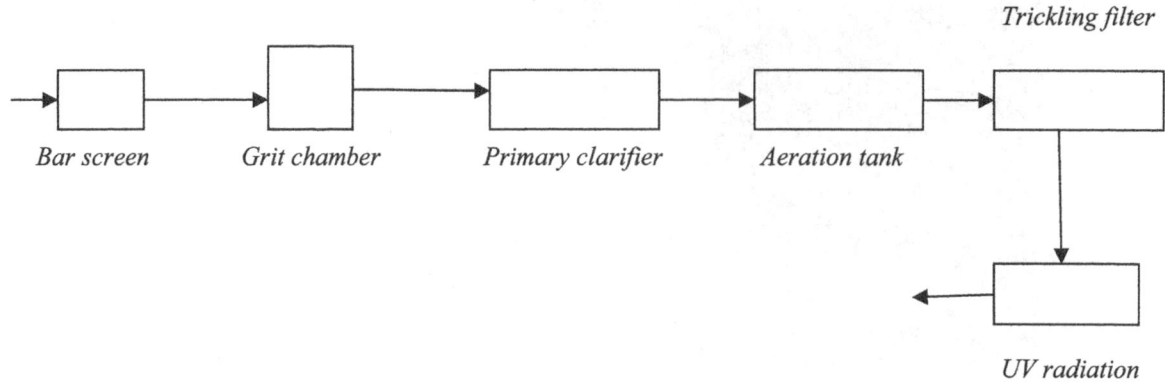

Above is a very simplistic representation of a wastewater plant. First large objects are removed using bar screens. Then sandy particles are removed. Primary clarifier removes significant amount of sludge and scum. Aeration tank pumps Oxygen to multiply Bacteria. Bacteria would digest organic matter and give out CO_2 and water. Trickling filter also removes organic matter using Bacteria. UV radiation would kill off any Bacteria in wastewater. Real life wastewater plant would look like as shown below.

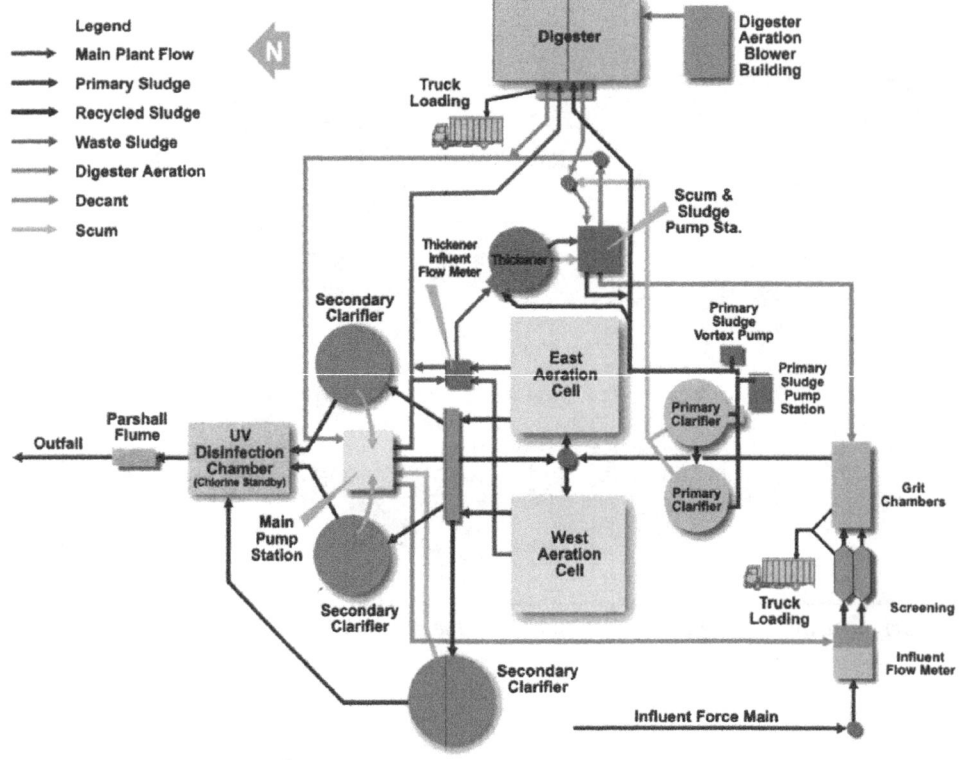

As you could see there are sludge thickeners, scum pumping stations, UV disinfection, grit chambers, screening, truck loading to remove grit and many other elements to do various activities.

BOD (Biological Oxygen Demand):

We are interested in knowing how much micro-organisms (Bacteria) are present in wastewater. Unfortunately it is not easy to measure the amount of Bacteria present in wastewater. Nevertheless, there are easier ways to measure the Oxygen level. Bacteria needs Oxygen to live. If there is less Oxygen in the wastewater, that means Bacteria has used up the Oxygen. Let us look at an example. Let us say there is a city and we cannot go inside to count the number of people. However, we can see the number of food trucks going in. By measuring the consumption of food by the city dwellers we can assess the population of the city. We are doing something similar here. We are unable to count Bacteria, but we can measure the Oxygen level.

Low level of dissolved Oxygen in wastewater ----------→   High level of Bacteria. Oxygen been used by Bacteria.

High level of dissolved Oxygen in wastewater   ----------→   Low level of Bacteria.

Dissolved Oxygen level is measured in mg/L.

Photo on left shows water with plenty of dissolved Oxygen. Photo on the right shows very little dissolved Oxygen. When dissolved Oxygen level is low, fish would start to die.

Practice Problem: Sewer pipe broke and sanitary waste was discharged into a river. The fish in the river started to die. What caused the death of fish?

Solution: Sanitary waste contains organic matter. Organic matter contains Bacteria. These Bacteria would consume organic matter in sanitary waste and multiply. During the process they also use dissolved Oxygen in the river water. Dissolved Oxygen level will go down and fish would die.

Dissolved Oxygen Levels Needed for Fish:

Different fish species need different dissolved Oxygen levels.

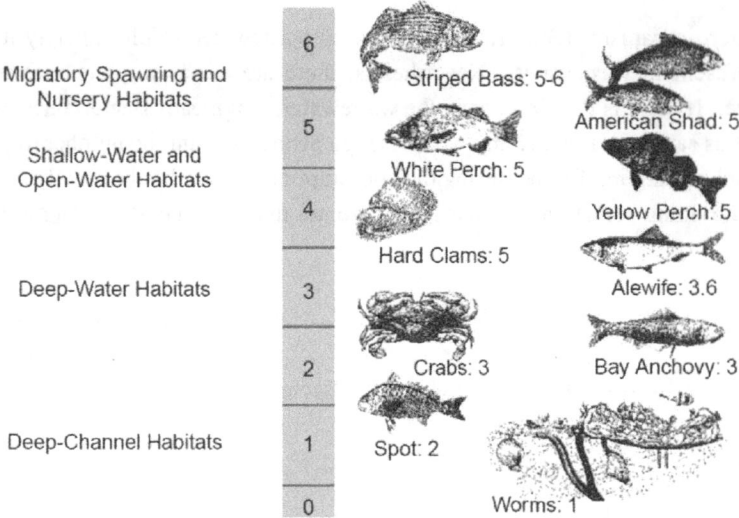

As per above figure;

Dissolved Oxygen level 0.2 mg/L or less = Very few worms may be able to live

Dissolved Oxygen level 1.0 mg/L = Some fish can live

Dissolved Oxygen level 3.0 mg/L = Crabs and anchovy can survive

According to above figure Stripped Bass need very high level of dissolved Oxygen to survive. (5.0 to 6.0 mg/L)

BOD$_5$ Test Procedure:

5 day biological oxygen demand is known as BOD$_5$. In other words, how much Oxygen was used up by Bacteria in 5 days. Why 5 days was selected. Why not 7 days or 3 days?

If we select 7 days that would delay the results. If we select 3 days, we may not a good reading. Hence 5 days is used that gives a representative reading and also has results in a reasonable time period.

STEP 1: Wastewater sample is taken and diluted with distilled water. Dilution is necessary for various practical reasons.

STEP 2: Measure the dissolved oxygen level of the sample. This is the DO$_i$ or dissolved oxygen initial value.

STEP 3: Seal the sample and wait for 5 days. One can wait two days or six days. However, typically 5 days are selected so that values can be compared.

What happens during this stage? Wastewater has bacteria. This bacteria would start to use up the oxygen in the wastewater sample. If there is more bacteria, more oxygen would be used up. If the sample has less bacteria, less oxygen would be used up.

STEP 4: Measure the dissolved oxygen level after 5 days, (DO$_f$), or dissolved oxygen final.

- If there is lot of bacteria, then DO$_f$ would be much lower than DO$_i$.

    Why? Bacteria will be using dissolved oxygen to survive.

- If there is less bacteria, then DO$_f$ value would be closer to DO$_i$.

- If there is no bacteria, then DO$_f$ value would be same as DO$_i$.

    Why? Since there is no bacteria, there is nobody there to use the dissolved oxygen.

Definition of Biological Oxygen Demand:

BOD_5 is defined as follows.

$$BOD_5 = \frac{DO_i - DO_f}{\frac{V_{sample}}{V_{sample} + V_{Dilution}}}$$

DO_i = Initial dissolved Oxygen level
DO_f = Final dissolved Oxygen level
V_{sample} = Volume of wastewater sample
$V_{Dialution}$ =   Volume of dilution

$$BOD_5 = \frac{DO_i - DO_f}{\frac{V_{sample}}{V_{sample} + V_{Dilution}}}$$

Practice Problem):   100 ml wastewater sample was diluted with 700 ml of water. Initial dissolved Oxygen level (DO_i) was found to be 12.8 mg/L. After 5 days of incubation, final dissolved Oxygen level (DO_f) was found to be 4.9 mg/L. What is the 5 day BOD of the sample?

A) 111.8 mg/L   B) 56.8 mg/L   C) 63.2 mg/L   D) 27.5 mg/L

Solution: BOD_5 is defined as follows.

$$BOD_5 = \frac{DO_i - DO_f}{\frac{V_{sample}}{V_{sample} + V_{Dilution}}}$$

DO_i = Initial dissolved Oxygen level = 12.8 mg/L.
DO_f = Final dissolved Oxygen level = 4.9 mg/L
V_{sample} = Volume of wastewater sample = 100 ml
$V_{Dialution}$ =   Volume of dilution = 700 ml

$$BOD_5 = \frac{DO_i - DO_f}{\frac{V_{sample}}{V_{sample} + V_{Dilution}}} \quad = \frac{12.8 - 4.9}{\frac{100}{100 + 700}} \quad = 63.2 \text{ mg/L}$$

(Ans C)

Practice Problem):   100 ml wastewater sample was diluted with X ml of water. Initial dissolved Oxygen level (DO_i) was found to be 9.0 mg/L. After 5 days of incubation, final dissolved Oxygen level (DO_f) was found to be 3.2 mg/L. 5 day BOD of the sample is 43.2 mg/L. What is volume of dilution?

A) 320 ml   B) 645 ml   C) 732 ml   D) 275 ml

Solution: BOD_5 is defined as follows.

$$BOD_5 = \frac{DO_i - DO_f}{\frac{V_{sample}}{V_{sample} + V_{Dilution}}}$$

DO_i = Initial dissolved Oxygen level = 9.0 mg/L.
DO_f = Final dissolved Oxygen level = 3.2 mg/L
V_{sample} = Volume of wastewater sample = 100 ml
$V_{Dialution}$ = Volume of dilution = X ml
BOD_5 = 43.2 mg/L.

$$BOD_5 = \frac{DO_i - DO_f}{\frac{V_{sample}}{Vsample + VDilution}} = \frac{9.0 - 3.2}{\frac{100}{(100 + X)}} = 43.2 \text{ mg/L}$$

5.8 = 43.2 x (100/100 + X)
5.8 x (100 + X) = 43.2 x 100
X = 4320/5.8 – 100 = 645 ml
(Ans B)

Ultimate BOD and the Deoxygenation Rate:

General BOD curve as shown in the figure below.

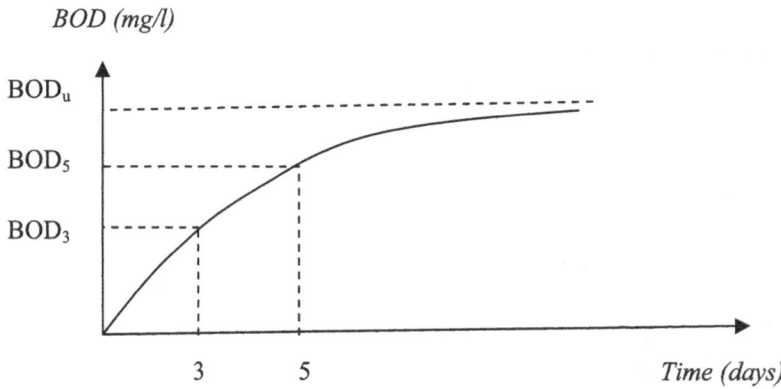

Above figure shows BOD vs. time graph. BOD value gradually tops off and that is known as ultimate BOD. If you know BOD_5 or BOD_3 then ultimate BOD can be found using the equation given below.

$$BOD_t = BOD_u \cdot (1 - e^{-k_1 t})$$

BOD_t = BOD value at time "t" in mg/L.
BOD_u = Ultimate BOD value in mg/L.
k_1 = Deoxygenation rate constant ($days^{-1}$)
t = time in days

Practice Problem):   Wastewater sample has a BOD_5 value of 82 mg/L. Deoxygenation rate constant of the sample is 0.25. What is the ultimate BOD of the wastewater sample?
A) 114.9 mg/L   B) 78.9 mg/L   C) 112.8 mg/L   D) 167.8 mg/L

Solution):
$BOD_t = BOD_u \cdot (1 - e^{-k_1 t})$

BOD_t = BOD value at time "t" in mg/L.
BOD_u = Ultimate BOD value in mg/L.
k_1 = Deoxygenation rate constant ($days^{-1}$)
t = time in days

STEP 1:   Write down known values;
In this problem 5 day BOD is given to be 82 mg/L.

Hence BOD_t = BOD_5 = 82 mg/L.
k_1 is given to be 0.25.
t = 5 days

STEP 2:   Apply the equation;

$BOD_t = BOD_u \cdot (1 - e^{-k_1 t})$

$82 = BOD_u \cdot (1 - e^{-0.25 \times 5})$
$82 = BOD_u \cdot (1 - 0.286505)$
$BOD_u = 114.93$ mg/L.
Ans A

Practice Problem):   Wastewater sample has a 5 day BOD value of 65 mg/L. The Deoxygenation rate constant is 0.28. Find the 10 day BOD value of this wastewater sample.
A) 100.2 mg/L   B) 68.2 mg/L   C) 78.9 mg/L   D) 81.0 mg/L

Solution): Same equation used for the previous problem can be used for this problem as well.

STEP 1:   Write down given values;
BOD_5 = 65 mg/L.
k_1 = 0.28.
t = 5 days

STEP 1:   Apply the equation between BOD_5 and ultimate BOD;

$BOD_t = BOD_u \cdot (1 - e^{-k_1 t})$
$65 = BOD_u \cdot (1 - e^{-0.28 \times 5})$
BOD_u = Ultimate BOD = $65/(1 - e^{-0.28 \times 5})$ = 86.27 mg/L

STEP 2:   Apply the equation from 0 to 10 days;
Since you have already found the ultimate BOD, now you can find BOD_{10}.
BOD_{10} is y_{10} in the equation.

$BOD_t = BOD_u \cdot (1 - e^{-k_1 t})$
$BOD_{10} = BOD_u \cdot (1 - e^{-k_1 t})$
$BOD_{10} = 86.27 \cdot (1 - e^{-0.28 \times 10})$
$BOD_{10} = 81.02$ mg/L.
Ans D

Practice Problem): In a sanitary sewer system, manholes are provided at

A) All abrupt grade changes to the sewer pipeline
B) Locations where pipe diameter changes
C) Terminal of a line
D) All of the above

Solution: Typically manholes are provided at

- All abrupt grade changes to the sewer pipeline
- Locations where pipe diameter changes
- Terminal of a line
- At intersection of main sewers
- Whenever there is a change to pipe materials

Ans D

5.3.3 Water Treatment;

People of ancient civilizations understood the need for clean water for drinking purposes. Sushruta of India recommended to boil and heat water under the sun and filtering with gravel and charcoal prior to drinking. (*Sushruta Samhita,* Arabic translation Kitab-i-Susrud). In Greece, Hippocrates advised to filter water with a cloth bag known as "*Hippocrates sleeve*". Ancient Chinese book *I-Ching* recommended to boil water prior to drinking. In addition, the author recommend to keep drinking water wells clean.

Sushruta wrote a book known as Sushruta Samhita, which described purifying water using heating in the sun and filtering using gravel and charcoal. Hippocrates used filters made of cloth and known as Hippocrates sleeve.

We can say that ancient people used boiling, heating in the sun and filtering as methods to purify water. Boiling is prohibitively expensive and heating in the sun is not very practical. Today following methods are used to purify water.

Coagulation
Sedimentation
Filtration
Chemical Treatment - (Chlorine)
Disinfection
Storage

Coagulation: Coagulation is the process of adding coagulants. Coagulation removes dirt and other particles suspended in water. Alum is the main chemical used to coagulate particles. Alum get attached to dirt particles and form tiny sticky particles called "floc". The combined weight of the dirt and the alum (floc) become heavy enough to sink to the bottom during sedimentation.

Sedimentation: The heavy particles (floc) settle to the bottom and the clear water moves to filtration.

Filtration: The water passes through filters, some made of layers of sand, gravel, and charcoal that help remove even smaller particles. Filters also remove Bacteria and various other micro-organisms.

Disinfection: One of the major advances in water purification is the discovery of Chlorine by a Swedish chemist Carl Wilhelm Scheele. Small amount of chlorine is added to kill any bacteria or microorganisms that may be in the water.

Hardness and Softness in Water Supply System:

Water "hardness" and "softness" is determined by calcium and magnesium concentration in water. High concentrations of these minerals would make the water "hard". It is difficult to use hard water for cleaning of cloths and dishes. Also high concentrations of minerals can be a health hazard.

pH Value in Drinking Water:

pH value less than 7.0 indicates acidic water. pH value greater than 7.0 indicates alkaline water. pH value of drinking water is maintained slightly above 7.0. Hence the drinking water is slightly alkaline. The main reason for this is due to the fact that pipes would corrode in an acidic environment. Corroding pipes could increase the lead and iron levels in drinking water.

Parameters tested in Drinking Water:

Following parameters are generally tested to establish the quality of drinking water

- **Microbial - (bacteria and algae)**: Microbes create diseases. Drinking water is constantly tested for microbes.

- **Turbidity**: Water contains suspended solids. Turbidity is a measure of suspended and colloidal particles. Some typical solid particles that could be in drinking water are clay, silt, organic and inorganic matter, algae, and microorganisms.

- **Corrosiveness (pH and alkalinity):**
  Acidic water has a pH value less than 7.0. Acidic water tends to corrode pipes and increase the lead content in drinking water. Hence it is important to keep the drinking water pH level above 7.0. Typically between 8 and 9. Water with pH value above 7.0 is known as alkaline water.

- **Chemical (inorganic and organic):**   Tests are done to find out the levels of inorganic and organic chemicals.

- **Radionuclides:**   Radioactive particles can create a health hazard. Hence concentration of radioactive particles also has to be tested.

5.3.4. Water Distribution Systems:

Water is of paramount importance for any civilization. Ancient man used canals to divert water from reservoirs and lakes. Most wonderful example of ancient Engineering come to us from the Roman civilization. Aqua ducts were used to carry water to cities.

Roman Aqua Ducts distributed water to cities.

Water Main Materials:
Steel, Ductile Iron, Reinforced concrete and Fiberglass are used for water mains. Smaller size pipes feeding houses are typically copper.

5.4.0 Hydraulics – Pressure Conduits:

Imagine a building. People in the building need cold and hot water. In addition, pipes are needed to carry water for fire suppression systems, drainage, wastewater and storm water. In some buildings steam also been circulated. Steam pipes are not part of civil engineering. Mechanical engineers deal with steam pipes.

Very complicated piping network in a large building;

Water in closed conduits flows under pressure.
Following properties are required of water pipes;

- Resistance to corrosion and aggressive chemicals
- Should not contaminate the water due to rust
- Ease of installation
- Little maintenance
- Light weight (Load on structure will be less. Hence cost of structural beams and columns will be less).
- Reduced risk of theft (Expensive material can be stolen by thieves)

Pipe Materials:
- Large Pipes and Mains; Cast iron, ductile iron, steel, reinforced concrete, PVC, HDPE
- Home Piping: Copper, galvanized steel, PEX

5.4.1. Energy Equation:

Energy Conservation Law: Today we know that energy cannot be created nor destroyed. This law did not come in a direct way.   Gottfried Wilhelm Leibniz was working with machines in 1670s. He noticed that moving parts of machines had a constant parameter "m. v^2". Nobody knew why m. v^2 parameter seems to be a constant in many situations. They called it "kinetic energy". In addition, researchers noticed that "m. v^2" parameter in moving bodies gradually decreased due to friction. If "m. v^2" parameter represents kinetic energy where does it go? During this time period, Count Rumford noticed that loss of kinetic energy is proportional to the heat gain. In other words if a machine loses more kinetic energy it would generate more heat. Gradually it was apparent to the researchers that heat, work and kinetic energy are interconnected. Kinetic energy can change to heat or vice versa. Energy conservation law states that energy cannot be created nor destroyed.

Daniel Bernoulli;

Daniel Bernoulli who lived from 1700 to 1782 was a Swiss mathematician. He was the first to apply the idea of energy conservation to fluids. He saw that moving fluids had three forms of energy.

- Pressure energy
- Kinetic energy
- Energy due to elevation (Datum head)

Pressure Energy: Pressure is also a form of energy. In typical pipes, water flows under pressure. Hence pressure energy also has to be considered. Pressure energy is also known as pressure head. Head is another term for energy.
Pressure head is given by the following equation

$$\text{Pressure head} = P/\gamma$$

P = Pressure in psf and γ is the density of water in pcf.

Pressure head units = P (psf)/γ (pcf) = ft

Kinetic Energy (Velocity Head):   Kinetic energy is due to velocity of water particles.   Kinetic head is given by the following equation.

$$\text{Kinetic head } = v^2/2g$$

v = velocity in ft/sec.
g = Gravitational constant. (32.2 ft/sec^2)

Energy due to High Elevation (Datum Head):

Datum head is due to height of water (Z ft).

Total Energy inside a Pipe = Pressure head + Velocity Head + Datum Head
$$= P/\gamma + V^2/2g + Z$$

$$\text{Total Energy} = P/\gamma + V^2/2g + Z$$

Practice Problem: A horizontal water pipe is shown below. Water is moving from point A to point B. Assuming there is no head loss due to friction, which point has a higher pressure head?.

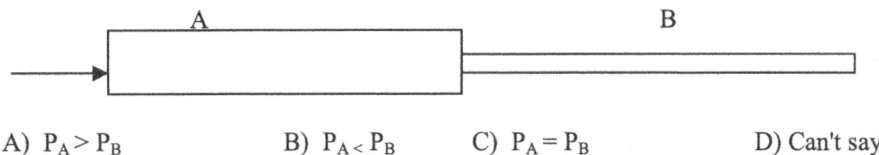

A) $P_A > P_B$ B) $P_A < P_B$ C) $P_A = P_B$ D) Can't say

Solution:

Large shown above has a low velocity. Hence large pipe will have a low kinetic energy. Smaller pipe has a larger velocity.
Let us apply the Bernoulli's energy equation;

$$P_A/\gamma + V_A^2/2g + Z = P_B/\gamma + V_B^2/2g + Z$$

Datum head "Z" is the same for both pipes. Hence,

$$P_A/\gamma + V_A^2/2g = P_B/\gamma + V_B^2/2g$$

We know that $V_A < V_B$
V_A is less than V_B since area of pipe A is larger.
Hence as per above equation, P_A has to be greater than P_B.   (Ans A)

Practice Problem: A pipe is sloped as shown. Pressure at point A is 5 psi. The internal diameter of the pipe is 6 inches. Find the pressure at point B. Assume there is no head loss due to friction. There is a 7 ft drop in elevation between two points.

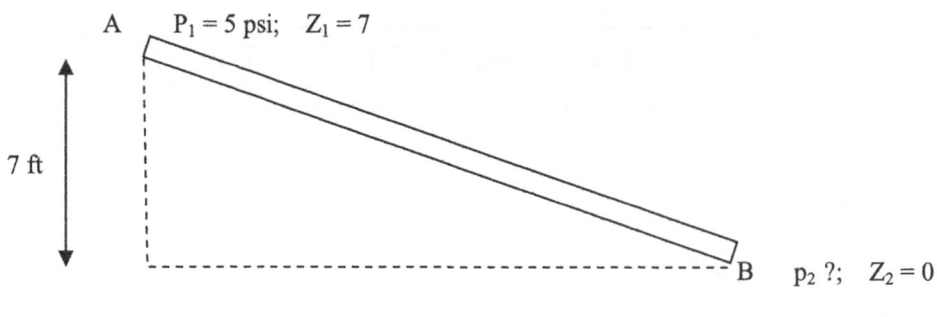

A) 5.00 psi   B) 6.03 psi   C) 3.54 psi   D) 12.34 psi

Solution:
STEP 1:   Write the energy equation;
Since there is no energy loss due to friction,

Energy at point A = Energy at point B

$P_1/\gamma + V_1^2/2g + Z_1 = P_2/\gamma + V_2^2/2g + Z_2$
Since the diameter of the pipe is same at point A and B, $V_1 = V_2$

STEP 2:   Find P_1;
Pressure at point A is given in psi. This has to be converted to psf.
$P_1 = 3 \text{ psi} = 3 \times 144 \text{ psf} = 432 \text{ psf}$

STEP 3:   Apply the energy equation;
$P_1/\gamma + V_1^2/2g + Z_1 = P_2/\gamma + V_2^2/2g + Z_2$
$432/62.4 + 7 = P_2/62.4 + 0$
Note the $V_1 = V_2$, $Z_1 = 7$ and $Z_2 = 0$
$13.92 = P_2/62.4$
$P_2 = 868.8 \text{ psf} = 868.8/144 \text{ psi} = 6.03 \text{ psi}$
Ans B

Discussion: Water pressure is higher at point B compared to pint A. When water is flowing from point A to point B, gravitational head converted to pressure head. Hence the pressure at point B is higher than point A.

5.4.2. Pressure Conduits:
Pipes are considered to be pressure conduits. Water is transported under pressure in pipes.

5.4.3 Continuity Equation:
Continuity equation is due to the fact that flow of water is same in all points unless water is removed from the pipes.

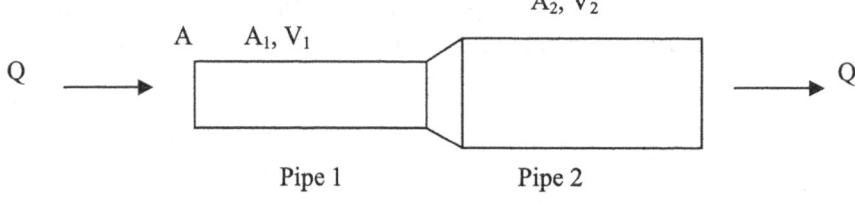

If a flow of Q, enters the pipe 1, flow of Q should exit pipe 2.

Flow = Velocity x Cross sectional Area
$Q = A_1 . V_1 = A_2 . V_2$

┌───┐
│   │
│   Continuity Equation;   │
│   │
│ $Q = A_1 . V_1 = A_2 . V_2$ │
│   │
└───┘

Practice Problem: Two pipes are shown below.

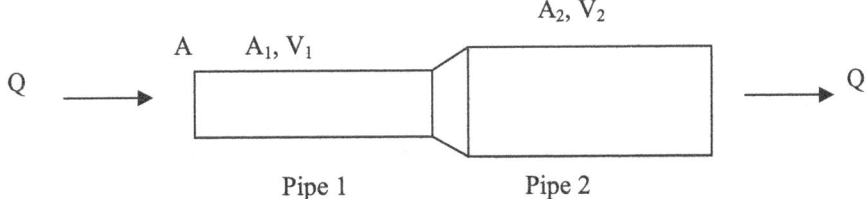

Pipe 1   Pipe 2

Diameter of pipe 1 = 1.5 ft
Diameter of pipe 2 = 2.5 sq. ft
Flow in the pipes is 2,300 gallons per minute. Find the velocity of water in pipes 1 and 2.

Solution:

STEP 1:   Write down the continuity equation.

$Q = A_1 . V_1 = A_2 . V_2$

Diameter of pipe 1 = 1.5 ft.
Hence $A_1 = \pi . D^2/4 = \pi . 1.5^2/4 = 1.77$ sq. ft

Diameter of pipe 2 = 2.5 ft.
Hence $A_2 = \pi . D^2/4 = \pi . 2.5^2/4 = 4.91$ sq. ft

Q = Flow = 2,300 gallons per minute. This needs to be converted to cu. ft/sec.
1 cu. ft = 7.48 gallons
1 gallon = 0.13369 cu. ft

Hence   Q = 2,300 x 0.13369 cu.ft/min
            Q = 2,300 x 0.13369/60 cu.ft/sec = 5.12 cu. ft/sec

STEP 2: Apply the continuity equation;

$Q = A_1 . V_1 = A_2 . V_2$

Let us take the first part;

$Q = A_1 . V_1$

$5.12 = 1.77 \times V_1$

Hence $V_1 = 2.89$ ft/sec

$Q = A_2 . V_2$

$5.12 = 4.91 \times V_2$

Hence $V_2 = 1.04$ ft/sec

Discussion: Note that V_2 is smaller than V_1, since pipe 2 is larger than pipe 1.

5.4.4 <u>Closed Pipe Flow Equations</u>;

Two types of flow encountered in pipes. They are;

- Laminar flow
- Turbulent flow

<u>Laminar Flow</u>; When the water flow is very slow, water seems to travel in layers. Hence this type of flow is known as laminar flow.

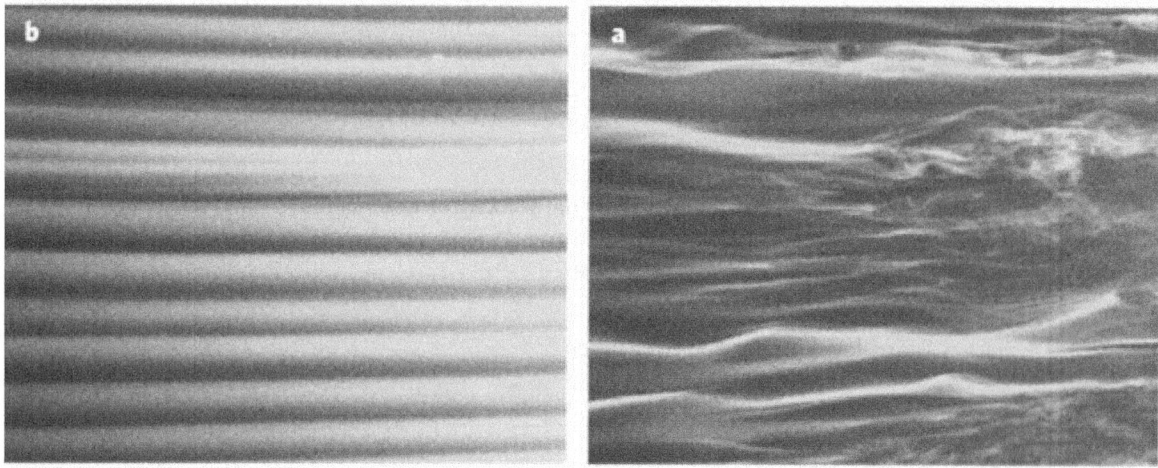

Left: Laminar flow is shown in the above left figure. The flow is slow and water travels in layers.

Right: Above photo on right shows turbulent flow.

When the water velocity starts to increase, layers of water will break up and turbulent flow will start.

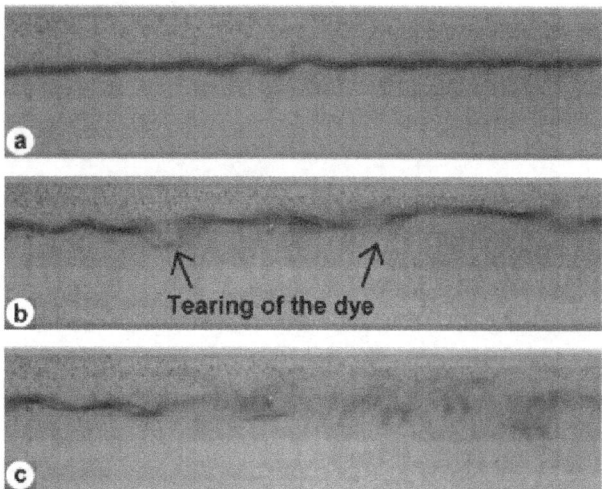

Above photograph marked "a" shows laminar flow. When the flow is increased, layers start to break up. When the flow is further increased, layers completely break up and turbulent flow will start.

Laminar Flow Equation in Pipes:

Consider a pipe with flowing water. The maximum velocity of water is found at the center of the pipe. The velocity of water reduces when moving towards the pipe walls.

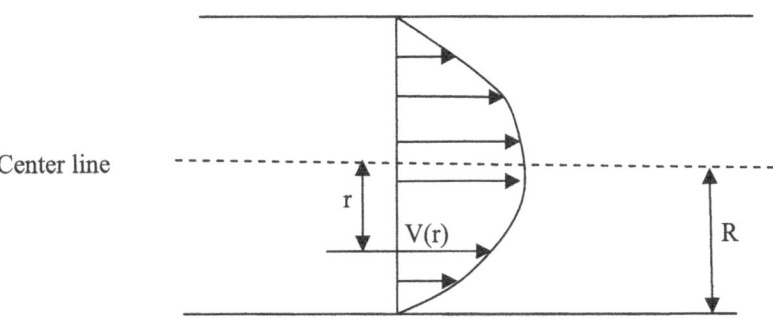

Center line

Above arrows represent flow velocity. Longer the arrow higher the velocity. R is the radius of the pipe. "r" is the distance measured from the center line of the pipe.

Following equation is used to compute laminar flow in pipes.

$$V(r) = V_{max} \left[1 - (r/R)^2\right]$$

$V(r)$ = Velocity of flow "r" distance from the centerline of the pipe
V_{max} = Maximum velocity. Maximum velocity occurs at the center line of the pipe.
$r = r$ distance from the centerline of the pipe
R = Radius of the pipe

When "r" = 0, $V(r) = V_{max}$
When "r" = R, $V(r) = V_{max}[1 - (R/R)^2] = 0$
Now let us complete a problem;

Practice Problem):   Laminar flow is seen in a pipe. The velocity at the center of the pipe is 4.5 ft/sec. The pipe is 5 ft in diameter. What is the velocity of flow 1 ft away from wall of the pipe?

A) 3.88 ft/sec   B) 2.88 ft/sec   C) 6.7 ft/sec   D) 4.1 ft/sec

Solution)

$$V(r) = V_{max} [1 - (r/R)^2]$$

$V(r)$ = Velocity of flow "r" distance from the centerline of the pipe
V_{max} = Maximum velocity. Maximum velocity occurs at the center line of the pipe.
r = "r" distance from the centerline of the pipe
R = Radius of the pipe

V_{max} = 4.5 ft/sec
R = 5/2 = 2.5 ft

Find the velocity 1 ft away from the wall.
The velocity needs to be found 1 ft away from the wall. This point is (2.5 – 1.0) ft away from the center of the pipe.

r = 2.5 – 1.0 = 1.5 ft

$$V(r) = V_{max} [1 - (r/R)^2]$$

$V(r) = 4.5 [1 - (1.5/2.5)^2] = 2.88$ ft/sec
Ans B

There is no similar equation for turbulent flow. It is not easy to find the velocity of water at a given point inside a pipe for turbulent flow due to turbulent nature of the flow.

5.4.5 Reynold's Number;

It is important to know whether a flow is laminar or turbulent. Reynold's number is given by the following equation.

$Re = v.D/v$

Re = Reynolds number
v = Velocity in ft/sec
D = Diameter in ft
v = Kinematic viscosity ft/sec.
If the Reynold's number is less than 2,100 the flow is laminar. If the Reynold's number is greater than 2,100 the flow is turbulent.

Re < 2,100   Laminar flow
Re > 2,100   Turbulent flow

Practice Problem):   Water is flowing thru a 3 inch pipe. Velocity of water flow is 0.15 ft/sec. Kinematic viscosity of water is 1.924×10^{-5} ft/sec. The flow is

A) Laminar flow   B) Turbulent flow   C) Transitional flow   D) None of the above

Solution:
   - Flow is laminar when the Reynolds number is less than 2,100.
   - Flow is turbulent when the Reynolds number is greater than 10,000.
   - Flow is transitional when the Reynolds number is between 2,100 and 10,000.

STEP 1:   Find the Reynolds number;

The equation for the Reynolds number is given below.

$$Re = v.D/v$$

Re = Reynolds number
v = Velocity in ft/sec
D = Diameter in ft
v = Kinematic viscosity ft/sec.

$$Re = [0.15 \times (3/12)]/1.924 \times 10^{-5}$$

$$Re = 1,949$$

The Reynolds number is less than 2,100. Hence the flow is laminar.
(Ans A)

5.4.6 Hazen-Williams Equation;

When water travels in a pipe, energy is lost due to friction. There are two equations available to find the energy loss in pipes due to friction.
They are;

- Hazen-Williams Equation;
- Darcy-Weisbach Equation

Hazen Williams Equation:

$$h_L = [4.73\ C^{-1.852}\ L\ .\ D^{-4.87}]\ .\ Q^{1.852}$$

h_L = Head loss due to friction in ft
C = Hazen Williams Coefficient
L = Length of the pipe (ft);
D = Diameter of the pipe (ft)
Q = Flow (cu. ft/sec)

Practice Problem: A pipe is shown below. Following parameters are known;
Diameter of the pipe = 6 in (0.5 ft).
Length of the pipe = 150 ft
p_1 = 11.2 psi, v_1 = 7 ft/sec; p_2= 13.4 psi, v_2 = 7 ft /sec
Head difference between point A and B is 10 ft.
Find the Hazen Williams coefficient.

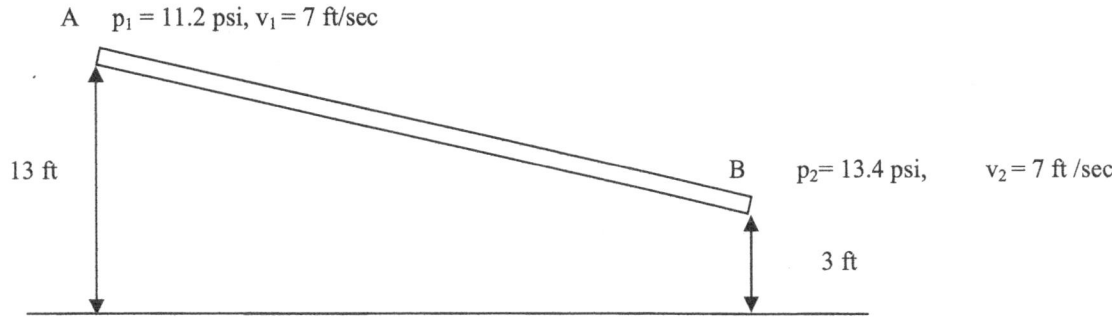

A) 45.2   B) 14.5   C) 231.2   D) 124.3

Solution:

Energy at point A = Energy at point B + Head Loss due to Friction

$$h_1 + v_1^2/2g + p_1/\gamma = h_2 + v_2^2/2g + p_2/\gamma + h_L$$

h_L = Head loss due to friction

Hazen William formula can be used to compute h_L (head loss due to friction).

Hazen Williams Equation:

$$h_L = [4.73 \; C^{-1.852} \; L \cdot D^{-4.87}] \cdot Q^{1.852}$$

h_L = Head loss due to friction in ft
C = Hazen Williams Coefficient
L = Length of the pipe (ft);
D = Diameter of the pipe (ft)
Q = Flow (cu. ft/sec)

Use the Bernoulli's equation with a term for head loss.

Energy at point A = Energy at point B   + Head loss

When water flows from A to B, there is a loss of head. The loss of head is given by h_L.

$$h_1 + v_1^2/2g + p_1/\gamma = h_2 + v_2^2/2g + p_2/\gamma + h_L.\text{------------------------}(1)$$

STEP 1:   Find Head Loss:

$$h_L = [4.73 \; C^{-1.852} \; L \cdot D^{-4.87}] \cdot Q^{1.852}$$

C = ?;
D = 0.5 ft;
L = 150 ft

$$h_L = [4.73 \times C^{-1.852} \times 150 \times 0.5^{-4.87}] \cdot Q^{1.852}$$

Since velocity is given, Q can be computed.
$$Q = \pi \cdot D^2/4 \times v = \pi \cdot 0.5^2/4 \times 7 = 1.374 \; cu.ft/sec$$

Insert the value of Q into above equation;

$$h_L = [4.73 \times C^{-1.852} \times 150 \times 0.5^{-4.87}] \, Q^{1.852}$$
$$h_L = [4.73 \times C^{-1.852} \times 150 \times 0.5^{-4.87}] \times 1.374^{1.852}$$
$$h_L = 37,369.73 \times C^{-1.852}$$

STEP 2:   Apply the energy equation;

$$h_1 + v_1^2/2g + p_1/\gamma = h_2 + v_2^2/2g + p_2/\gamma + h_L$$

$h_1 = 13$ ft;
$h_2 = 3$ ft
$v_1 = 7$ ft/sec
$v_1 = v_2$ (Since the diameter is the same at two points)
$p_1 = 11.2$ psi
p_1 needs to be converted to psf.
$p_1 = 11.2$ x 144 psf
$\gamma = 62.4$ pcf (Density of water)
$p_2 = 13.4$ x 144 psf

$h_1 + v_1^2/2g + p_1/\gamma = h_2 + v_2^2/2g + p_2/\gamma + h_L$
$13 + 7^2/2g + (11.2 \times 144)/62.4 = 3 + 7^2/2g + (13.4 \times 144)/62.4 + 37{,}369.73 \cdot C^{-1.852}$
$10 + (11.2 - 13.4) * 144/62.4 = 37{,}369.73 \cdot C^{-1.852}$
$4.92 = 37{,}369.73 \cdot C^{-1.852}$
$0.000132 = C^{-1.852}$

From arithmetic;

$$C^{-1.852} = \frac{1}{C^{1.852}}.$$

Hence;
$$0.000132 = \frac{1}{C^{1.852}}.$$

Invert both sides;

$$\frac{1}{0.000132} = C^{1.852}$$

$$7575.75 = C^{1.852}$$

From arithmetic;
$$(7575.75)^{1/1.852} = C$$

$1/1.852 = 0.5399$
$(7{,}575.75)^{0.5399} = C$
$C = 124.31$ Ans D

5.4.7 Darcy-Weisbach Equation;
When water flows thru a pipe, energy is lost due to friction. Energy loss is high if the velocity of flow is high. Energy loss is commonly known as head loss. Head loss in a pipe is dependent on following parameters;
- Velocity of flow – (Head loss is larger for high velocity flow)
- Pipe wall surface roughness
- Pipe diameter
- Length of the pipe

After considering all these parameters, Darcy Weishback equation was developed.

Henry Darcy (1,803 – 1,858) – French engineer who was in charge of the water supply of town of Dijon in France came up with an equation to calculate the head loss due to friction. This equation was further refined by Julius Weisbach and today known as Darcy - Weisbach Formula

Darcy - Weisbach Equation: Head loss due to friction in pipes can be computed using either Hazen William formula or Darcy Weishbach formula. We encountered Hazen William formula before.

$$h_L = \frac{f\,L\,v^2}{2gD}$$

h_L = Head loss due to friction,
f = Darcy friction coefficient
L = Length of the pipe
D = Diameter of the pipe

In the exam either "C" or "f" would be given. "C" is the Hazen William coefficient. If "C" is given, use the Hazen William formula. If "f" is given use the Darcy formula.

Practice Problem: 10 inch diameter pipe is carrying water at a velocity of 3.5 ft/sec and at a pressure of 12 psi. After travelling 100 ft, elevation of the pipe has dropped by 13 ft. Darcy friction coefficient of the pipe is 0.008. What is the pressure in the pipe at the end of the pipe?

$Z1 = 13\ ft,\ p_1 = 12\ psi,\ v_1 = 3.5\ ft/sec$

A ———————————————————————————————————
$D = 10\ in$
$100\ ft$ $Z2 = 0,\ p_2 = ?,\ v_2 = 3.5\ ft/sec$
B

A) 3.4 psi   B) 12.3 psi   C) 17.6 psi   D) 9.8 psi

Solution):

Assume the datum head at point A to be 1.3 and point B to be 0.0. Velocity of water given to be 3.5 ft/sec. Since the diameter of the pipe did not change, velocity f water at point B also would be 3.5 ft/sec.

Darcy - Weisbach Formula: Head loss due to friction in pipes can be computed using either Hazen William formula or Darcy Weishbach formula. In this case, Darcy friction coefficient is given.

$$h_L = \frac{f L v^2}{2gD}$$

h_L = Head loss due to friction,
f = Darcy friction coefficient
L = Length of the pipe
D = Diameter of the pipe

STEP 1: Apply the Bernoulli's equation from A to B.

$h_1 + v_1^2/2g + p_1/\gamma = h_2 + v_2^2/2g + p_2/\gamma + h_L.$------------------------(1)

$h_1 = 13$ ft
$h_2 = 0$ ft
$v_1 = v_2 = 3.5$ ft/sec;
h_L = Head loss due to friction

Velocity at two points has to be equal since flow coming from point A has to come out of point B.

$p_1 = 12$ psi = 12 x 144 psf

From (1) $13 + 3.5^2/(2 \times 32.2) + (12 \times 144)/62.4 = 0 + 3.5^2/(2 \times 32.2) + p_2/\gamma + h_L$

$40.7 = p_2/\gamma + h_L$ ---(2)

STEP 2:   Find the head loss. Head loss is given by the Darcy equation.

$$h_L = \frac{f L v^2}{2gD}$$

L = Length of the pipe = 100 ft;   D = Diameter = 10 in = 10/12 ft = 0.833 ft

$h_L = \frac{f L v^2}{2gD} = \frac{0.008 \times 100 \times (3.5)^2}{2 \times 32.2 \times 0.833} = 0.183$

Hence from equation (2) $40.7 = p_2/\gamma + 0.183$ ($\gamma = 62.4$ pcf)

$p_2 = 2528.3$ psf
$p_2 = 2528/144 = 17.6$ psi   (Ans C)

Note that pressure at point B is higher than point A. In this case some of the datum head got transferred to pressure head. In addition, there is a loss of total energy due to friction when water is travelling from point A to point B.

Discussion:

How to Find the Darcy Friction Coefficient?

In this problem, Darcy friction coefficient is given. Moody diagrams are used to find the Darcy friction coefficient. You need to know the Reynold's number and relative roughness to use the Moody diagrams.

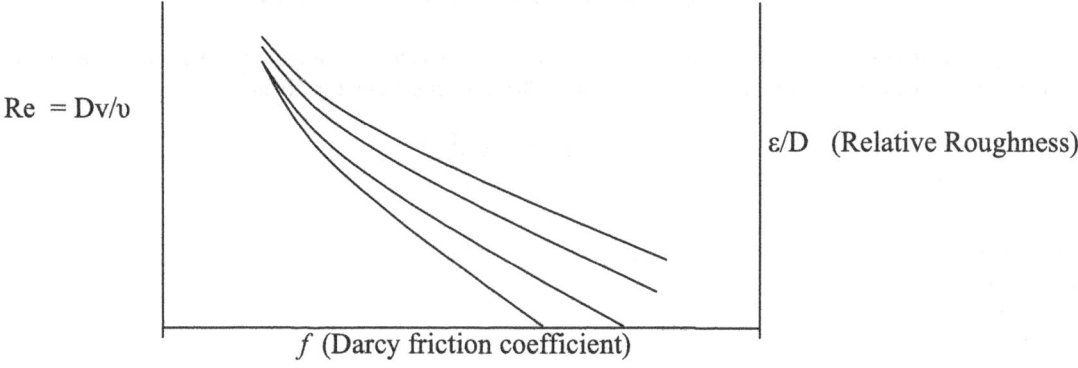

Moody diagrams are not provided in this book but available in many text books.

Using Moody Diagrams:

STEP 1:   Find the Reynold's Number (Re)

Reynold's number is given by Re = D. v/ υ

          D = Diameter   v = velocity of water   υ = Kinetic viscosity of the fluid

Usually "υ' is provided in the morning exam.

STEP 2:   Find the Relative Roughness: (ε/D)
          ε = Roughness of the pipe surface   D = Diameter

Roughness (ε) is usually provided in the morning exam.

Once you know the Reynold's number and relative roughness you can find Darcy's friction coefficient using the Moody diagram.

5.4.8 Minor Losses:
Energy or head is lost due to friction.   In addition, head is lost due to bends, contractions in pipes, valves and various other devices. These losses are known as minor losses.
Head loss reasons;
- Bends
- T- Junctions
- Valves   (Ball valves, Gate valves)

- Pipe contractions
- Pipe enlargements
- Pressure gauges and flow meters
- Various irregularities
- Inlets and outlets

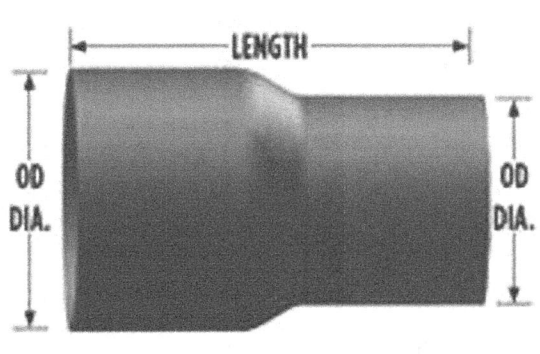 

Left: Pipe reducer or contraction
Right: T – joint

Minor losses are typically given by the following equation.

Minor losses = K. $v^2/2.g$

K = Coefficient for minor losses.;   v = Velocity of water
Minor losses can occur at valves.

Practice Problem: Water flows in a 6 inch pipe from point A to point B. There are four bends and a 2 gate valves in the pipe. Minor loss coefficient (K) for one bend is 0.03.   (Minor loss is given by equation K. $v^2/2.g$). Minor loss coefficient (K) for one valve is 0.04.   Length of the pipe is 120 ft and the Darcy friction coefficient of pipe material is 0.007. Pressure at point A is 7 psi and pressure at point B is 4 psi. Datum head at point A is 30 ft and datum head at point B is 15 ft. Find the velocity of water in the pipe.

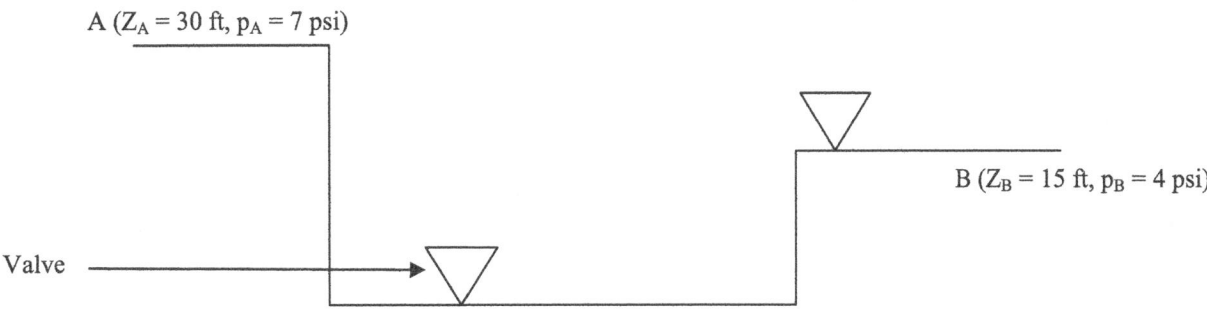

A (Z$_A$ = 30 ft, p$_A$ = 7 psi)

B (Z$_B$ = 15 ft, p$_B$ = 4 psi)

Valve

A) 21.43 ft/sec   B) 27.44 ft/sec   C) 18.91 ft/sec   D) 4.87 ft/sec

Solution: Water is travelling from point A to point B. When water travel from point A to point B, there is an energy loss or head loss. Hence we can write the following equation.

STEP 1:   Write the energy equation;

Energy at point A = Energy at point B + Head loss -----------------------------(1)

STEP 2:   Write the equation for head loss;

Head loss = Head loss due to friction + Head loss due to bends + Head loss due to valves --------------(2)

Head loss due to valves = $2 \times K. v^2/2.g = 2 \times 0.04. v^2/(2 \times 32.2) = 0.0012. v^2$
There are two valves. The K value for one valve is given to be 0.04.

Head loss due to bends = $4 \times K. v^2/2.g = 4 \times 0.03. v^2/(2 \times 32.2) = 0.0019. v^2$
There are four bends. The K value for one bend is given to be 0.03.

Head loss due to friction in pipes = $h_L = \dfrac{fL v^2}{2g.D} = \dfrac{0.007 \times 120 \times v^2}{2 \times 32.2 \times 0.5} = 0.026. v^2$

$f = 0.007$; $L = 120$ ft; $D = 0.5$
Total head loss = $0.0012. v^2 + 0.0019. v^2 + 0.026. v^2$
Total head loss = $0.0291. v^2$

STEP 3:   Find the energy at point A;
Now let us find the energy at point A;
Energy at point A = $Z_A + p_A/\gamma + v^2/2.g = 30 + 7 \times 144/62.4 + v^2/2.g = 46.15 + v^2/2.g$
Pressure at point A is 7 psi. This has to be converted to psf

STEP 4:   Find the energy at point B;
Energy at point B = $Z_B + p_B/\gamma + v^2/2.g = 15 + 4 \times 144/62.4 + v^2/2.g = 24.23 + v^2/2.g$
Since the pipe diameter is the same "v" at point A and B are the same.

STEP 5:   Apply the energy equation;
Energy at point A = Energy at point B + Head loss -----------------------------(1)

$46.15 + v^2/2.g = 24.23 + v^2/2.g + 0.0291. v^2$

$21.92 = 0.0291. v^2$
$v = 27.44$ ft/sec

5.4.9 Pipes in Series;
Two or more pipes can be placed in series. In such situations, head loss in pipes are added together. Following problem illustrates pipes in series situation.

Practice Problem: Two pipes are connected in series as shown in the figure below. Find the pressure at point C.

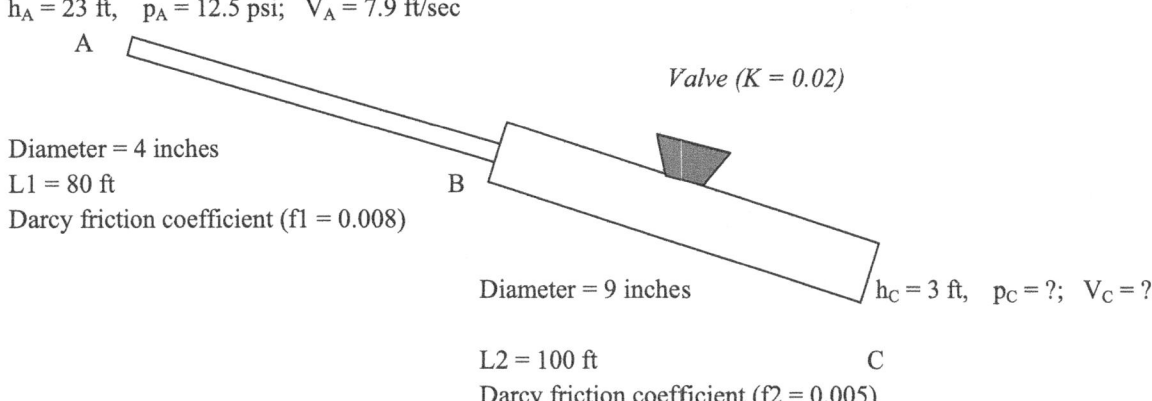

$h_A = 23$ ft, $p_A = 12.5$ psi; $V_A = 7.9$ ft/sec

A

Valve (K = 0.02)

Diameter = 4 inches
L1 = 80 ft
Darcy friction coefficient (f1 = 0.008)

B

Diameter = 9 inches

$h_C = 3$ ft, $p_C = ?$; $V_C = ?$

L2 = 100 ft   C
Darcy friction coefficient (f2 = 0.005)

K value for expansion near point B = 0.04
Find the pressure at point C

Solution:

STEP 1: Write the energy equation;

Head at point A = Head at point C + Head loss from A to B + Head loss from B to C + Minor losses -------(1)

Head at point A = $h_A + v_A^2/2g + p_A/\gamma$ = 23 + 7.9²/(2 x 32.2) + (12.5 x144)/62.4

STEP 1A   Head at point A = 52.82 ft

STEP 1B   Head at point C = $h_C + v_C^2/2g + p_C/\gamma$ = 3 + $v_C^2/2g$ + p_C/γ
Velocity at point C can be calculated.
Flow at point A and flow at point C should be the same.

$Q_A = Q_C$
$Q_A = \pi. D^2/4. V_A$ = π x (4/12)²/4 x 7.9 = 0.689 cu.ft/sec
$Q_C = \pi. D^2/4. V_C$ = π x (9/12)²/4 x V_C = 0.4417 V_C
 0.4417 V_C = 0.689
V_C =1.56 ft/sec
Hence head at point C = $h_C + v_C^2/2g + p_C/\gamma$ = 3 + $v_C^2/2g$ + p_C/γ
Head at point C = 3 + 1.56²/2g + p_C/γ

STEP 2: Find the head loss from point A to point B;

$h_L = \dfrac{f\,L\,v^2}{2g.D} = \dfrac{0.008 \times 80 \times 7.9^2}{2 \times 32.2 \times (4/12)} = 1.861$ ft

STEP 3: Find the head loss from point B to point C;

$h_L = \dfrac{f\,L\,v^2}{2g.D} = \dfrac{0.005 \times 100 \times 1.56^2}{2 \times 32.2 \times (9/12)} = 0.025$ ft

STEP 4: Find minor losses;

There are minor losses due to the valve and the expansion.

Minor loss due to the valve $= \frac{K. v^2}{2.g} = \frac{0.02 \times 1.56^2}{2 \times 32.2} = 0.00076$ ft

Minor loss due to expansion $= \frac{K. v^2}{2.g} = \frac{0.04 \times 7.9^2}{2 \times 32.2} = 0.039$ ft

Total minor losses $= 0.00076 + 0.039$ ft $= 0.0397$ ft

STEP 5:   Apply the energy equation;

From (1)
Head at point A = Head at point C + Head loss from A to B + Head loss from B to C + Minor losses -------(1)
$52.82 = 3 + 1.56^2/2.g + p_C/\gamma + 1.861 + 0.025 + 0.0397$
$52.82 = 3 + 0.0377 + p_C/\gamma + 1.926$
$p_C/\gamma = 47.86$
$p_C = 47.86 \times 62.4$ psf $= 47.86 \times 62.4/144$ psi $= 20.74$ psi

5.4.10   Parallel Pipes;
Head loss across a parallel pipe system is the same regardless of the path taken.

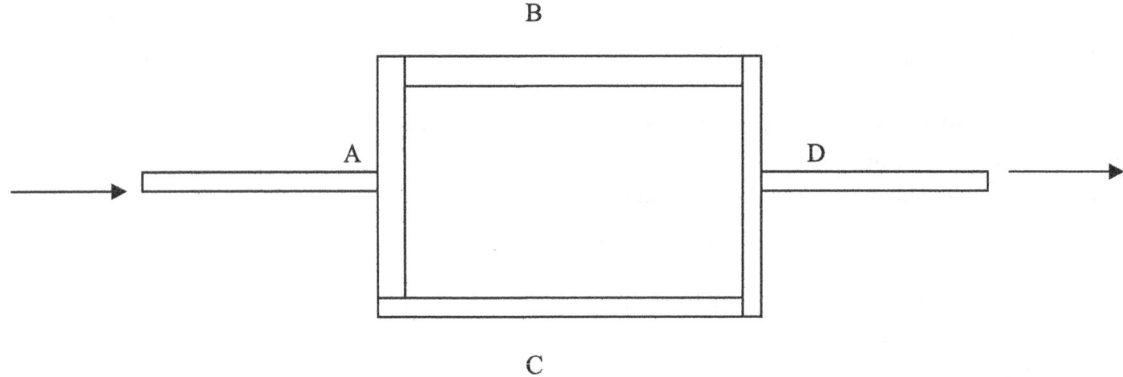

Head loss from A to D is same in both paths. Both paths ABC and ADC would give the same value for head loss. This can be explained by looking at individual water particles. Let us say there are 100 water particles at point D. All water particles have the same head or energy. Assume energy included in one particle to be X. Also assume there are 100 water particles at point A as well. All water particles at point A has the same energy. Assume energy included in one particle to be Y.

Energy of one particle at point D = X
Energy of one particle at point A = Y
Energy loss in one particle travelling from point A to D = Y – X

Some particles come via route ABD. Some other particles come via route ACD. It does not matter how the particles get to point D. As soon as they arrive at point D, all particles would have the same energy.
Hence energy loss from travelling thru route ABD = Energy loss from travelling thru route ACD.

Practice Problem: Parallel pipe system is shown in the figure below.

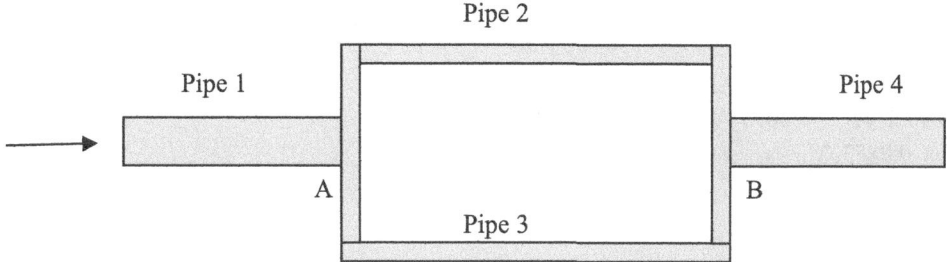

Properties of pipes 1, 2, 3 and 4 are given below.

	Diameter (in)	Length (ft)	Darcy Friction Coefficient
Pipe 1	12	50	0.008
Pipe 2	6	100	0.006
Pipe 3	4	120	0.004
Pipe 4	9	100	0.009

Flow in pipe 1 is 6.7 cu.ft/sec. Find the velocity of water in pipes 2 and 3.

Solution:
STEP 1: Apply the continuity equation;

There is no water leakage. Hence the flow in pipe 1 should be equal to the addition of flows in pipes 2 and 3.

Flow in pipe 1 = Flow in pipe 2 + Flow in pipe 3

Assume diameter and velocity of pipe 2 is D_2 and V_2.
Also assume diameter and velocity of pipe 3 is D_3 and V_3.

$6.7 = (\pi. D_2^2/4). V_2 + (\pi. D_3^2/4). V_3$
$6.7 = [\pi. (6/12)^2/4]. V_2 + [\pi. (4/12)^2/4]. V_3$
$6.7 = [\pi. (6/12)^2/4]. V_2 + [\pi. (4/12)^2/4]. V_3$
$6.7 = 0.1963. V_2 + 0.0873. V_3$ _____(1)

STEP 2: Equate the head losses in pipes 2 and 3;

As we discussed earlier head loss in pipe 2 and 3 are the same.

Head loss is pipe 2 = $\dfrac{f. L. V^2}{(2.g.D)}$ = $\dfrac{0.006 \times 100. V_2^2}{2 \times 32.2 \times 0.5}$ = $0.0186. V_2^2$

Head loss is pipe 3 = $\dfrac{f. L. V^2}{(2.g.D)}$ = $\dfrac{0.004 \times 120. V_3^2}{2 \times 32.2 \times (4/12)}$ = $0.0224. V_3^2$

Head loss in pipe 2 = Head loss in pipe 3

$0.0186. V_2^2 = 0.0224. V_3^2$
$V_2^2 = 1.20 \times V_3^2$

$V_2 = 1.095 \times V_3$ ―――――――――――――――――――――――(2)

Substitute V_2 in equation (1)

$6.7 = 0.1963 . V_2 + 0.0873 . V_3$ ――――――――――――――(1)
$6.7 = 0.1963 . (1.095 \times V_3) + 0.0873 . V_3$
$6.7 = 0.215 \times V_3 + 0.0873 . V_3$
$6.7 = 0.3023 V_3$
$V_3 = 22.16$
and from (2); $V_2 = 24.27$

Practice Problem: Parallel pipe system is shown in the figure below.

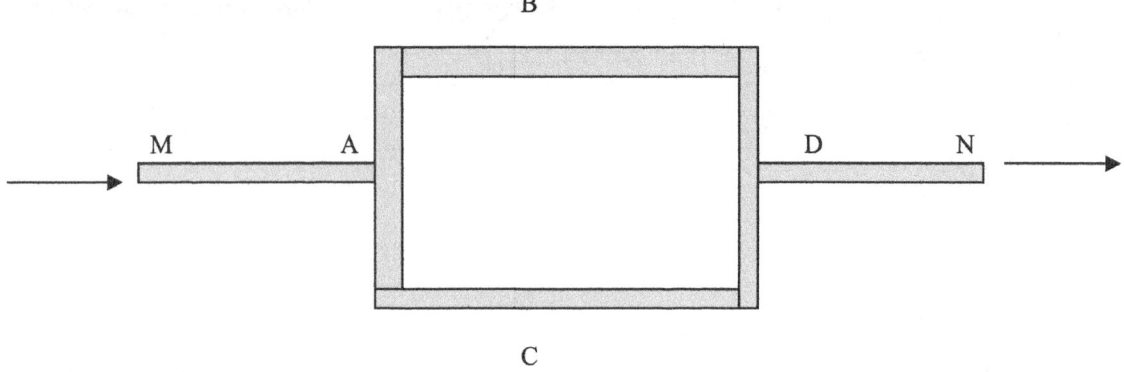

Following information is given;
Pressure at point A = 10 psi
Velocity at point A = 12.9 ft/sec
Diameter of pipe MA = 8 inches
Datum head at point A and D are the same.

Darcy friction coefficient of pipe ABD = 0.007
Diameter of pipe ABD = 6 inches
Length of pipe ABD = 50 ft

Darcy friction coefficient of pipe ACD = 0.009
Diameter of pipe ACD = 4 inches
Length of pipe ACD = 90 ft

Diameter of pipe DN = 12 inches
Find the pressure at point D.
A) 5.67 psi   B) 9.04 psi   C) 12.87 psi   D) 23.09 psi

Solution:

STEP 1:   Find the flow at point A;

Flow $= (\pi . D^2/4) \times$ Velocity
Flow $= [\pi . (8/12)^2/4] \times 12.9 = 4.50$ cu.ft/sec

STEP 2:   Find the flow in pipe ABD;

Flow in pipe ABD = $(\pi . D^2/4) \times V1$
Flow in pipe ABD = $[\pi . (6/12)^2/4] \times V1 = 0.196\ V1$
Assume velocity in pipe ABD to be V1.

STEP 3:   Find the flow in pipe ACD;
Flow in pipe ACD = $(\pi . D^2/4) \times V2$
Flow in pipe ACD = $[\pi . (4/12)^2/4] \times V2 = 0.087\ V2$
Assume velocity in pipe ACD to be V2.

STEP 4:   Find the total flow;

Total flow = Flow in pipe ABD + Flow in pipe ACD
Total flow was found in step 1 to be 4.50 cu.ft/sec
Hence;
$4.5 = 0.196\ V1 + 0.087\ V2$ --(1)

STEP 4:   Find the head loss in pipe ABD;

$$h_L = \frac{f L v^2}{2g.D} = \frac{0.007 \times 60 \times V1^2}{2 \times 32.2 \times (6/12)} = 0.013\ V1^2$$

STEP 5:   Find the head loss in pipe ACD;

$$h_L = \frac{f L v^2}{2g.D} = \frac{0.009 \times 90 \times V2^2}{2 \times 32.2 \times (4/12)} = 0.0377\ V2^2$$

STEP 6:   Equate the two head losses;

$h_L = 0.013\ V1^2$
$h_L = 0.0377\ V2^2$
As we discussed earlier, in parallel pipes head loss in different routes are the same.

Hence;
 $0.013\ V1^2 = 0.0377\ V2^2$
$V1^2 = 2.90 . V2^2$--(2)

$V1 = (2.90)^{1/2} . V2$
$V1 = 1.703\ V2$ --(3)

Substitute V1 in equation (1)

$4.5 = 0.196\ V1 + 0.087\ V2$ ---(1)
$4.5 = 0.196 \times 1.703\ V2 + 0.087\ V2$
$4.5 = 0.334\ V2 + 0.087\ V2$
$4.5 = 0.421\ V2$
$V2 = 10.69$ ft/sec

From (3) we can find V1;
V1 = 1.703 V2 = 18.21 ft/sec

STEP 7:   Apply energy equation from A to D;

$h_A + v_A^2/2g + p_A/\gamma = h_D + v_D^2/2g + p_D/\gamma + H_L$ (Head loss)

Datum head at points A and D are the same.
$v_A^2/2g + p_A/\gamma = v_D^2/2g + p_D/\gamma + H_L$
$12.9^2/2 \times 32.2 + 10 \times 144/62.4 = v_D^2/2g + p_D/\gamma + H_L$ --(4)

H_L is the head loss from point A to point D.
$H_L = 0.013 V1^2$ (See step 6)
$H_L = 0.013 \times 18.21^2 = 4.31$ ft

V_D is the velocity of water at point D. We know the full flow is 4.5 cu.ft/sec. Diameter of pipe DN is given to be 12 inches.

Flow $= (\pi. D^2/4) \times$ Velocity
$4.5 = [\pi. (12/12)^2/4] \times V_D$
$V_D = 5.73$ ft/sec

Substitute known values in above (4)

$12.9^2/2 \times 32.2 + 10 \times 144/62.4 = v_D^2/2g + p_D/\gamma + H_L$ --(4)
$2.58 + 23.1 = 5.73^2/(2 \times 32.2) + p_D/\gamma + 4.31$
$p_D/\gamma = 20.86$
$p_D = 20.86 \times 62.4 = 1,301.6$ psf $= 1,301.6/144$ psi
$p_D = 9.04$ psi   Ans B

5.5   Pump Application and Analysis

Pumps:

Pumps are used for many purposes in civil engineering and construction work. Some of the uses of pumps are to remove water from excavations, trenches and drains. Also, pumps are used for the purpose of lowering groundwater level. Centrifugal pumps are the most commonly used pumps in civil engineering work.

Centrifugal Pumps:

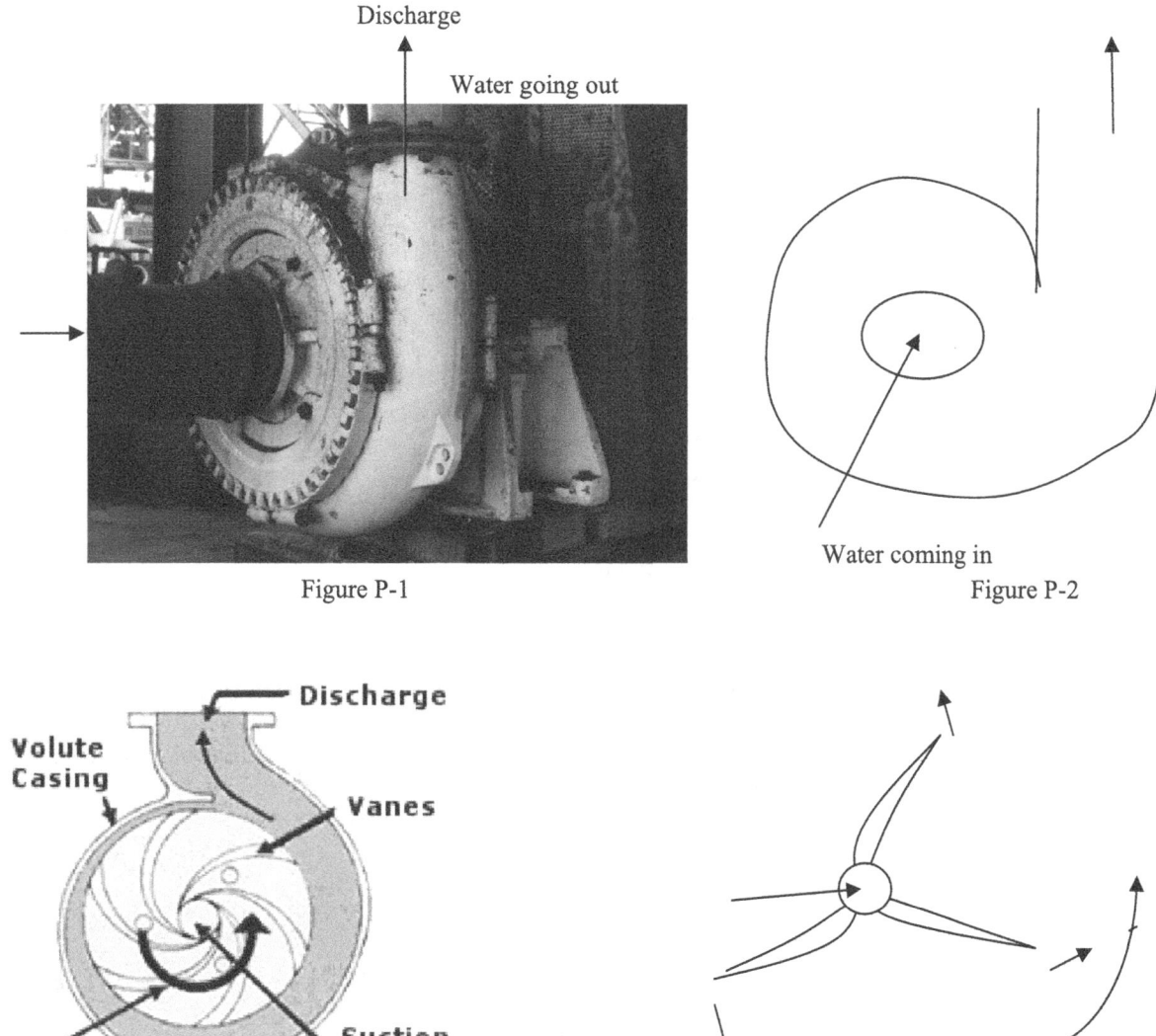

Figure P-1

Figure P-2

Figure P-3

Figure P-4

Above figure P-1 shows a centrifugal pump. The pipe coming in from left of the picture brings in water. Figure P-2 is a schematic of a centrifugal pump. Water inflow and outflow is shown. Figure P-3 shows the vanes (rotating blades) inside a centrifugal pump. Figure P-4 shows the rotating vanes. Let us concentrate on figure P-4. When the vanes rotate, it is natural to throw the water away from the outer edges of the vanes. Small arrows in the figure show vanes expelling water from outer edges. This would cause suction at the center. The center of the pump is known as suction eye. Water in the pipe get sucked in and thrown out or discharged from discharge pipe.

In centrifugal pumps, water inflow is perpendicular to water discharge.

Positive Displacement Pumps: In the case of positive displacement pumps, water is pushed without the centrifugal action. Two screw pumps are shown below. They belong to positive displacement category. Screw pumps do utilize the centrifugal action.

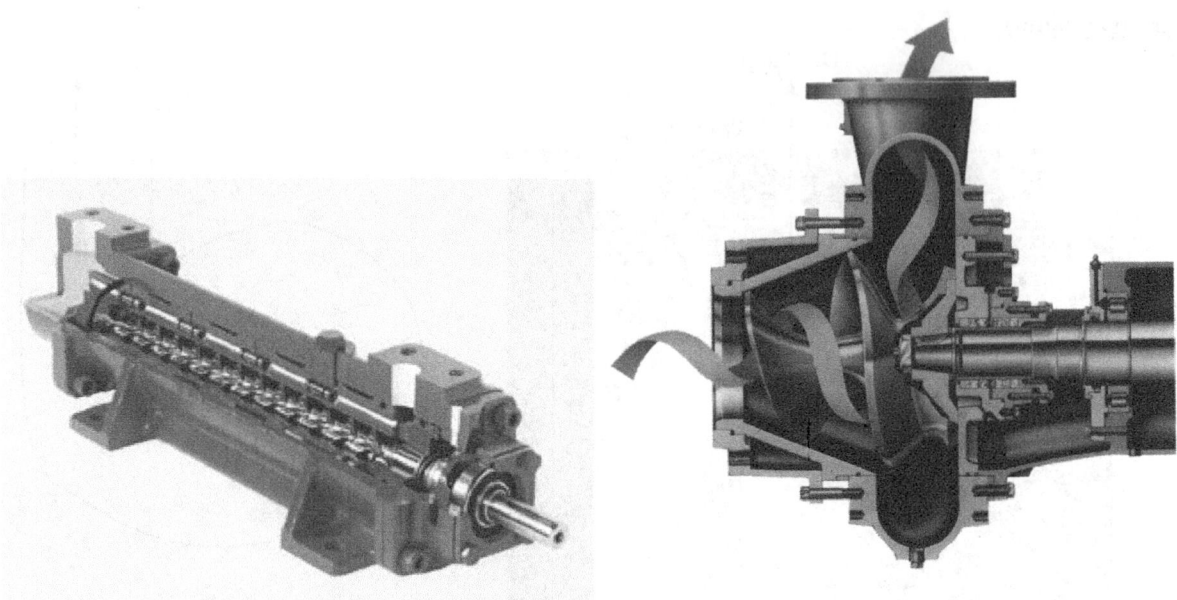

Screw Pumps

<u>Head Added by Pumps:</u>   All pumps basically increases the pressure in the flow. Pressure is a form of energy. Hence pumps add energy to the water flow in pipes.

Suction Head and Discharge Head: Pumps typically pump water from a low point to a higher point. In such situations, suction head and discharge head are shown in the figure below.

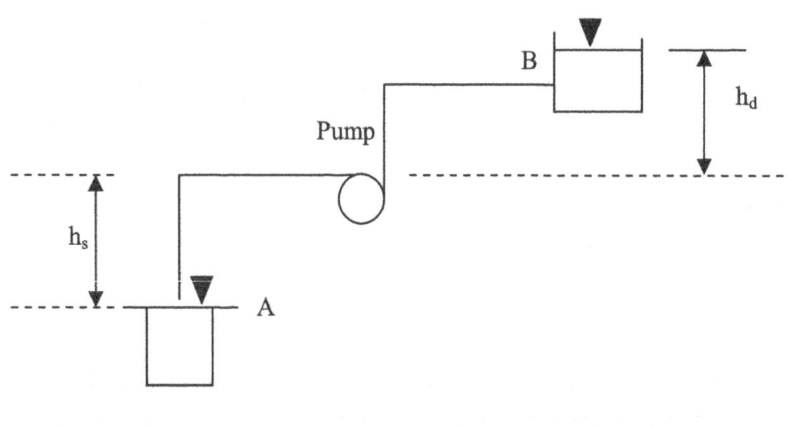

h$_s$ = Suction head   *h$_d$ = Discharge head*

A pump is used to pump water from reservoir A to reservoir B. Pump has to lift water from reservoir A to pump level. This head is known as suction head (h$_s$). Then the pump has to lift the water from pump level to reservoir B. This head is known as discharge head (h$_d$).

Total head added by the pump = h$_d$ + h$_s$
Units of head is ft for US units and meters for SI units.

Practice Problem: Find the total head added by the pump in the figure below.

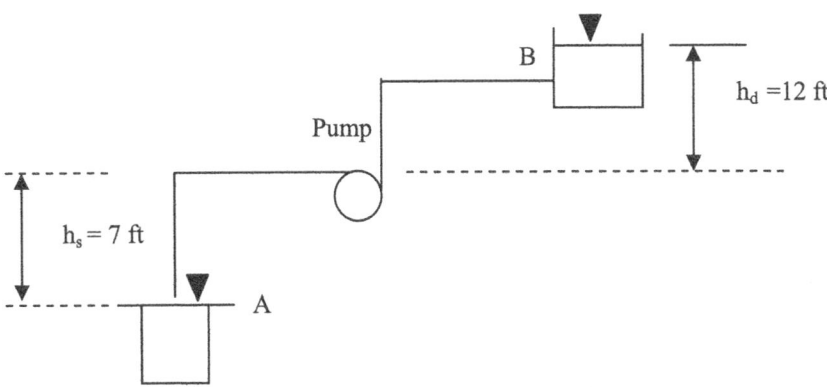

h_s = Suction head = 7 ft h_d = Discharge head = 12 ft

Solution: Total head added by the pump = 7 + 12 = 19 ft.

Practice Problem: Find the total head added by the pump in the figure below.

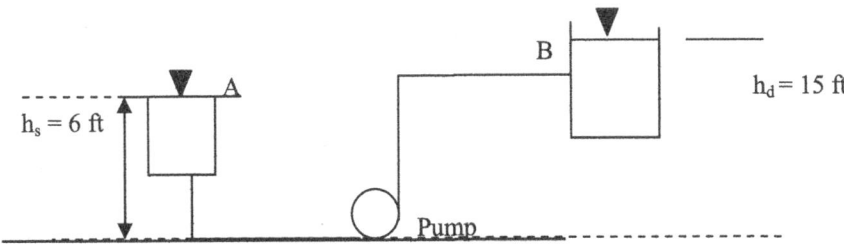

Solution: Total head added by the pump = 15 - 6 = 9 ft.

Water will flow freely from reservoir A to the pump. Hence pump has to add only additional head from A to B, which is 9 ft.

Pump Horse Power (HP):   Pump horse power is given by the following equation.

$$\text{Pump horse power (HP)} = h \times m'/550$$

h = Head added by the pump.   m' = Flow in lbs/sec.
Typically flow is given in cu.ft per sec or gallons/minute. This has to be converted to lbs/sec.

Practice Problem: Suction head of a pump is 8 ft and the discharge head is 9 ft. The flow is 300 gal/min. Find the horse power of the pump.

Solution: head added by the pump = Suction head + discharge head = 8 + 9 = 17 ft

Pump horse power (HP) = h x m'/550

Flow is given in gallons/minute. This has to be converted to lbs/sec.

Flow = 300 gal/min = 300/60 gal/sec = 5 gal/sec

1 gallon = 8.342 lbs;   Hence 5 gal/sec = 5 x 8.342 lbs/sec.= 41.71 lbs/sec
h = 17 ft;   m' = 41.71 lbs/sec
Pump horse power = h x m'/550 = 17 x 41.71/550 = 1.29 HP

Pump Horse Power with Friction Head Considered: When water flows thru pipes, there is considerable friction occurs. Pumps need to provide enough head to overcome friction.

Practice Problem: a)   Find the total head added by the pump in the figure below. Head loss due to friction in suction pipes = 4 ft.   Head loss due to friction in discharge pipes = 5 ft.

b)   If the flow is 3 cu. ft/min, what is the horse power of the pump?

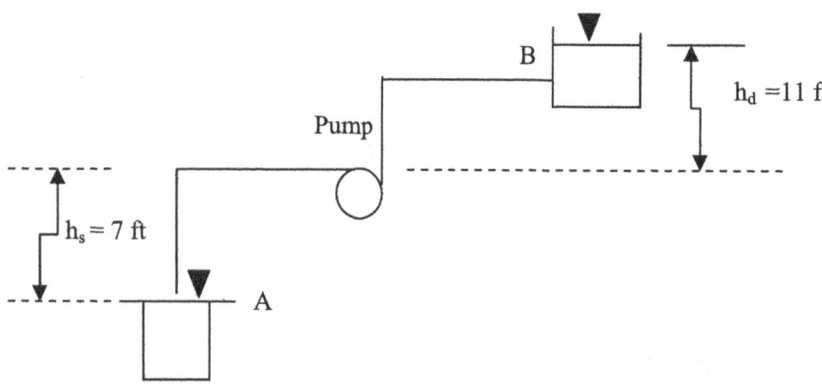

Solution: a)   Total *datum* head added by the pump = 7 + 11 = 18 ft.

Head loss in suction pipes due to friction = 4 ft
Head loss in discharge pipes due to friction = 5 ft
Total head loss due to friction = 9 ft
Note that total datum head added by the pump is 18 ft. (11 + 7 = 18). But this is not enough to lift water from point A to point B. The pump has to over come the friction as well. Hence;
Total head added by the pump = 18 + 9 = 27 ft

b)   Horse power of the pump = h x m'/550
h = 27 ft
Flow = 3 cu. ft/min
Flow has to be converted to lbs/sec.
Flow = 3 cu. ft/min = 3/60 cu. ft/sec = 0.05 cu. ft/sec
1 cu. ft of water = 62.4 lbs
Flow = 0.05 cu. ft/sec = 0.05 x 62.4 lbs/sec = 3.12 lbs/sec
Horse power of the pump = h x m'/550 = 27 x 3.12/550 = 0.153 HP.

5.5.1 Pump Performance Curve, Pump Efficiency Curve and System Curve:

As a civil engineer, you may have to pick the correct pump for the job. Let us say in one project you have to pump water to an elevation of 25 ft at a rate of 100 gallons per minute. In another project you need to pump water to an elevation of 15 ft at a rate of 200 gallons per minute. Same pump may not be the best choice for both cases.

Following figure shows pump performance curve and the pump efficiency curve. To understand the pump performance curve, let us look at a truck. When a truck is empty and not loaded it can travel fast. let us say maximum speed it can travel when the truck is empty is 60 mph. It is common sense that the same speed cannot be achieved when the truck is fully loaded. When the truck is fully loaded the speed could be 40 mph. Hence we can draw a graph between load on the truck vs. speed. Higher the load, lesser the speed.

Similarly, when a pump is pumping a large volume of water, it may not be able to lift to a higher elevation. Larger the volume of water been pumped lower the height of operation. Take a look at the pump performance curves below. Higher the flow, lower the head. Every pump has its own pump performance curve as in the case of trucks where each truck would have its own performance curve.

On the other hand, pump efficiency curve tells us what head and what flow would give the best efficiency for a given pump. Pump A has its highest efficiency at a flow of 100 gpm at a head of around 25 ft. Similarly pump B has its highest efficiency when the flow is 200 gpm and head is 12 ft.

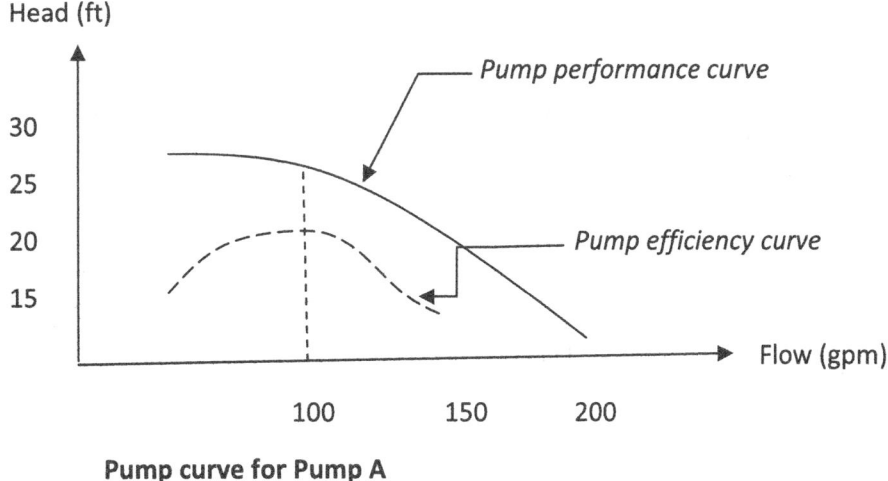

Pump curve for Pump A

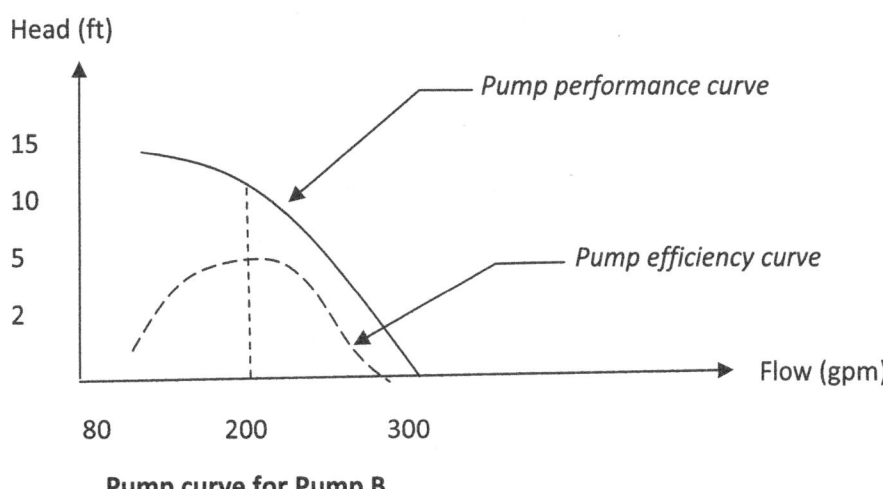

Pump curve for Pump B

When a pump is operating at its highest efficiency, energy bill would be lowest. Hence you need to find a pump that would operate at its highest efficiency. Highest efficiency of pump A is attained when pumping at 100 gpm. The highest effieciency of pump B is attained when pumping at 200 gpm.

Let us say you need to pump at a rate of 100 gpm to an elevation of 25 ft, then pump A should be your best choice. But if you want to pump at a rate of 150 gpm to an elevation of 20 ft, pump A is not very efficient. Yet pump A can do the job.

Now if you want to pump at a rate of 150 gpm to an elevation of 50 ft, pump A may not be able to do the job. (See the pump performance curve).

If you look at pumps A and B, pump A can pump to higher elevations but at a slower rate. On the other hand, pump B can pump at a faster rate, but it cannot lift the water to higher elevations.

System Curve:

System curve has nothing to do with pump curve. System curve depends on the friction head loss in pipes in the system. Let us assme a pipe line. The pipe line has certain characteristics. When water is flowing at a certain velocity, it would have a certain friction. Friction head loss is given by;

Friction head loss = $H_f = f.L. V^2/(2g.d)$

f = Friction coefficient

L = Length of pipe

V = Velocity of water

d = Diameter of the pipe

Length of pipes, friction coefficient and diameter of pipe are dependant on the specific system in consideration. Also you would notice that higher the flow, higher the friction head loss.

Let us look at a typical dewatering scheme.

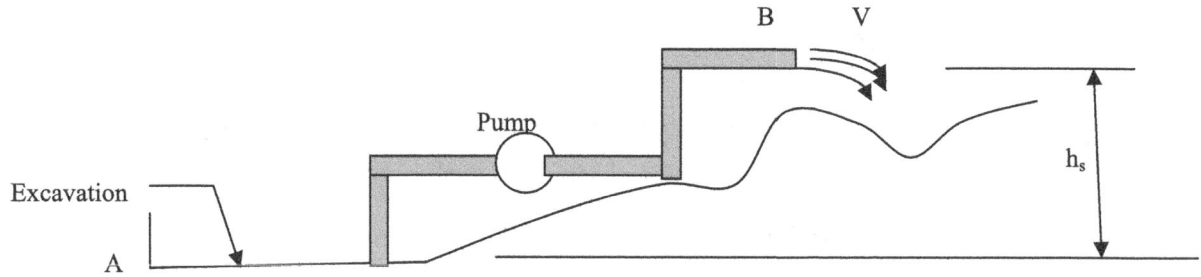

Water in the excavation is exposed to atmosphere. Hence the pressure head is zero. Also water does not move. Hence kinetic head is zero. We are required to pump water at velocity V.

Energy at point A = 0

Energy at point B = $h_s + V^2/2g + P_B/\gamma$

h_s = Static head;   V= Velocity; P_B = Pressure at point B

In this case, water at point B is exposed to atmosphere. Hence P_B is zero.

H_R = Head needed to be added by the pump = $H_f + h_s + V^2/2g$

H_f is the friction head loss.

Above H_f and $V^2/2g$, depend on velocity of water. Higher the velocity, higher the head loss due to friction.

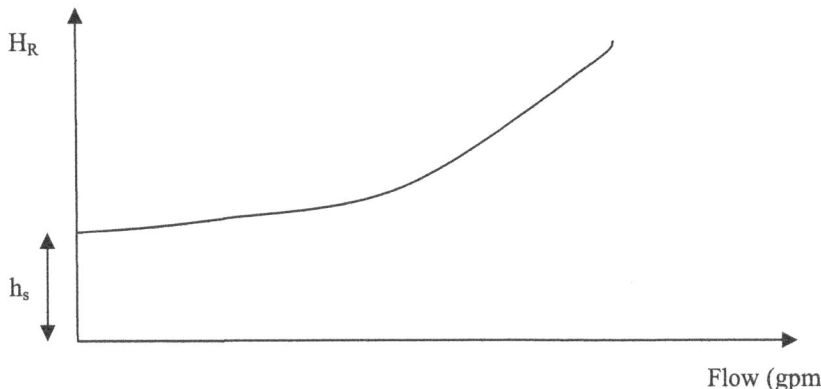

Figure: System Curve

Higher the flow, higher the head required. Above curve depends on length of pipes, friction coefficient and static head. All these parameters are dependant on the system.

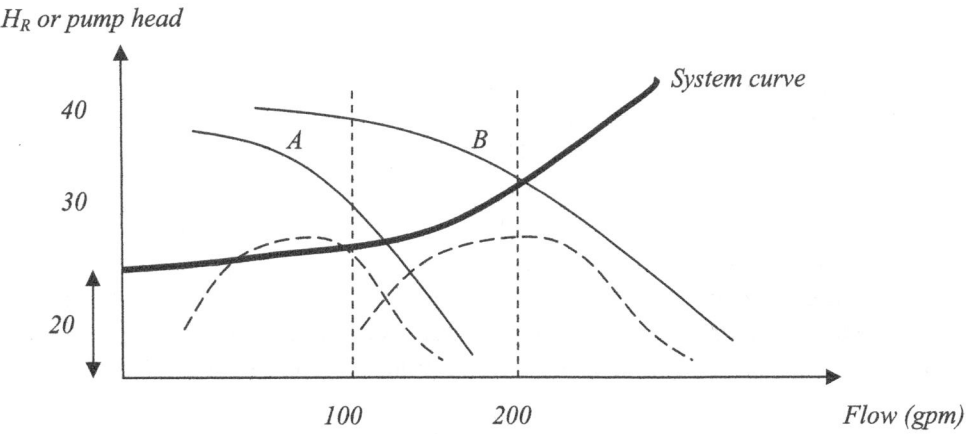

Now let us look at some pump curves drawn along side with the system curve.

Two pump performance curves (A and B) and their efficincy curves are shown.
Let us assume that we need to pump at 100 gpm. To pump 100 gpm, we need a pump that can deliver a head of around 22 ft as per the system curve given. (See the system curve). The system requires at least 22 ft of head to pump at 100 gpm to provide the static head required and to overcome the friction. Both pumps A and B are capable of doing so. But pump A is better suited for the job since it has a higher efficiency than pump B at 100 gpm. Now if we want to pump at 200 gpm, pump A is not suitable. Pump A is not capable of providing a flow of 200 gpm. Then we find pump B to be a better candidate. In this case only candidate.

System curve when pumping horizontally;

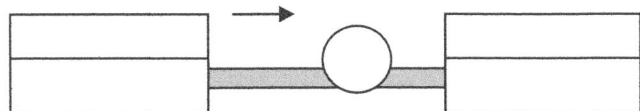

Water is pumped from one reservoir to the other. Both are at same elevation. System curve for this situation is shown below.

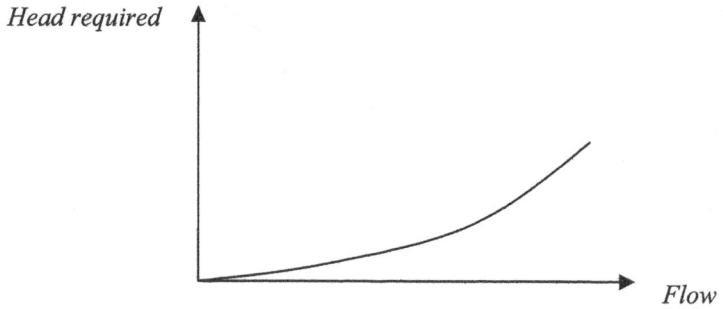

When the flow is increased head required goes up since friction head loss and velocity head depends on flow.

5.5.2 Net Positive Suction Head:

When selecting pumps, one has to look for the net positive suction head of the pump as well. Now let us look what this means.

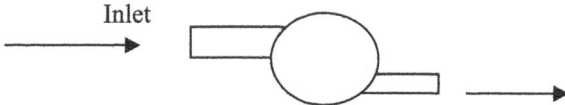

In the above pump, water comes in from the inlet. For the impellers to work properly, certain amount of head is required. If the head available at the inlet is very low, there would be air pockets inside the pump. This is known as cavitation. Cavitation means cavities of air has developed. In such situations, pump will not operate as designed. Hence every pump would indicate the net positive suction head (NPSH) required for the pump to operate properly. Some pumps may require a larger NPSH while some other pumps may require a smaller NPSH.

On the other hand, head available in the system has nothing to do with the pump. Available NPSH at the inlet of the pump depends on the suction head, friction head loss and vapor pressure.

If water is pumped from an open reservoir, available NPSH is given by the following equation.

$$NPSH_{Available} = Atmospheric\ head - Suction\ head - Friction\ loss - Vapor\ pressure.$$

As you could see pump characteristics are not in the equation.
Let us look at an example;

Example: Water is pumped from an excavation 12 ft deep. Friction head loss in suction pipes is 10 ft. Vapor pressure of water is 0.6 ft. Atmospheric pressure is 35 ft. What is the available NPSH at the inlet of the pump?

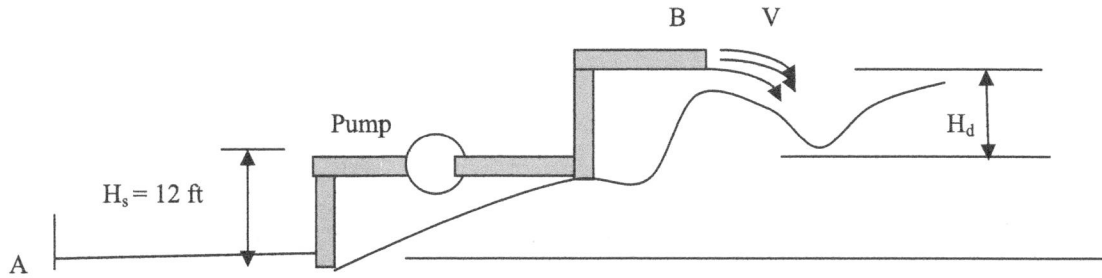

H_s = Suction head

$NPSH_{Available}$ = Atmospheric head - Suction head - Friction loss - Vapor pressure.
$NPSH_{Available} = 35 - 12 - 10 - 0.6 = 12.4$

$NPSH_{Available}$ is dependent on the friction head loss and the suction head. These two parameters have nothing to do with the pump.

Once you have calculated the $NPSH_{Available}$, you need to find a pump that requires a NPSH less than what is available. If $NPSH_{required}$ of the pump is greater than the $NPSH_{Available}$, then the pump will not work.
For the pump to work;

$NPSH_{Available} > NPSH_{required}$

Example: Two pumps are available at the job site mentioned in the previous example.
Pump A ----> $NPSH_{required}$ = 15 ft
Pump B ----> $NPSH_{required}$ = 8 ft
Which pump is best suited for the previous example?
$NPSH_{Available}$ = 12.4   (Computed earlier)

Pump A requires a NPSH of 15 ft. What is available is 12.4. This pump will not perform properly. When the available NPSH is less than required NPSH, water vapor and air pockets will form inside the pump. This is known as cavitation. This is not a desirable situation.
On the other hand, NPSH required for pump B is only 8 ft of water. For this application, pump B should be selected.

Practice Problem: Dewatering from an excavation is shown below.

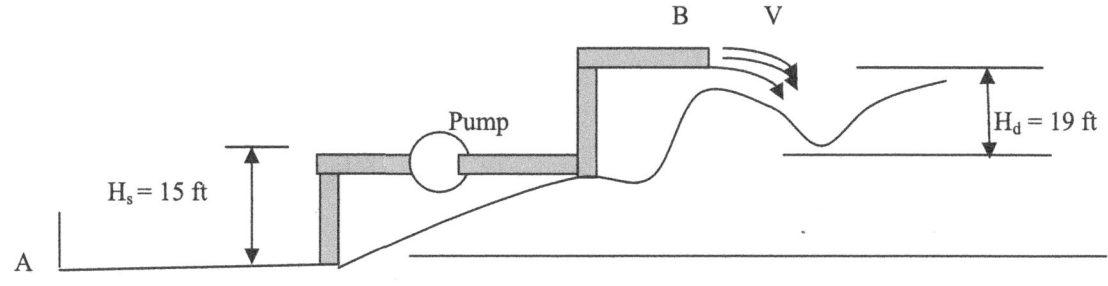

Vapor pressure = 0.6 ft.   Atmospheric head = 35 ft
Pump rate required = 400 gpm.
Ignore the velocity head.

System curve is shown below;

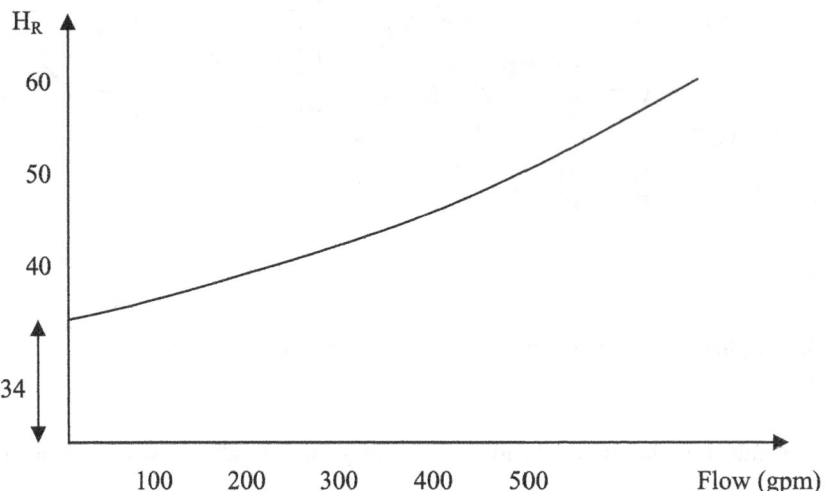

Performance curves for two pumps are given below.

<u>Pump A:</u>   (NPSH = 7 ft)

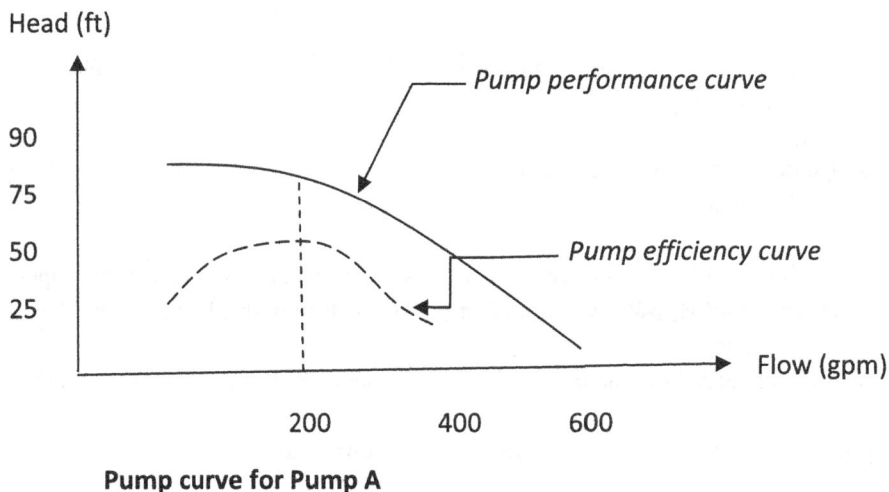

Pump curve for Pump A

Pump B: (NPSH = 15 ft)

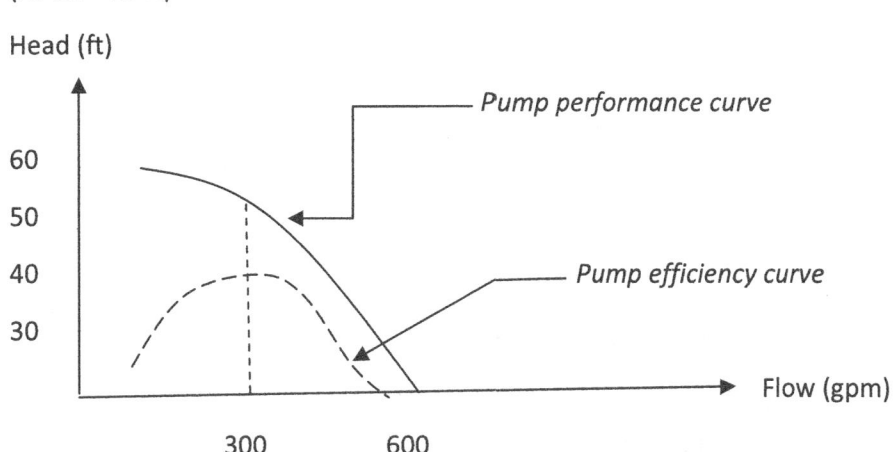

Pump curve for Pump B

Friction head loss in suction pipes is given by the following table;

Flow (gpm)	Friction head loss in suction pipes
100	8
150	10
200	11
400	12
500	19

Which pump is best suited for the project?

Solution:

If you look at the system curve, to pump at a rate of 400 gpm, you need a head of 45 ft.
Now let us look at the pump curve of pump A.

Can pump A, pump at a rate of 400 gpm and at the same time provide a head of 45 ft?
The answer is yes. But there is a problem. The efficiency of the pump at 400 gpm is very low.

Now let us look at pump B. We will ask the same question.
Can pump B, pump at a rate of 400 gpm and at the same time provide a head of 45 ft?
The answer is yes. The efficiency of the pump B at 400 gpm seems to be reasonable.

Now let us look whether the pumps would operate properly.

Find available NPSH;

$NPSH_{Available}$ = Atmospheric head - suction head - friction head in suction pipes - vapor pressure

$NPSH_{Available}$ = 35 - 15 - 12 - 0.6 = 7.4 ft

A table has been provided to find the friction head loss in suction pipes. For a rate of 400 gpm, friction head loss in suction pipes is 12 ft.

NPSH required by pump A is 7 ft. Available NPSH is larger than required NPSH.

On the other hand, NPSH required for pump B is 15 ft. Required NPSH is much larger than what is available. Hence pump B cannot be used.

Pump A has to be selected. This pump has a low efficiency level and high energy bill to be expected. But there is no choice. Pump B cannot be used since available NPSH is not enough for this pump to operate.

Example: What can be done to increase the $NPSH_{avialable}$?

$NPSH_{Available}$ = Atmospheric head - suction head - friction head in suction pipes - vapor pressure

One can replace the old piping with new piping. This would decrease the friction head. Also suction head can be reduced. This can be done by placing the pump at a lower elevation. This option may not be feasible depending on the conditions of the site.

5.5.3 Practice Problems:

Problem 1.6):   Find the velocity at point B in the pipe shown. Ignore the head loss due to friction.

A)   32.8 ft/sec   B) 28.8 ft/sec   C) 13.9 ft/sec   D) 65.8 ft/sec

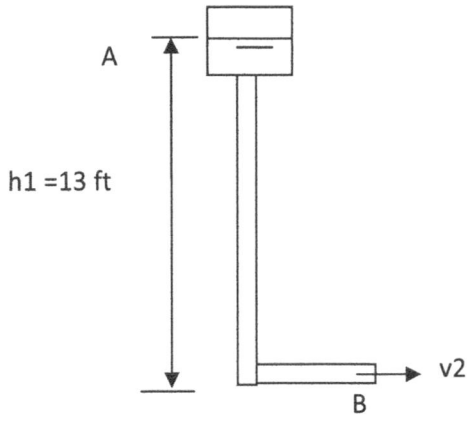

Problem 1.7):   Pressure inside a pipe is 1 psi. What is the pressure head in ft of water?
A) 3.8 ft   B) 2.3 ft   C) 5.8 ft   D) 9.7 ft

Problem 2.4):   Find the pressure at point B in the pipe shown below.   Assume there is no energy loss due to friction.

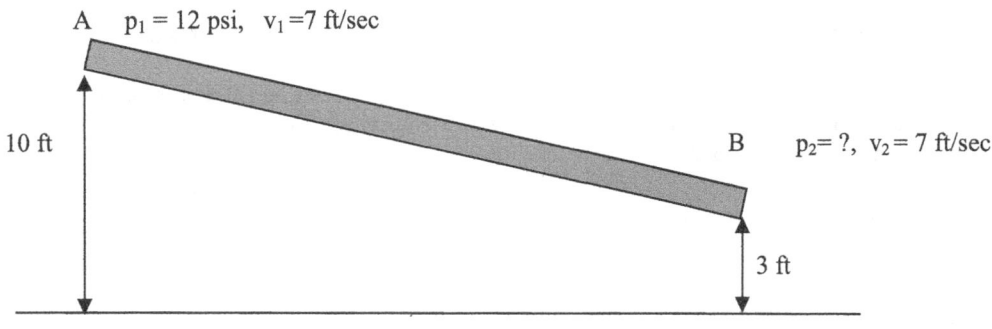

A)   15.03 psi   B) 13.03 psi   C) 2.89 psi   D) 17.98 psi

Problem 2.5:   A pipe is shown below. Find the pressure at point B.
(Use Hazen Williams formula to calculate the head loss due to friction).

Hazen Williams coefficient = 120
Diameter of the pipe = 6 in (0.5 ft)
Length of the pipe = 35 ft

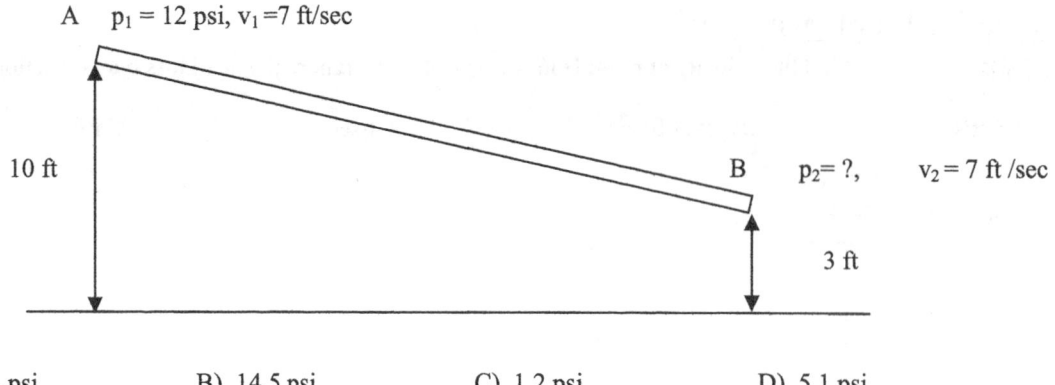

A) 18.2 psi   B) 14.5 psi   C) 1.2 psi   D) 5.1 psi

Problem 2.6:   Pitot tube is shown in the figure. Find the flow velocity.

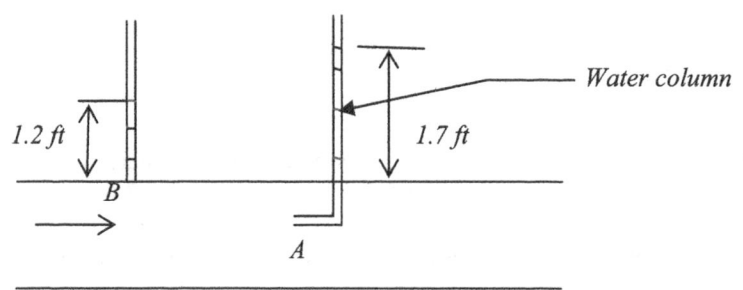

A) 12.3 ft/sec   B) 5.67 ft/sec   C) 3.45 ft/sec   D) 9.4 ft/sec

Problem 3.7:   1.2 ft diameter pipe is carrying water at a velocity of 2.1 ft/sec and at a pressure of 12 psi. After travelling 900 ft, elevation of the pipe has dropped by 9 ft. Darcy friction coefficient of the pipe is 0.008. What is the pressure in the pipe at the end of the pipe?

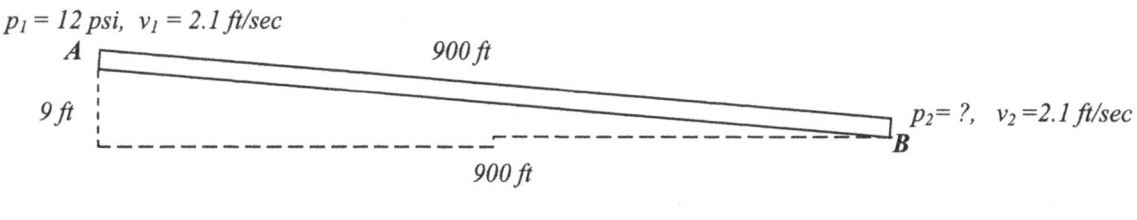

A) 3.4 psi   B) 12.3 psi   C) 15.7 psi   D) 9.8 psi

Problem 3.8:   Coefficient of the Pitot tube shown in the figure is 0.9. If the flow velocity is 6 ft/sec, find "y":

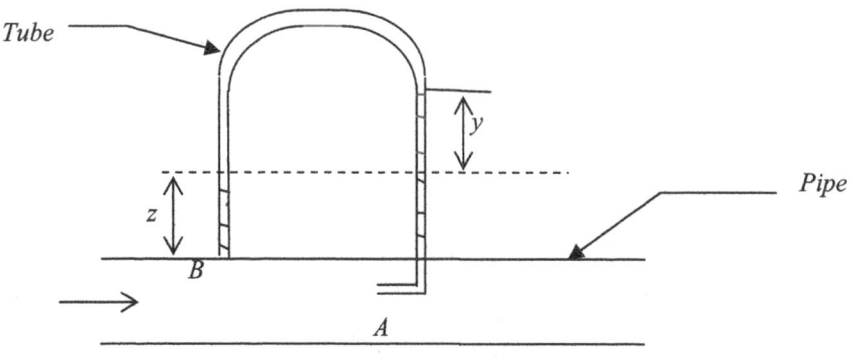

A) 1.3 ft   B) 6.7 ft   C) 0.69 ft   D) 2.1 ft

Problem 3.9:   A Venturi meter is shown (with a water column) in the figure. Assuming a discharge coefficient of 1.0 find the flow in the pipe.

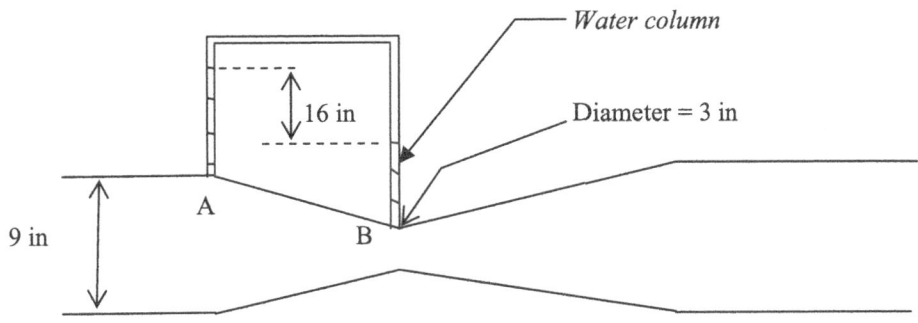

A) 0.01 cfs   B) 0.45 cfs   C) 0.69 cfs   D) 2.1 cfs

Answers:

1.6) B   1.7) B   2.4) A   2.5) B   2.6) B   3.7) C
3.8) C   3.9) B

Step by step solutions to above problems provided in "Four Sample Exams for the Civil PE Exam", Second Edition.

Problem 1.8):   Net precipitation (after deducting for infiltration) of a watershed area was found to be 2.45 inches. The main culvert located at the most downstream end of the watershed area had 40,210 cu. ft of water passing thru it during the storm. Find the total acreage of the watershed area.

A) 2.34 Acres   B) 12.76 Acres   C) 4.52 Acres   D) 8.76 Acres

4.52 acres   (Ans C)

Problem 1.9):   20 year and 50 year storms are shown in the figure.

Rainfall Intensity (in/hr)

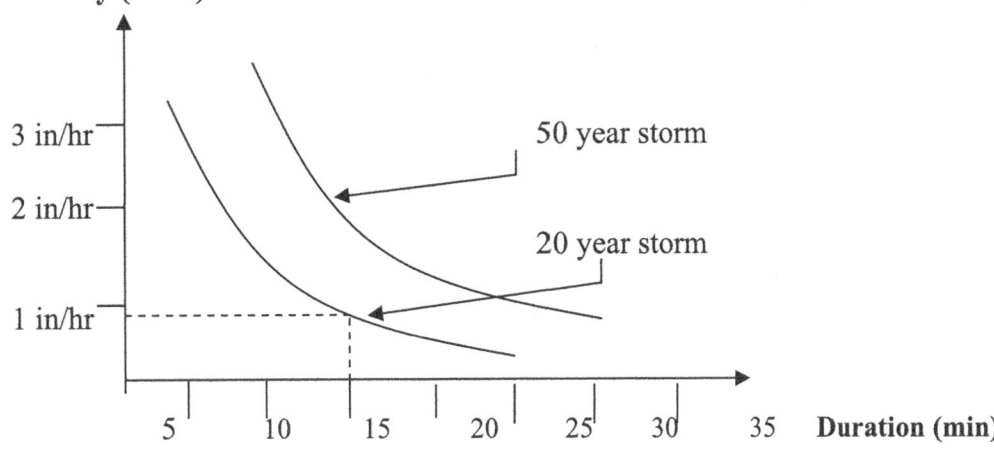

Drainage basin has an area of 13 acres. Time of concentration for the drainage basin was found to be 15 min. What is the design flow if the decision was made to design for a 20 year rainfall? Runoff coefficient (C) of the drainage area is 0.73.

A) 4.56 cu.ft/sec   B) 8.61 cu.ft/sec   C) 6.90 cu.ft/sec   D) 12.78 cu.ft/sec

8.61 cu.ft/sec   (Ans B)

Problem 1.10):   Hydrograph of a drain pipe is shown below. Find the maximum flow in the drain pipe (cu.ft/sec) if the net precipitation is 3.5 inches. Area of the drainage basin is 4 acres.

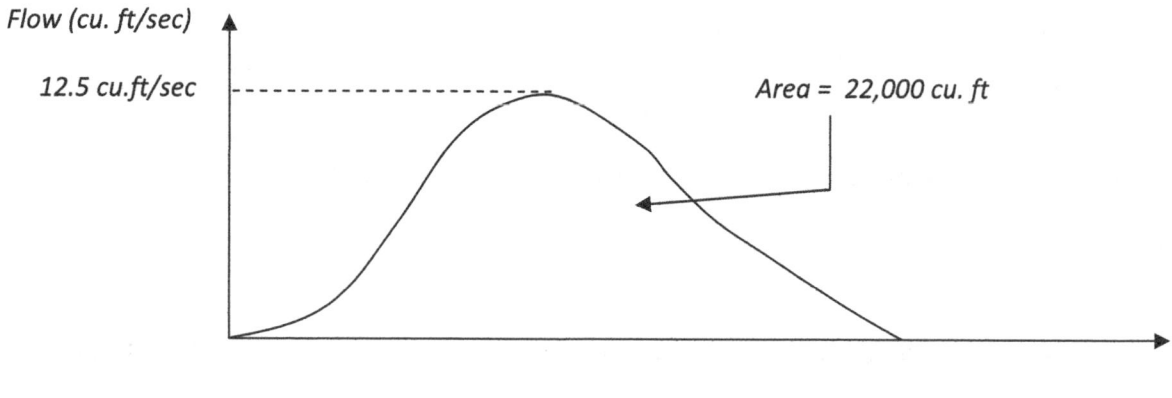

A) 12.6 cu.ft/sec   B) 3.6 cu.ft/sec   C) 28.9 cu.ft/sec   D) 23.7 cu.ft/sec

28.9 cu.ft/sec   (Ans C)

Problem 1.39):   What is the correct order of processes in a wastewater plant?

A) Screens, grit chamber, primary clarifier, secondary clarifier, trickling filter, Chlorine and metal removal system, UV radiation unit
B) Screens, grit chamber, primary clarifier, trickling filter, secondary clarifier, Chlorine and metal removal system, UV radiation unit
C) Screens, grit chamber, primary clarifier, trickling filter, Chlorine and metal removal system, secondary clarifier, UV radiation unit
D) primary clarifier, trickling filter, Screens, grit chamber, Chlorine and metal removal system, secondary clarifier, UV radiation unit

Ans B

Problem 1.40)   Primary clarifier (Primary sedimentation tank) which has a diameter of 40 ft and a height of 8 ft receives 0.7 MGD of wastewater. What is the retention time of the clarifier?
A) 2.13 hrs   B) 2.57 hrs   C) 3.81hrs   D) 5.67 hrs
2.57 hrs   (Ans B).

Problem 3.1)   Channel cross section is shown in the figure. Manning coefficient (n) of the left section of the channel is 0.13 and Manning coefficient (n) for the right section of the channel is 0.09.   If the flow in the channel is 400 cu.ft/sec, what is the slope?

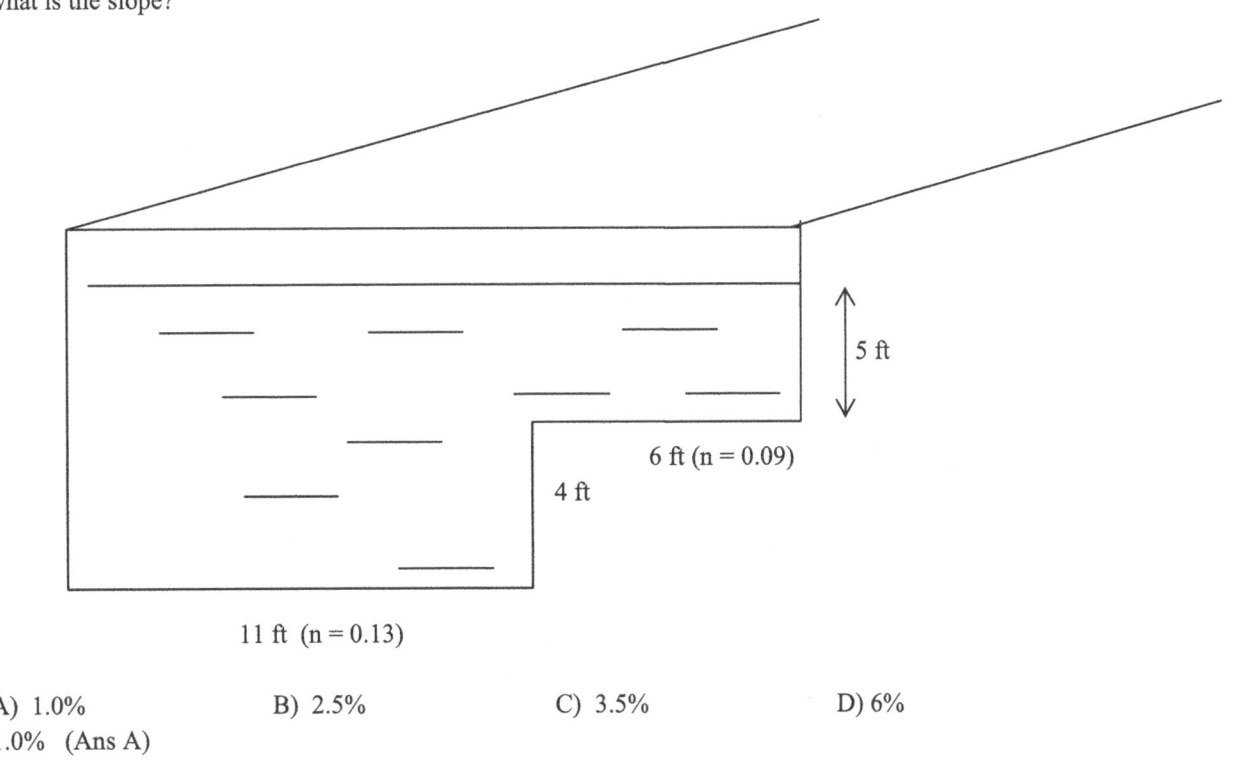

A) 1.0%   B) 2.5%   C) 3.5%   D) 6%

1.0% (Ans A)

Problem 3.2)   Find which section has the highest velocity? All sections have the same Manning roughness coefficient and same slope.

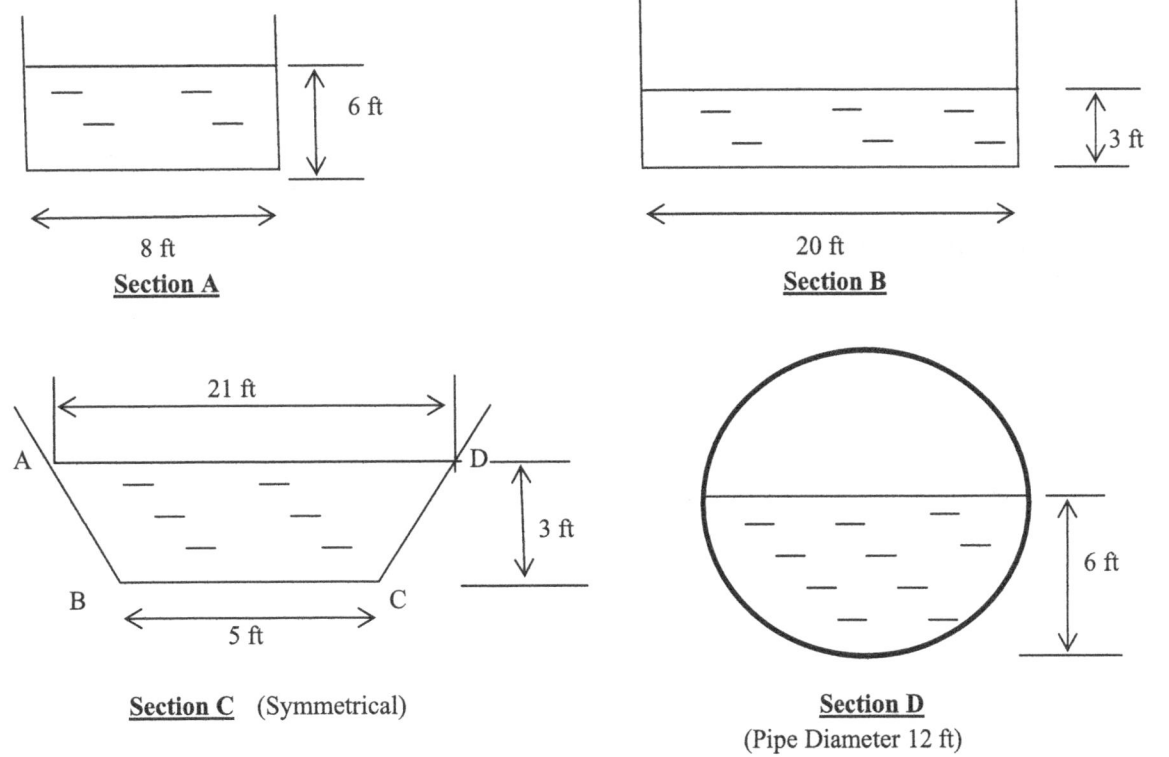

A) Section A;   B) Section B   C) Section C   D) Section D
(Ans D)

Problem 3.3)   Find the Froude number of the trapezoidal channel shown if the flow is 35 cu.ft/sec. Channel width at the bottom is 8 ft and channel depth is 6 ft. Channel slopes are 1 to 2 as shown.

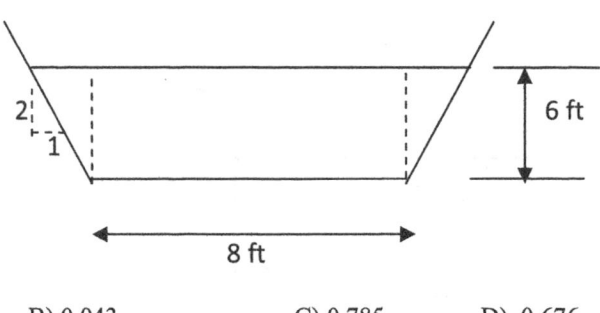

A) 0.034   B) 0.043   C) 0.785   D) 0.676
0.043   (Ans B)

Problem 3.4)   Find the critical depth of channel shown if the flow is 35 cu.ft/sec. Channel width at the bottom is 8 ft. Channel slopes are 1 to 2 as shown.

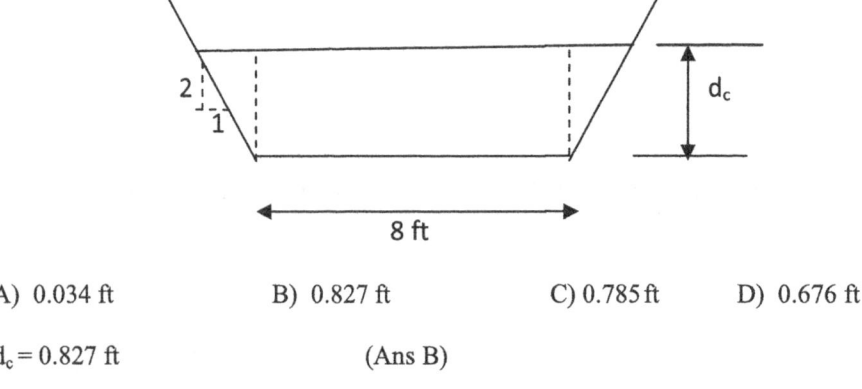

A) 0.034 ft   B) 0.827 ft   C) 0.785 ft   D) 0.676 ft

$d_c = 0.827$ ft   (Ans B)

Problem 3.5)   Hydraulic jump in a rectangular channel is shown in the figure. Upstream depth before the jump is 1.1 ft and the velocity is 15 fps. Width of the rectangular channel is 5.5 ft. What can you say of the flow after the jump?

A)   Flow is subcritical after the jump
B)   Flow is super critical after the jump
C)   Flow is critical after the jump
D)   Not enough data to answer the question.

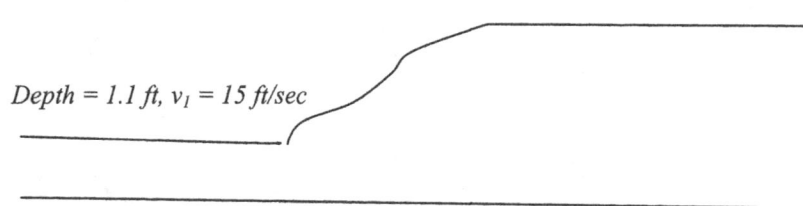

Depth = 1.1 ft, v_1 = 15 ft/sec

$Fr = 0.46$ (Flow is subcritical).   (Ans A)
Problem 3.10)   Two areas are draining into a drain pipe as shown. Information of the two areas is given below.

Area 1 = Area is 9 acres, time of concentration 31 minutes, runoff coefficient 0.51
Area 2 = Area is 3 acres, time of concentration 30 minutes, runoff coefficient 0.40

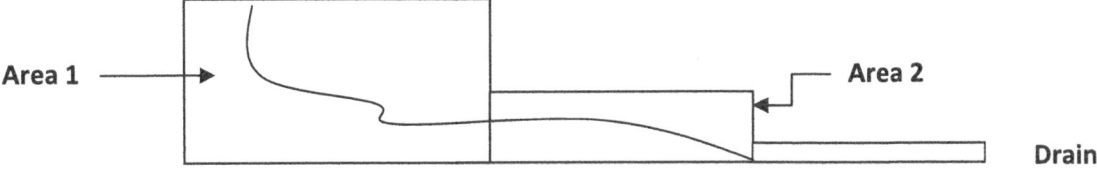

Rainfall intensity for 10 year, 15 year and 30 year rainfalls are given by following graphs.

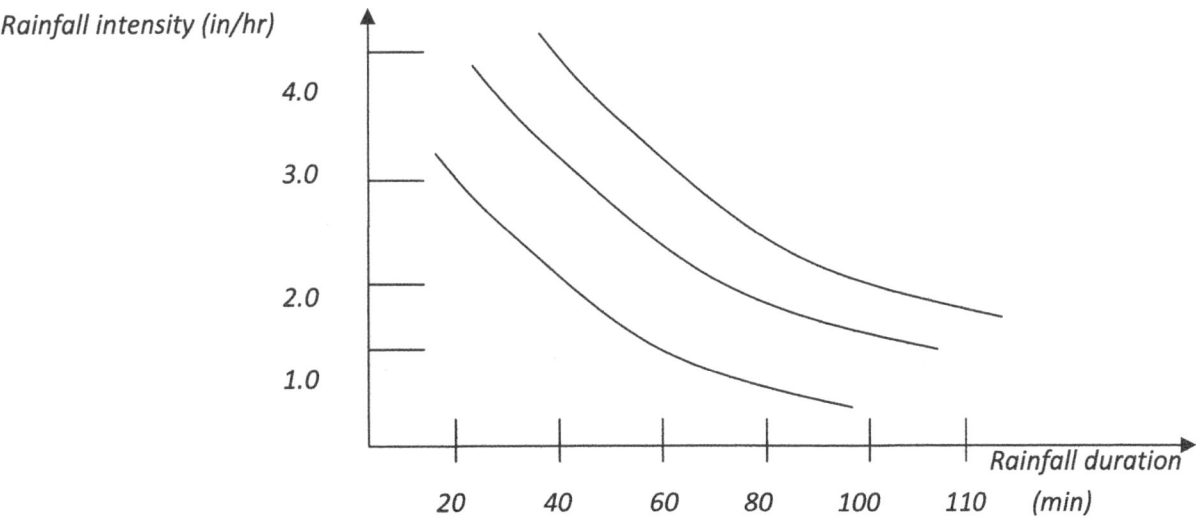

Find the maximum flow in the drain pipe for a 30 year rainfall.

A) 78.9 cu.ft/sec   B) 16.84 cu.ft/sec   C) 61.3 cu.ft/sec   D) 23.5 cu.ft/sec

16.84 cu.ft/sec (Ans B)

Problem 3.11)   Area 1 and area 2 are discharging into a pipeline as shown. Following information is provided.

	Area	Runoff coefficient	Slope	Distance to hydrologically most distant point
Area 1	23 acres	0.70	8%	D1 = 0.6 miles
Area 2	35 acres	0.55	18%	D2 = 0.4 miles

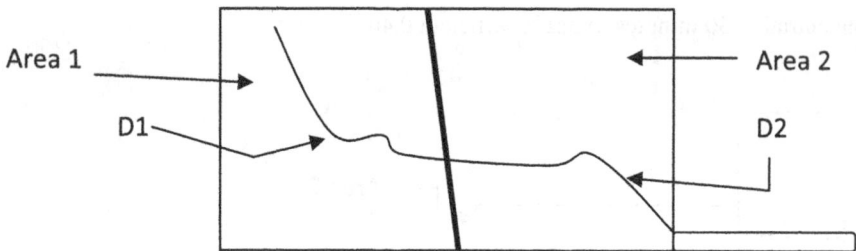

D1 = 0.6 miles
D2 = 0.4 miles

Rainfall data for the region as shown below are for 10 year, 20 year and 100 year rainfalls.

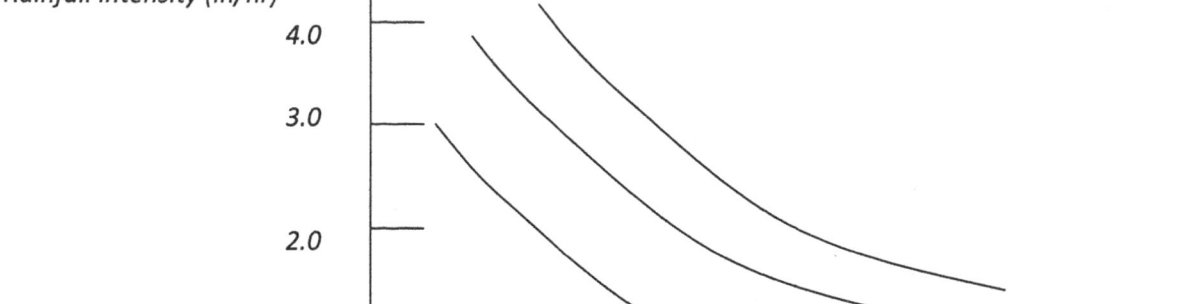

Slope	Velocity
5%	2 ft/sec
10%	10 ft/sec
15%	13 ft/sec
20%	20 ft/sec

Find the flow in the pipeline for a 20 year rainfall in cu.ft/sec.
A) 78.9 cu.ft/sec   B) 71.6 cu.ft/sec   C) 141.3 cu.ft/sec   D) 110.6 cu.ft/sec

Q = 110.6 cu.ft/sec   (Ans D)

Problem 3.12)   Wastewater engineer designed a circular primary sedimentation tank with a detention time of 2.5 hrs. Height of the tank is 5.6 ft. What is the diameter of the tank if the flow is 1.2 MGD?

A) 12.3 ft   B) 98.7 ft   C) 61.6 ft   D) 24.8 ft

D = 61.6 ft   Ans (C)

Problem 4.1):   Hydraulic jump is observed in a rectangular channel with an upstream depth of 3 ft. Froude number of the channel upstream section is 5.2. What is the downstream depth of the channel after the jump.

A) 12.34 ft   B) 20.61 ft   C) 5.67 ft   D) 9.08 ft

y_2 = 20.61 ft   (Ans B)

Problem 4.2)   Broad crested weir has a crest length of 300 ft and a discharge of 646 MGD. What is the height of water at the weir if the weir coefficient is 0.8.

A) 2.59 ft   B) 5.45 ft   C) 12.78 ft   D) 4.12 ft

H = 2.59 ft   (Ans A)

Problem 4.3)   A circular channel has a diameter of 12 ft.   The channel depth is 8.5 ft as shown. Find the flow in the channel if the slope is 0.5% and Manning coefficient is 0.02.

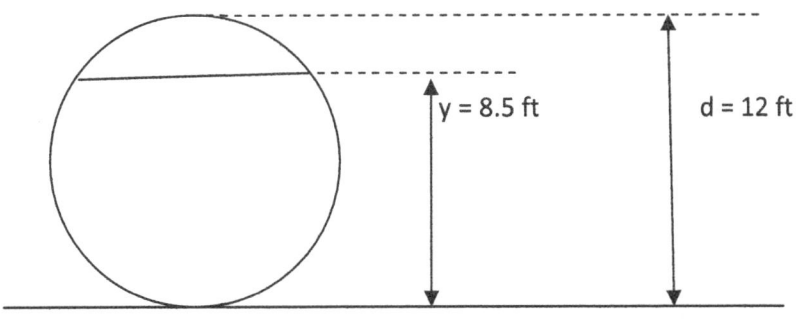

A) 1,421 cu.ft/sec   B) 1,053 cu.ft/sec   C) 672 cu.ft/sec   D) 418 cu.ft/sec

Q = 1,053.3 cu.ft/sec   (Ans B)

Problem 4.4)   Triangular channel is shown below. The flow rate was found to be 1,200 gallons/minute. Manning coefficient was found to be 0.003 and slope was found to be 0.0025. Find the depth of the channel.

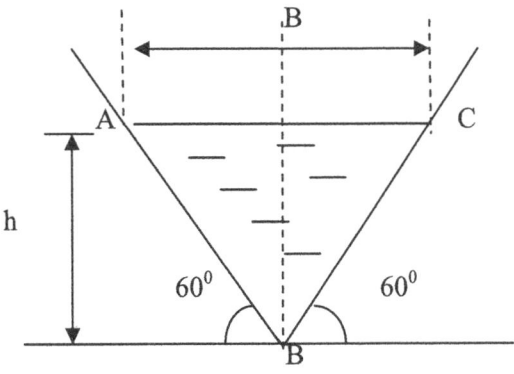

A) 0.75 ft   B) 1.28 ft   C) 0.82 ft   D) 0.55 ft

h = 0.753 ft   (Ans A)

Problem 4.5):   Find the total head added by the pump in the figure below.
Assume the following;
- Friction head loss in pipes is zero.
- Pressure at point B is 3 psi
- Velocity head at point B is negligible.

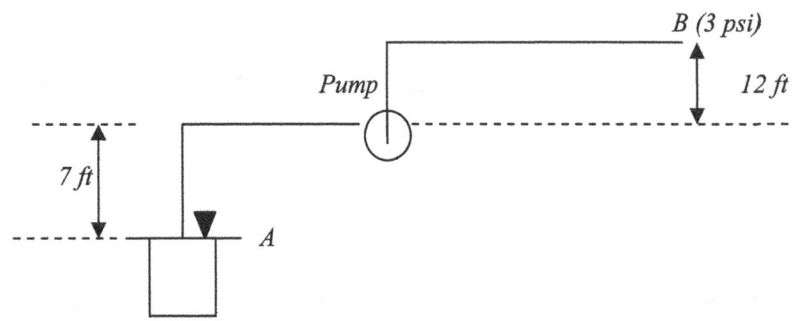

A) 5.0 ft   B) 25.9 ft   C) 7.1 ft   D) 12.2 ft

Total head added by the pump = 25.92 ft   (Ans B)

Problem 4.6):   Parallal pipe system is shown in the figure.

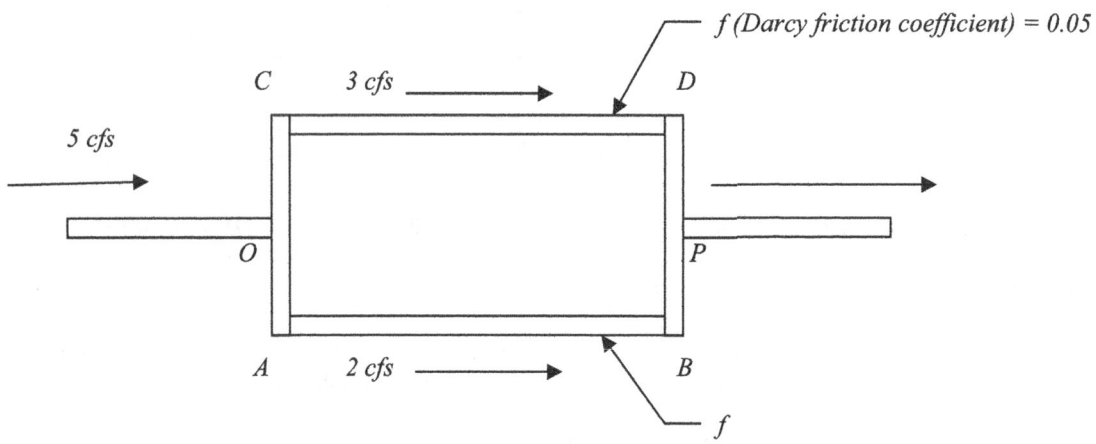

Length of pipe along OABP = 120 ft
Length of pipe along OCDP = 170 ft
Diameter of all pipes are 6 inches.
Find the Darcy friction coefficient of pipe AB.

A) 0.022   B) 0.114   C) 0.003   D) 0.159

f = 0.159   (Ans D)

Problem 4.7):   Suction head of a pump is 8 ft and the discharge head is 9 ft. The flow is 300 gal/min. Find the horse power of the pump. Ignore the head loss due to friction and velocity head in pipes.

A) 5.3 HP   B) 1.2 HP   C) 0.3 HP   D) 4.1 HP

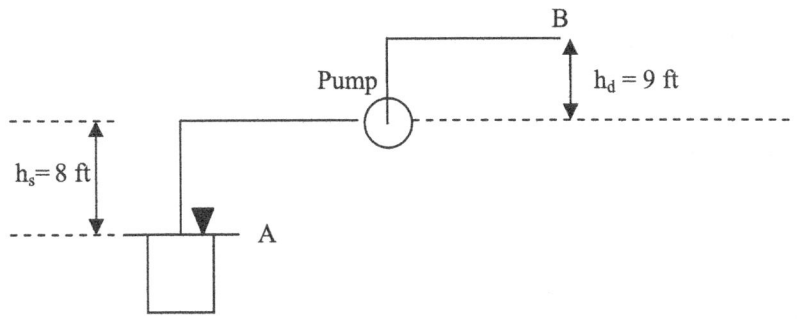

Pump horse power = h x m'/550 = 17 x 41.71/550 = 1.29 HP   (Ans B)

Problem 4.9):   Approximate hydrograph for a stream is shown below.

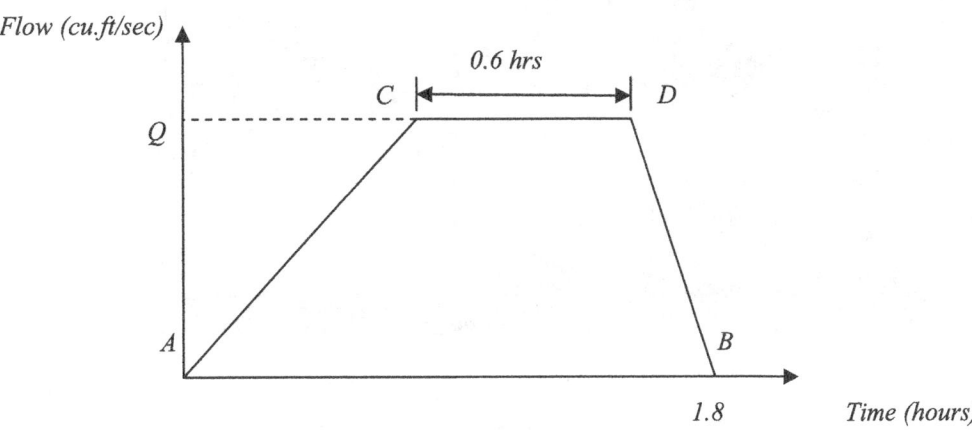

Area of the drainage basin was found to be 15 acres and the net precipitation was 1.3 inches.

Find the maximum flow (Q) in cu.ft/sec.

A) 16.4 cu.ft/sec   B) 22.9 cu.ft/sec   C) 76.7 cu.ft/sec   D) 42.2 cu.ft/sec

= 54,450 cu.ft   (Ans A)

Step by step solutions are given in "Four Sample Exams for the Civil PE Exam, Second Edition"

6.0 <u>Geometrics</u>   -   3 Questions

It is the dream of highway designers to build straight roads. When you travel thru Nevada, Texas or some parts of California one would encounter long straight roads.

In many instances, straight roads are not possible due to mountains, depressions, hills, towns, wildlife, cemeteries and bad soil conditions. Horizontal curves need to be designed so that the speed of the roadway will be maintained. Sharper the curve, lower the speed limit.

Sharp curves have low radius. Larger radius would need a longer curve and that would cost more to build.

Prior to venture into horizontal curves it is important to spend little time learning trigonometry.

<u>Trigonometry Refresher:</u>   Knowledge of trigonometry is essential for the civil PE exam.
It is highly unlikely that you would require any more trigonometry than sin, cos and tan functions.

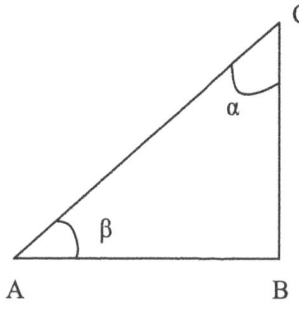

$\text{Sin } \beta = BC/AC$
$\text{Sin } \alpha = AB/AC$
$\text{Cos } \beta = AB/AC$
$\text{Cos } \alpha = BC/AC$

$\beta = 90 - \alpha$
$\text{Sin } \beta = \text{Sin } (90 - \alpha) = \text{Cos } \alpha$
$\text{Cos } \beta = \text{Cos } (90 - \alpha) = \text{Sin } \alpha$

$\text{Tan } \alpha = AB/BC$
$\text{Tan } \beta = BC/AB$
$\alpha = 90 - \beta$

BC = AC Sin β
BC = AB Tan β
AB = AC Sin α
AB = BC Tan α

AC = BC/Cos α
AC = AB/Cos β
AC = BC/Sin β
AC = AB/Sin α
BC = AB/Tan α
AB = BC/Tan β

Practice problem: Find α, β, γ, AD, DC and BC

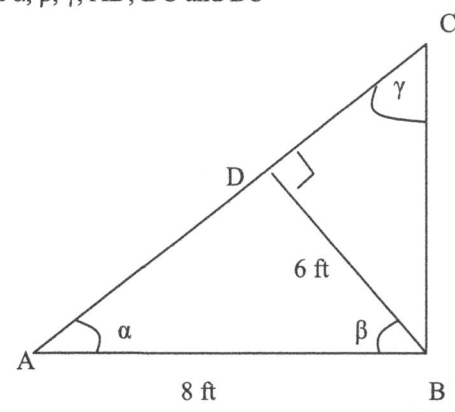

Solution:

Sin α = 6/8;   α = 48.6^0
Hence β = 90 – α = 41.4^0.
γ = 90 – α = 41.4
Note that β = γ in this case.
AD = 8 Cos α = 8 Cos (48.6) = 5.29
DC = AC – AD
AC = 8/Cos α = 8/ Cos (48.6) = 12.1 ft
Hence DC = 12.1 – 5.29 = 6.81
BC = 8 Tan α = 8 Tan (48.6) = 9.1

Radians: Angles can be measured with radians as well.
360^0 = 2π radians

2π. r = Circumference of a circle
The length of an arc is given by multiplication of the angle measured in radians by the radius.

Length of an arc = Angle measured in radians x radius

Practice Problem: Find the length of an arc that projects an angle of 45^0 at the center of the circle. The radius of the circle is 2.5 m.

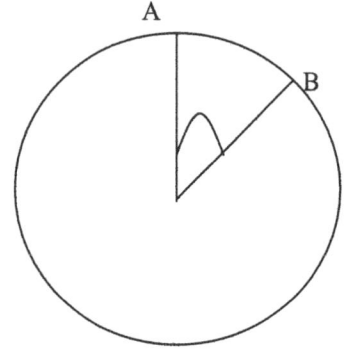

Solution:

Length of an arc = Angle measured in radians x radius
45 degrees has to be converted to radians.

Use the following equation to convert degrees to radians.

$$X_{radians} = X_{degrees} \text{ x } \pi/180$$

Length of arc AB = [45 x π/180] x 2.5 = 1.96 m

Practice Problem: Curve length between PC and PT is 121.5 ft. The radius of the horizontal curve is 300 ft. What is the intersection angle (I) in degrees.

A) 23.20   B) 34.90   C) 126.70   D) 50.20

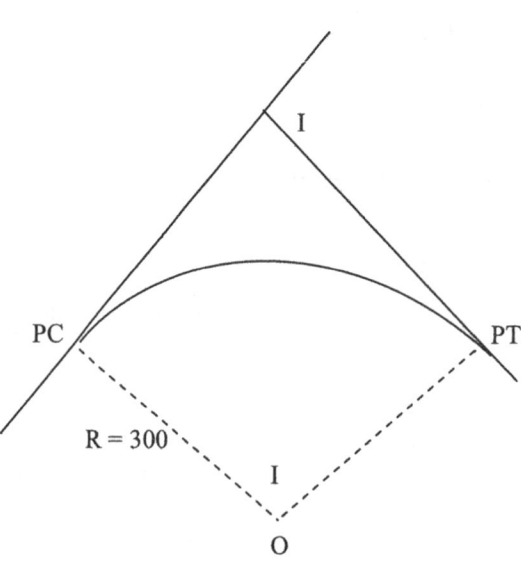

Solution:

Length of an arc = Angle measured in radians x radius

Length of the arc = Curve distance from PC to PT = 121.5 (given)
Radius = 300 ft (given)

Length of an arc = Angle measured in radians x radius
121.5 = Angle measured in radians x 300
Angle measured in radians = 121.5/300 = 0.405

Convert the angle to degrees using the following equation;

$I_{radians} = I_{degrees}$ x $\pi/180$
$0.405 = I_{degrees}$ x $\pi/180$
$I_{degrees}$ = 23.20 degrees   Ans A

Practice Problem: Find the length AB of the figure shown. O is the center of the circle and radius of the circle is 50 m.

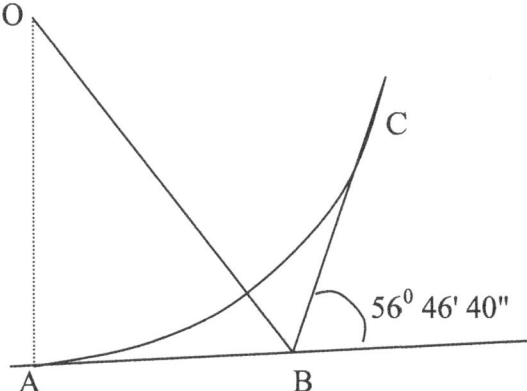

Solution: Angle ABC = 180^0 - 56^0 46' 40"

It is easy to work with decimals rather than minutes and seconds.

56^0 46' 40" = 56 + 46/60 + 40/3600 = 56.78^0
Angle ABC = 180 - 56.78 = 123.22^0

Angle OBA = =123.22/2 = 61.61^0

Tan (angle OBA) = OA/AB   (OA = Radius)
Tan (61.61^0) = radius/AB
Since radius is given to be 50 m,

AB = 50/Tan (61.61^0) = 50/1.85 = 27.03 m

6.1   Horizontal Curves:

Horizontal Curve

Straight road would start to curve at the point known as "*Point of curvature*" or PC. (See the figure below). The curve ends and straight road starts again at a point known as PT or *point of tangent*. Tangents drawn at each point intersects at PI or point of intersection. Following figure shows a horizontal curve with typical station markings.

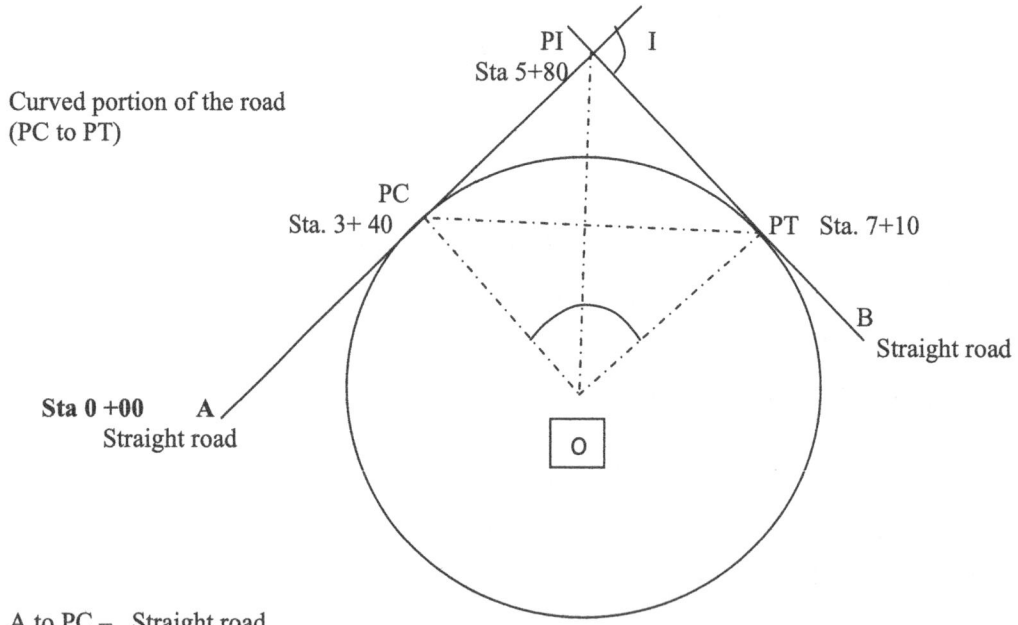

A to PC –   Straight road
PC to PT – Curved portion of the road
PT to B –   Straight road again

Station Markings: Station at PC is 3+40 (or 340 ft from the origin of stations). Station at PI is 5+80. The length between PC and PI = 580 – 340 =   240 ft
Station at PT = 7+10.   The radius of the circle is 850 ft.
This station is measured along the curved portion of the road from PC to PT.
Length of the curved portion of the road from PC to PT = 710 – 340 =   370 ft

<u>Angle at the Center of the Circle</u>:   Curved length of the road from PC to PT is 370 ft. If the radius is known, angle at the center of the circle can be calculated.

The radius is given to be 850 ft.

Length of an arc = Angle measured in radians x radius

Curve distance from PC to PT = 370 (Found earlier)

Radius = 850 ft (given)

Length of an arc = Angle measured in radians x radius

370 = Angle measured in radians x 850

Angle measured in radians = 370/850 = 0.4353

Convert the angle to degrees using the following equation;

$I_{radians} = I_{degrees}$ x $\pi/180$

$0.4353 = I_{degrees}$ x $\pi/180$

$I_{degrees}$ = 24.94 degrees

Practice problem: What is the length of an arc generated by an angle of 23.5 degrees. The radius of the circle is 575 ft.

<u>Solution</u>:

Length of an arc = Angle measured in radians x radius

Radius = 575 ft (given)

Length of an arc = Angle measured in radians x radius

Length of an arc = Angle measured in radians x 575

The angle is given in degrees. Convert the angle to radians using the following equation;

$I_{radians} = I_{degrees}$ x $\pi/180$

$I_{radians}$ = 23.5 x $\pi/180$ = 0.4102 radians

Length of an arc = Angle measured in radians x 575

Length of an arc = 0.4102 x 575

= 235.8 ft

Angle Computation:

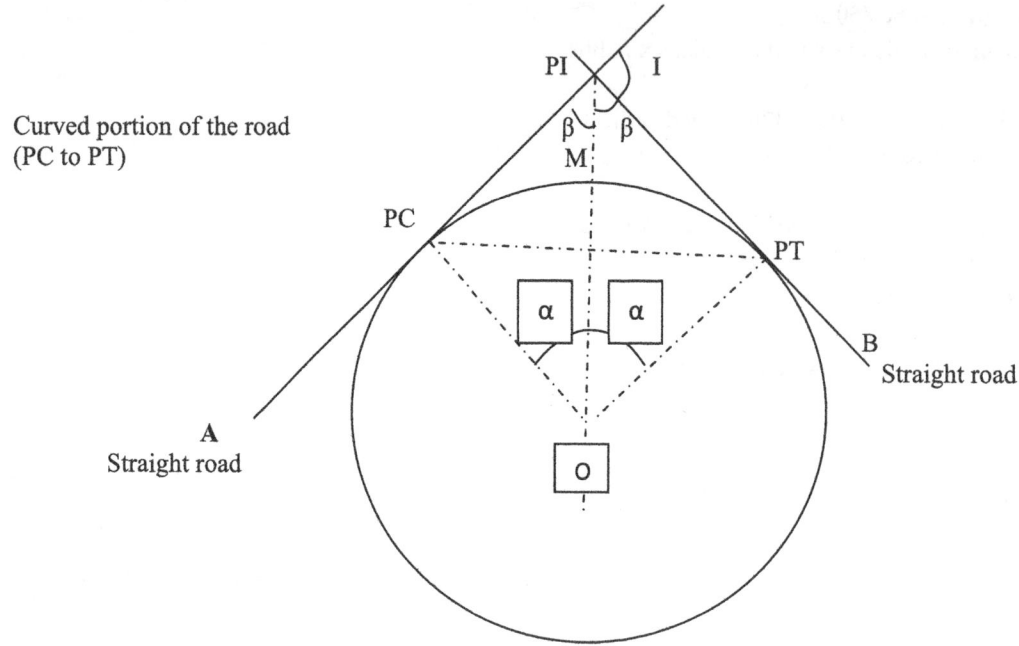

Curved portion of the road
(PC to PT)

A to PC – Straight road

PC to PT – Curved portion of the road

PT to B – Straight road again

Following properties are easily determined.

Intersection angle at point PI is known as I.

Consider two legs, PCPI and PTPI.

Angle PC PI PT = 180^0 - I   (Angle at point PI, between two legs PCPI and PIPT)

From symmetry, angle PC PI O = β = (180 – I)/2 = 90 – I/2

Hence α = 90 – β = I/2

If the radius of the circle is R and I are known, lengths PIPC, PCO and PIO can be deduced.

1)   PI PC = PCO tan α = R Tan α = R Tan I/2

2)   PCO = PIO Cos α
     PCO = R
     Hence PIO = PCO/Cos α = R/Cos α = R/Cos (I/2)

3)   PIM = PIO – R = R/Cos α – R = R/Cos (I/2) - R

4)   Length of the curved portion PC to PI can be found as follows.

Total perimeter of the circle = 2π R

Total perimeter of the circle generates an angle of 360^0 at the center.

360^0 extends a curve of $2\pi R$

Hence 1^0 extends a curve of $2\pi R/360$

α^0 extends a curve of $2\pi R/360$ x α

$2\alpha^0$ extends a curve of $2\pi R/360$ x 2α

Length of the curved portion of the road from PC to PT = $2\pi R/360$ x 2α

Practice Problem: Station of a road at PC is 5 + 30. Station at PI is 8 + 20. The angle of intersection (I) is 40^0.
  a)   Find the radius. (Refer to below figure)
  b)   Find the station at PT (measured along the curved portion of the road).
Solution:

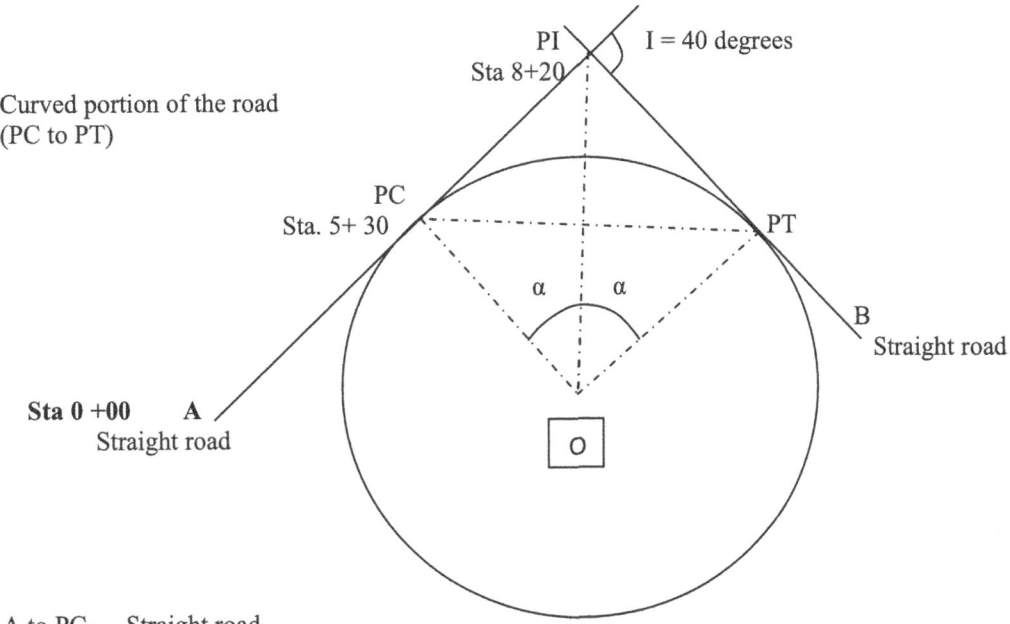

A to PC – Straight road
PC to PT – Curved portion of the road
PT to B – Straight road again

a) Length between PC and PI is 820 – 530 = 290 ft
          PCPI = 290 = R tan α
          α = I/2 = 20^0 (Showed earlier)
          820- 530 = 290 = R tan 20^0.
          R = 290/Tan 20^0 = 796.8 ft

  b)   Length of the curved portion of the road from PC to PT = 2πR/360 x 2α
  = 2π x 796.8/360 x 2 x 20 = 555.7 ft

  Station at PC = 5 + 30 = 530
  Station at PT = 530 + 555.7 = 1,085.7
  Station at PT = 10 + 85.7
Note: Stations can be marked along a curved road or a straight line. Station of PT is obtained from station of PC. Station of PT is **never** obtained from station of PI. Station at PT should be obtained using the station at PC. Following equation is used to obtain the station at PT

Station at PT = Station at PC + Curved length from PC to PT

Practice Problem)   Intersection angle of a horizontal curve is 118^0 12' 14". Radius of the curve is 332.5 ft. Find the area of the hatched section.

A) 2.62 acres   B) 1.59 acres   C) 9.45 acres   D) 6.53 acres

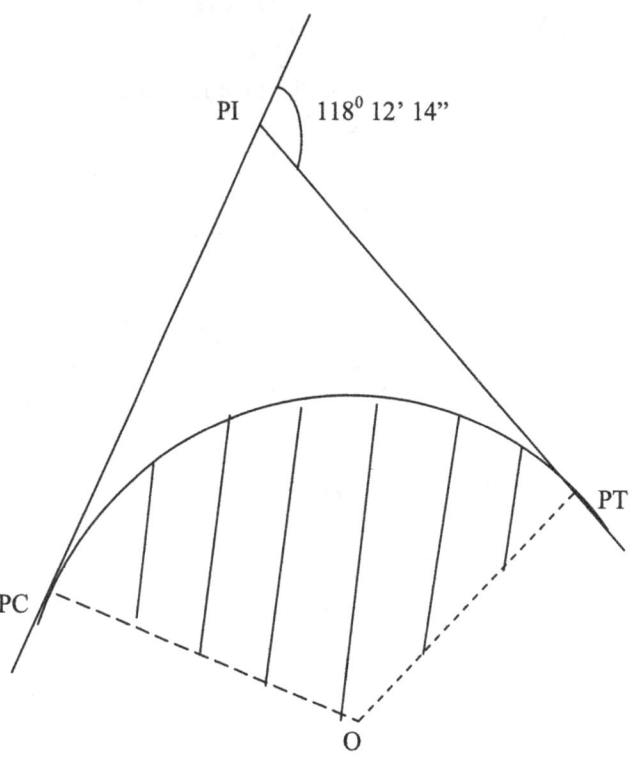

Solution):
STEP 1:   Find the angle PC O PT
Angle PC O PT is same as the intersection angle.
Angle PC O PT = 118^0 12' 14" = 118.20^0

STEP 2:   Find the area;

Area of a full circle = $\pi. R^2$
Full circle has 360 degrees.
Area inside one degree arc = $\pi. R^2/360$
Area inside 118.20^0 degree arc = $\pi. R^2/360$ x 118.20

R is given to be 332.5
Hence area of the hatched section = $\pi. (332.5)^2/360$ x 118.20 = 114,037.6 sq. ft
= 114,037.6/43,560 acres = 2.62 acres
Note that one acre is equal to 43,560 sq. ft.
Ans A

Practice Problem)   Station at point of curvature (PC) is given to be 8 +13 and the radius of the curve is 239 ft. Find the station at PI. (Angle of intersection is 50 degrees).

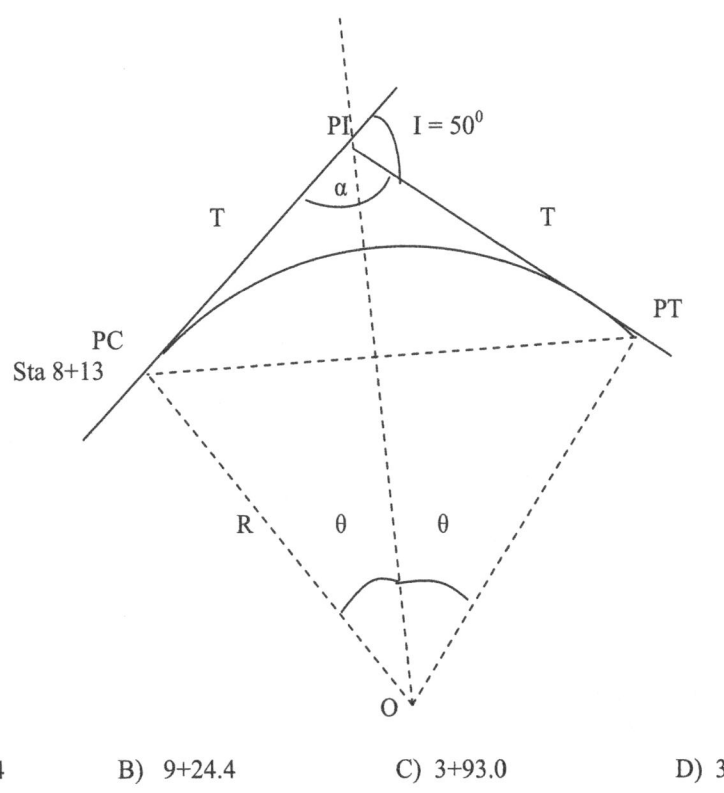

A)   8+16.4   B) 9+24.4   C) 3+93.0   D) 3+11.9

Solution):

Following nomenclature is used for horizontal curves.
Length between PC and PI is normally represented by "T".
In this problem, station at PC, radius (R) and intersection angle are given.

STEP 1:   Find the distance T
From trigonometry, $T = R \tan \theta$

STEP 1: Find θ
Consider the angle at PT PI PC.
Angle at PT PI PC = α = 180 - I

Hence angle O PI PC should be half of that.
Angle O PI PC   = α /2 = (180 - I)/2
                = 90 - I/2

Now consider the triangle PI PC O.
Since α = 180 - I
θ = 90 - (180 - I)/2 = 90 - (90 - I/2) = I/2

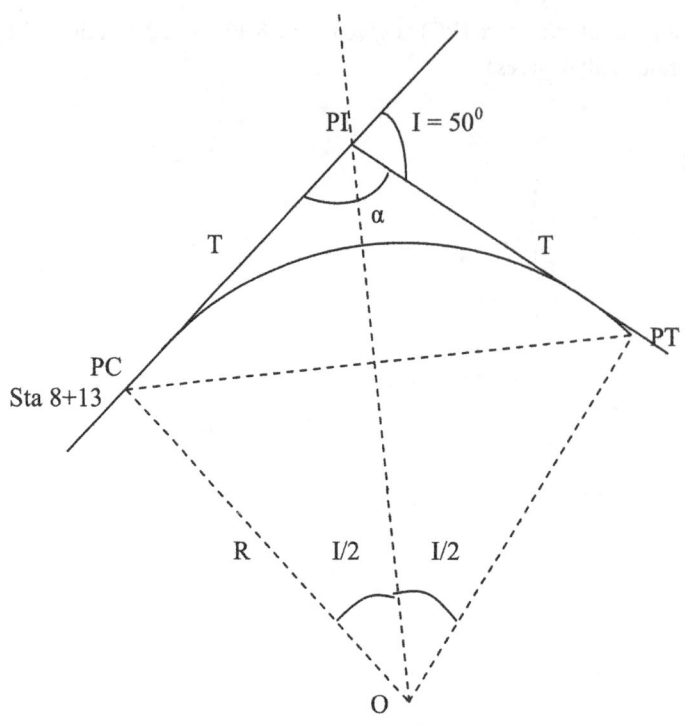

$T = R \times Tan\ I/2$
$T = R \times Tan\ (50/2) = 239 \times Tan\ 25 = 111.4\ ft$
Station at PI = 813 + 111.4 = 924.4 = 9 + 24.4   Ans B

Discussion:
Distance is represented with stations. Some examples are shown below.

100 ft = 1+ 00 Station
410 ft = 4 + 10 station
17,500 ft = 175 + 00 station

Practice Problem: Following information available for a horizontal curve.
PC station = 12+30
PT station is = 16+40
PI station = 17 + 30
Find the intersection angle (I).

A) 132.8   B) 143.9   C) 73.4   D) 23.9

Solution:

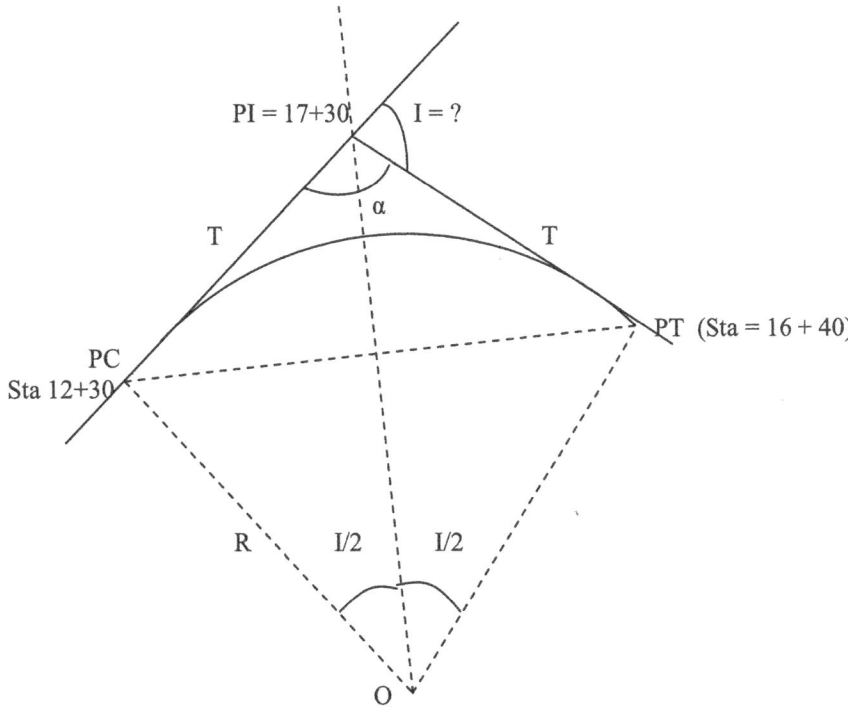

STEP 1:   Find the distance between PC and PI (T)
T = PI station – PC station = 1,730 – 1,230 = 500 ft

T = R. Tan (I/2)
500 = R. Tan (I/2) ------------------------------(1)
If we can find R, we can find I
Here angle I is in degrees.

STEP 2:   Find the curve distance between PT and PC;
Curve distance between PT and PC = 1,640 – 1,230 = 410 ft

Curve distance between PC and PT is equal to Radius times the angle (I) measured in radians.

$410 = R \cdot I_{radians}$

STEP 3:   Convert radians to degrees;
$I_{radians} = I_{degrees}$ x $\pi/180$

$410 = R \cdot I_{radians}$
Hence $410 = R \cdot (I_{degrees}$ x $\pi/180)$
410 x $180/\pi = R \cdot I_{degrees}$
$R = 23,491.3/I_{degrees}$

Insert R in eq 1
500 = R. Tan (I/2) ------------------------------(1)
$500 = (23,491.3/I_{degrees}$) x Tan $(I_{degrees}/2)$
0.0213 x $I_{degrees} =$ Tan $(I_{degrees}/2)$

This has to be solved using trial and error.
I = 143.9
Ans B

Practice Problem: A surveyor set up a station at 20 + 00 and measures 2,800 ft to obtain the PI station. The bearing of the PC PI line is N 50^0 43' 23" E. The bearing of the PI PT line is N 130^0 22' 14" E. The radius of the horizontal curve is 345 ft.
a) Find the intersection angle
b) Find the station at PC
c) Find the station at PT

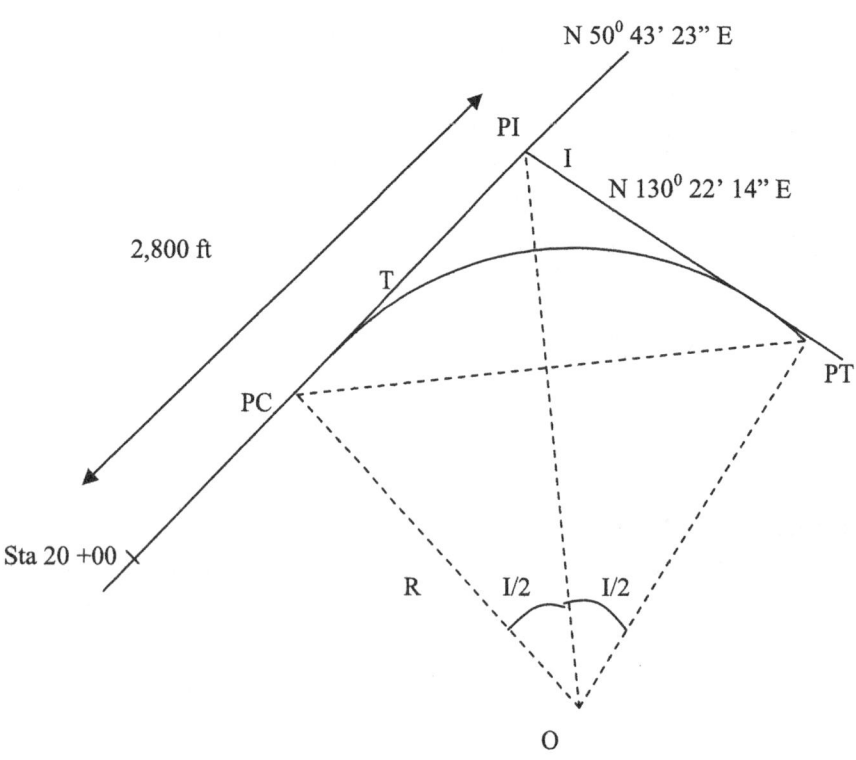

Solution:
a) Find the intersection angle;

The bearing of PC PI line is N 50^0 43' 23" E.

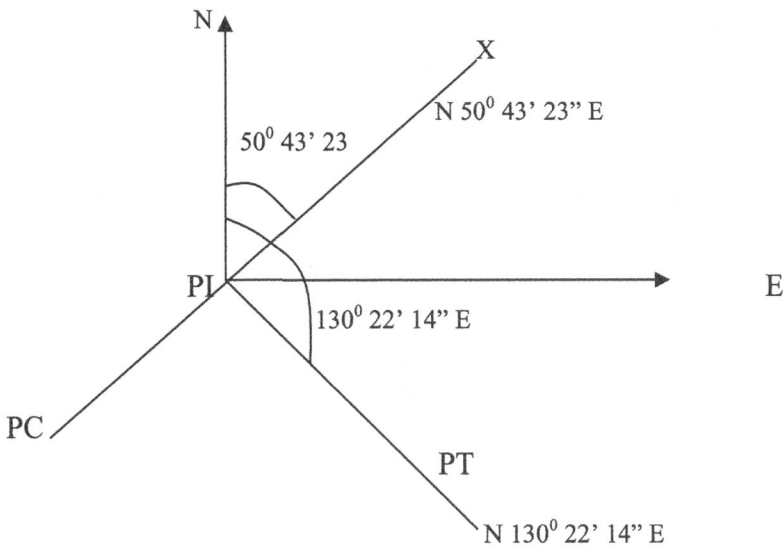

The angle N PI PT = 130^0 22' 14" = 130.370
The angle N PI X = 50^0 43' 23" = 50.723
The angle X PI PT = Intersection angle (I) = 130^0 22' 14" - 50^0 43' 23" = 130.370 - 50.723 = 79.647

b) Find the station at PC;

Station at PI can be found by adding 2,800 ft to station 20 + 00.
Hence station at PI = 2,000 + 2,800 = 4,800 = 48 +00

T = R tan (I/2) = 345. tan (79.647/2) = 287.68

Station at PC = Station at PI – 287.68
Station at PC = 4,800 – 287.68 = 4,512.32 = Sta. 45 + 12.32

c) Find the station at PT;

Find the curve distance from PC to PT;

The curve distance from PC to PT = R. $I_{radians}$ = R. $I_{degrees}$ x $\pi/180$
The curve distance from PC to PT = 345 x 79.647 x $\pi/180$ = 479.56 ft

The station at PT = Station at PC + Curved distance from PC to PT
The station at PT = 4,512.32 + 479.56 = 4,991.88 ft
The station at PT = 49 + 91.88

Obstructions in Horizontal Curves:

"28 percent of the 41,059 people who died on our highways in 2007 — more than 11,000 of our family members, friends, and neighbors — died in horizontal curve-related crashes. They make up a 1.0 percentage of the road miles but account for one-quarter of all highway fatalities." Division Administrator Tom Smith, of the Federal Highway Administration's (FHWA).

When designing horizontal curves one has to be mindful of obstructions. Horizontal curves has to be designed so that drivers have enough time to see the traffic coming towards them.

Two horizontal curves are shown above. One on the left is visible and safe. One on the right has trees and poles obstructing the view of the driver.

6.2 Stopping Sight Distance:

Stopping sight distance is the length required to stop a vehicle just after seeing a stopped vehicle, fallen tree or any other obstruction. Just imagine driving in a curvy road. One would always wonder there is a broken vehicle hiding behind the curve. On the other hand if the speed limit is low, motorists may be able to stop it as soon as they see the other vehicle. As per AASHTO Stopping sight distance consists of two distances:

Brake Reaction Distance : The distance travelled by the vehicle from the instant the driver sights an object necessitating a stop to the instant the brakes are applied

Braking Distance: The distance needed to stop the vehicle from the instant brake application begins.

> Stopping Sight Distance = Brake Reaction Distance + Braking Distance

In computing and measuring stopping sight distances, the height of the driver's eye is estimated to be 3.5 ft and the height of the object to be seen by the driver is 2.0 ft, equivalent to the taillight height of the passenger car.

AASHTO A policy on Geometric Design for Highways and Streets provides the following table for stopping sight distance;

Stopping Sight Distance (AASHTO) *A policy on Geometric Design for Highways and Streets*				
			Stopping sight distance	
Design Speed (mph)	Brake reaction distance (ft)	Braking distance on level ground (ft)	Calculated (ft)	Design (ft)
15	55.1	21.6	76.7 (=55.7 + 21.6)	80
20	73.5	38.4	111.9 (=73.5 + 38.4)	115
25	91.9	60.0	151.9	155
30	110.3	86.4	196.7	200
35	128.6	117.6	246.2	250
40	147.0	153.6	300.6	305
45	165.4	194.4	359.8	360
50	183.8	240.0	423.8	425
55	202.1	290.3	492.4	495
60	220.5	345.5	566.0	570
65	238.9	405.5	644.4	645
70	257.3	470.3	727.6	730
75	275.6	539.9	815.5	820
80	294.0	614.3	908.3	910

For an example, if the vehicle is travelling at 50 mph, the design stopping sight distance is 425 ft.

(Note: AASHTO is not required for the morning session of the exam. Necessary tables will be provided with the problem statement). In the above table, AASHTO gives the brake reaction distance. This is the distance that a vehicle would travel while the driver decides to brake or not. Once the brakes are applied, vehicle would travel forward. Distance that a vehicle would travel after brakes are applied is known as breaking distance. Stopping sight distance is the addition of these two.)

Stopping sight distance = Brake reaction distance + Braking distance on level ground

Design stopping sight distance is a rounded off value of the calculated value.

Horizontal Sight Offset: Horizontal sight offset is shown in the figure below.

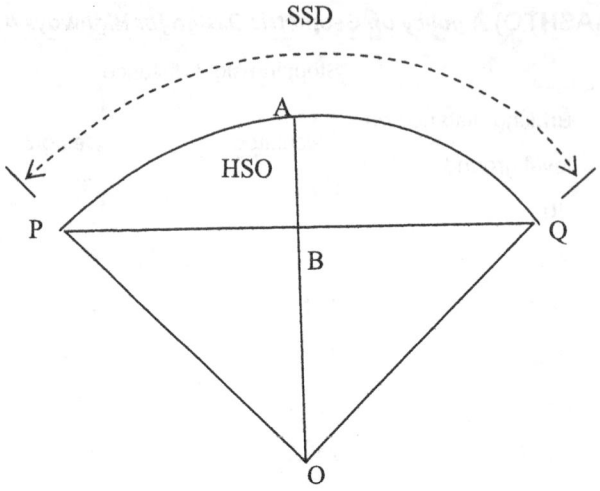

AB = HSO (Horizontal sight offset);   SSD = Stopping Sight Distance (Curved length from P to Q)

A driver at point P should be able to see an obstruction in the road such as accident at point Q and should be able to stop the car if necessary.

Below figure shows two cars and a line connecting two drivers. One driver should be able to see the other car.

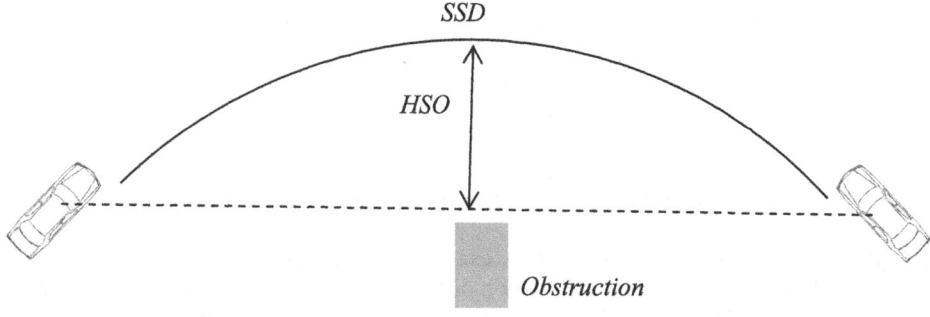

HSO is the horizontal sight offset. All obstructions should be placed outside of HSO as shown in the figure.

Practice Problem: Engineers have designed a horizontal curve so that traffic at PC station will be able to see the traffic at PT station. Property owner would like to build a structure as shown. What's the closest distance to the road that the structure can be built?.   (Find distance AB)

Radius of the curve = 900 ft

Angle of intersection is 18^0.

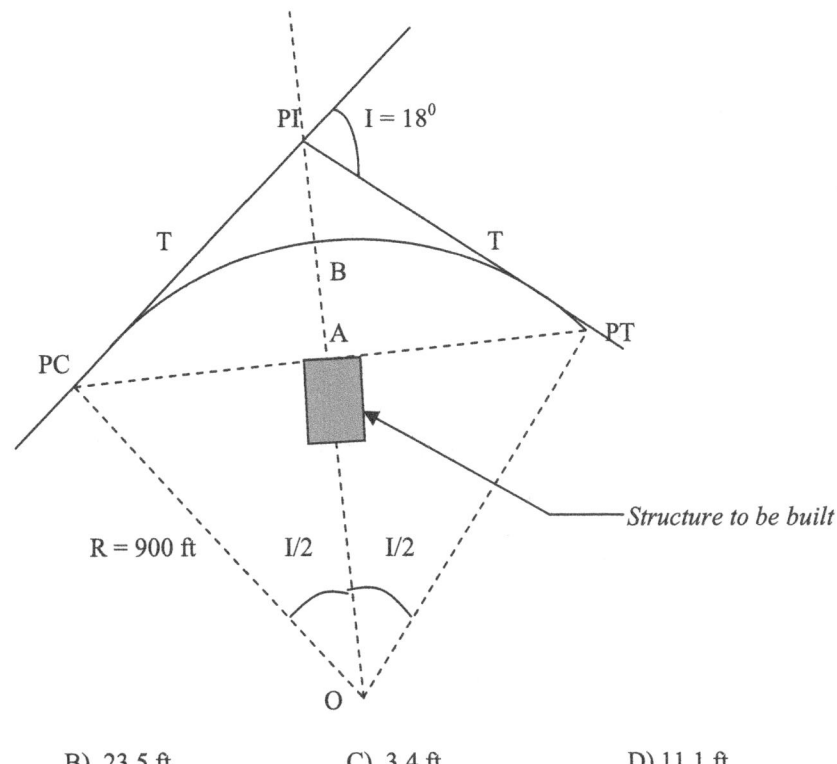

A) 13.4 ft   B) 23.5 ft   C) 3.4 ft   D) 11.1 ft

Solution: OB = Radius of the curve = 900 ft
AB = OB – OA;   OA = R Cos I/2
Intersection angle (I) is given to be 18^0.
OA = 900 Cos (18/2) = 888.9 ft
HSO = AB = OB – OA = 900 - 888.9 = 11.1 ft   (Ans D)

6.3   HSO and SSD Relationship;

HSO (Horizontal sight offset) and SSD (stopping sight distance) are related. Higher the posted vehicle speed of a curve, higher the stopping sight distance (SSD). Higher the SSD, higher the HSO (Horizontal sight offset).   Distance AB = HSO. Let us look at a horizontal curve to understand this.

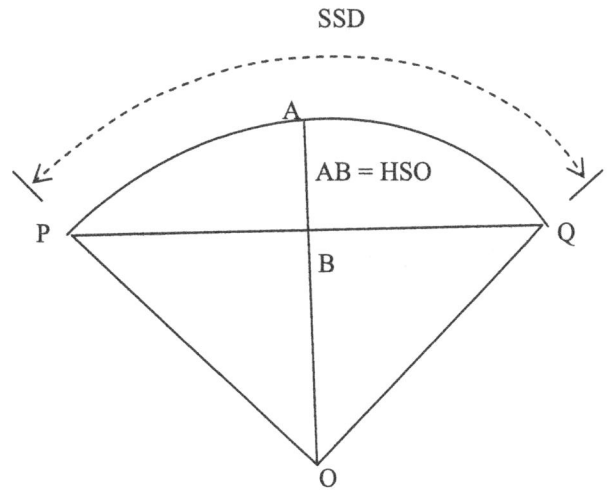

Assume there is a driver at point P. If he happens to see a stopped disabled car at point Q, then he should be able to stop the car prior to reaching point Q. Stopping sight distance is curve length from P to Q.

What is HSO? The length AB is considered to be HSO.
HSO is the horizontal sight offset. In other words, no object should be covering the AB line. If there is a house or tree in HSO, the driver may not be able to see the disabled vehicle at point Q.

Let us assume there is a house as shown below.

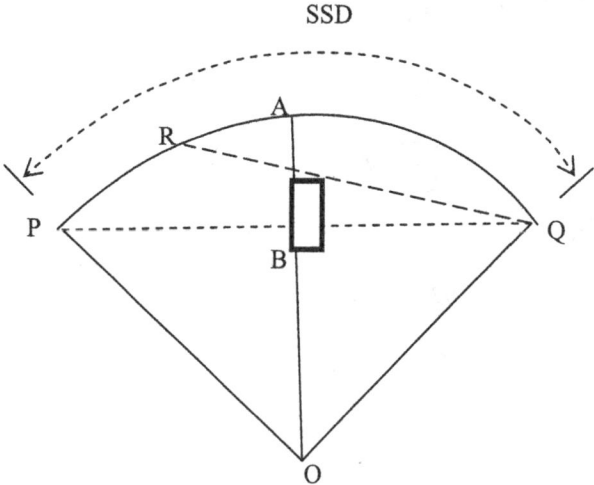

If there is a house located as shown, the driver will not see the disabled vehicle until he reaches the point "R". Then he may not be able to stop the car prior to reaching point Q since stopping sight distance is "SSD" which is longer than curve length from point R to Q. You would have an accident.

Hence no trees, houses or any other object should be along HSO. That line should be kept clear.
Now let us develop a relationship between HSO and radius.

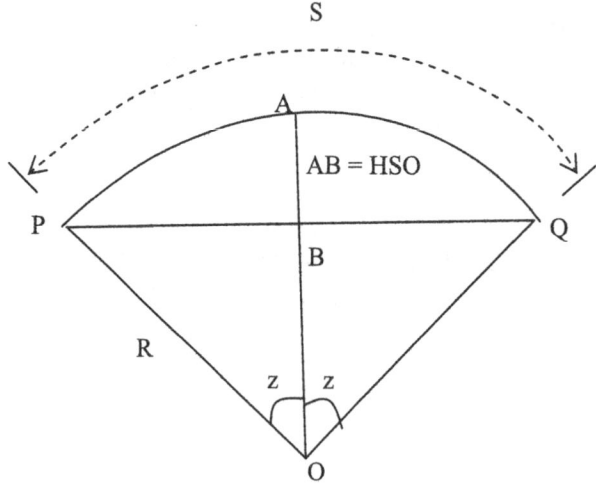

Let us call angle POA as z.
Then QOA = z as well.

AO = R = Radius
BO = R cos z
AB = AO − BO = R − R cos z

AB = R (1 – cos z)
AB = HSO
HSO = R (1 – cos z)   ----------------------------(1)

There is a relationship between the curve length of a circle and angle at the center of the circle. That relationship is;

Curve length = Radius x Angle at the center of the circle measured in radians

Hence;

$$SSD = R. (2\ z_{radians})   \text{-------(2)}   (z \text{ should be measured in radians}).$$

Convert z to degrees.

180 degrees = π radians
1 degree = $\pi/180$ radians
$z_{radians} = z_{degrees}$ x $(\pi/180)$
From above (2)
SSD = R. (2 $z_{radians}$)
Substitute $z_{radians}$.

SSD = R. (2.$z_{degrees}$ x $\pi/180$)
$z_{degrees}$ = (SSD/R) x $(90/\pi)$
Insert $z_{degrees}$ in above equation (1).

HSO = R (1 – cos z)   ----------------------------------(1)
HSO = R [1 – cos {(SSD/R) x $(90/\pi)$}]
$90/\pi$ = 28.65
HSO = R [1 – cos (28.65 SSD/R)]

Practice Problem)   Stopping sight distance of a road is 460 ft. Radius of the horizontal curve is 600 ft. Find the horizontal sight offset required for the horizontal curve.
A) 123.56 ft   B) 25.89 ft   C) 91.00 ft   D) 43.55 ft

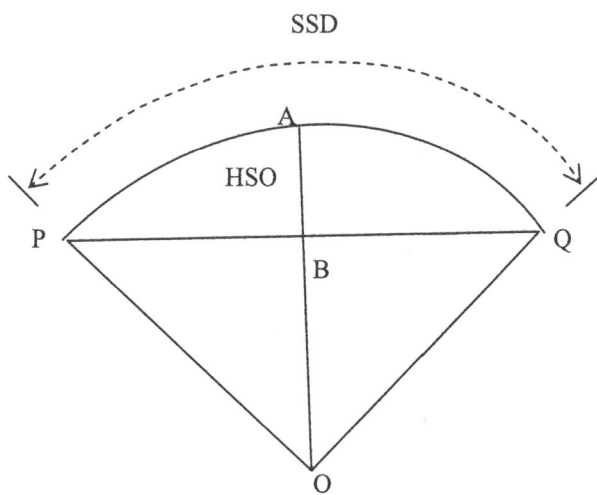

AB = HSO (Horizontal sight offset)
Solution):   HSO is given by the following equation.

$$HSO = R\,[1 - \cos\,(28.65.\ SSD/R)]$$

HSO = Horizontal sight offset (ft)
R = Radius of the horizontal curve (ft)
SSD = Stopping sight distance (ft)

STEP 1:   Write down the parameters given;
R = Radius = 600 ft
SSD = Stopping sight distance = 460 ft

STEP 2:   Apply the above equation;
HSO = R [1 – cos (28.65. SSD/R)]
HSO = 600 [1 – cos (28.65 x 460 /600)]
IISO = 43.55 ft
Ans D

Practice Problem: Horizontal curve has a radius of 210 ft and has an obstruction as shown. AB distance (HSO) is given to be 70 ft. What is the maximum speed as per AASHTO table for stopping sight distance?
(AASHTO table was given earlier in this book).

A) 35 mph   B) 45 mph   C) 40 mph   D) 60 mph

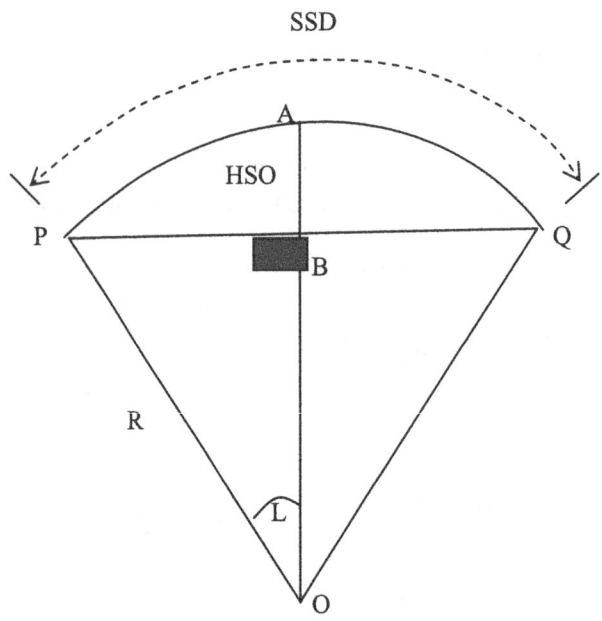

AB = HSO (Horizontal sight offset);   SSD = Stopping Sight Distance (Curved length from P to Q)

AB = 70 (Given)
OA = Radius = 210 (Given)
OB = OA – AB = 210 – 70 = 140
Angle BOP = L
OB = R. Cos L
140 = 210. Cos L
Cos L = 140/210 = 0.6667
L = Cos^{-1} (0.6667)

L = 48.19

$L_{radians} = 48.19 \times \pi/180 = 0.8411$

SSD = PAQ (Measured along the curve).

$SSD = R \times (2.L_{radians})$

$SSD = 210 \times (2 \times 0.8411)$

SSD = 353.25 ft

Go to the AASHTO table for stopping sight distance. (AASHTO table was given earlier in this book). Look at the column marked "Design stopping distance". 353 ft is between 305 and 360.

Speed 40 mph ------------SSD = 305 ft

Speed 45 mph ------------SSD = 360 ft

A vehicle travelling at 45 mph needs a stopping sight distance of 360 ft. However, we have only 353 ft. Hence the vehicles have to travel slower than 45 mph. The correct answer is 40 mph.   (More accurate answer can be found by interpolation. However, posted speed limits are rounded off to nearest 5 mph).

Ans C

6.4   Degree of Curve;

Horizontal curves have a radius. Shorter the radius, higher the degree of curvature.

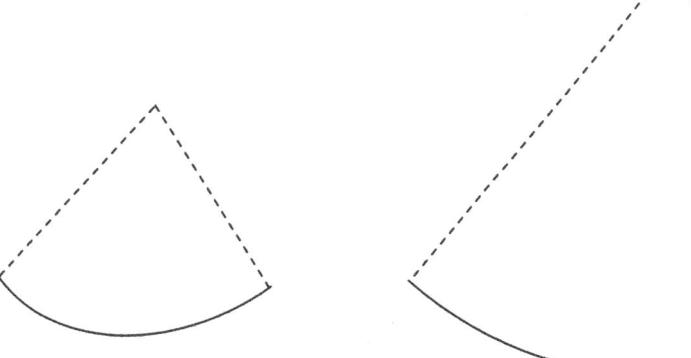

Figure on left: Figure on left is a curve with a small radius. This curve has a high degree of curvature

Figure on right; Figure on right is a curve with a large radius. This curve has a low degree of curvature.

It is correct to say that curves with large radius has a low degree of curvature and vice versa.

How to find the degree of curvature?   There are two types of degree of curvature.

- Degree of curve (arc method)
- Degree of curve (chord method)

In arc method, angle subtended at the center of a circle with an arc of 100 ft is known as degree of curve (D).

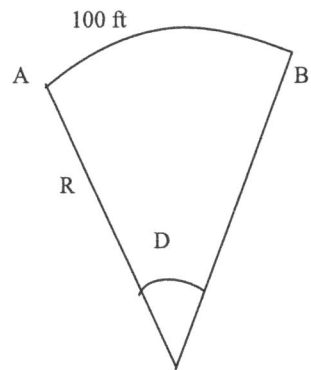

Arc Method (Arc AB = 100 ft)

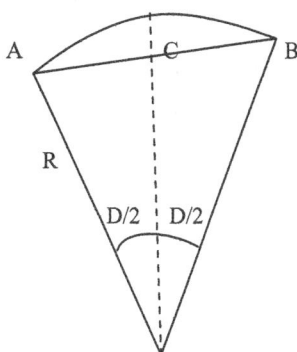

Chord Method (Chord AB = 100 ft)

Arc Method;

As per definition of degree of curve, D degrees will produce an arc of 100 ft.
Full circle is 360 degrees. Full circle produces $2 \pi R$.

360 degrees ----→ $2 \pi R$
1 degree --------→ $2 \pi R / 360$
D degrees ------→ $2 \pi R.D / 360$

As per definition, D degrees would produce an arc of 100 ft.
D degrees ------→ $2 \pi R.D / 360 = 100$ ft

$$D = \frac{100 \times 360}{2 \pi R} = \frac{5,729.5}{R}$$

Degree of curve (arc method)

$$R = \frac{5,729.5}{D} \quad \text{----------------------(1)}$$

D = Degree of curve;   R = Radius

If "D" degrees is known, R can be computed.

Chord Method:

Chord AB (Straight line) = 100 ft   (See the figure given above).

$AC = R \sin D/2$
$AB = 2 AC = 2 R \sin D/2$

In chord method 100 ft chord is used.
Hence

$100 = 2 R \sin D/2$

$R = 50/\sin D/2$ ---(2)

Degree of curve (chord method)

$$R = 50/(\sin D/2) \text{----------------------(2)}$$

D = Degree of curve;   R = Radius

Practice Problem: Degree of curve of a horizontal curve (arc method) is given to be 6.1 degrees. What is the radius of the curve?

   A) 939.3 ft   B) 212.8 ft   C) 621.1 ft   D) 1,801.8 ft

Solution

Degree of curve in this problem = 6.1 degrees.
Since it is arc method, use equation 1.

$$R = \frac{5,729.5}{D} = \frac{5,729.5}{6.1}$$

R = 939.3ft   (Ans A)

Practice Problem: Degree of curve of a horizontal curve (chord method) is given to be 6.1 degrees. What is the radius of the curve?

   A) 939.7 ft   B) 212.8 ft   C) 621.1 ft   D) 1,801.8 ft

Solution
Degree of curve in this problem = 6.1 degrees.
Since it is chord method, use equation 2.

$$R = 50/(\operatorname{Sin} D/2)$$

R = 939.7ft   (Ans A)
Note: For the most part, both methods give very similar answers.

6.5   Vertical Curves:
Vertical curves are needed where there are hills or valleys. There are two types of vertical curves;
- Crest curves
- Sag curves

Left: Crest curve
Right: Sag curve

Crest Curves: Crest curve is shown below.

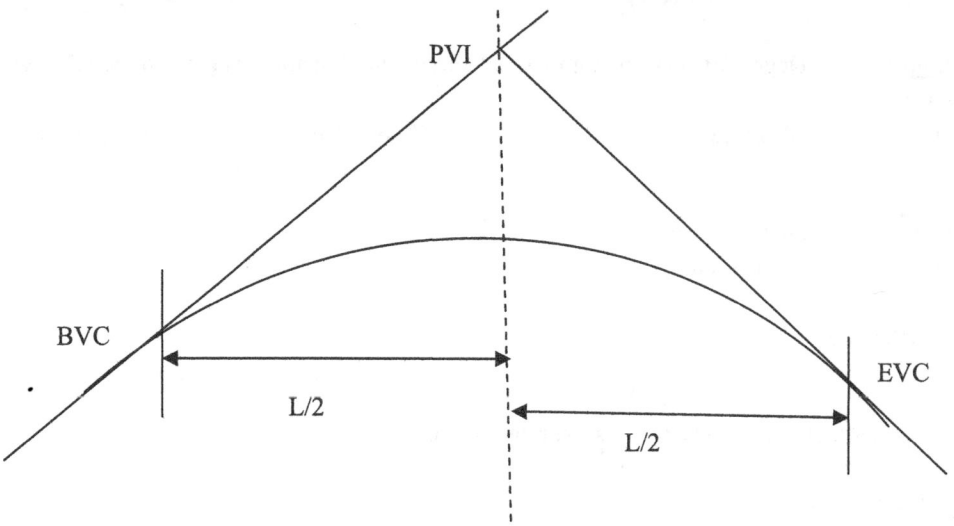

BVC = Beginning of vertical curve
EVC = End of vertical curve
PVI = Point of vertical intersection
In vertical curves, PVI is located at the center of BVC and EVC. See the figure above.

Vertical Curve Equation (Valid for both crest curves and sag curves);

$$y = \frac{(G2 - G1)}{2L} . x^2 + G1 . x$$

y = Vertical height to point in the curve from BVC in ft
x = Horizontal distance between two points
G1 and G2 are gradients in decimals.
Upward gradient is taken to be positive. (+).
Downward gradient is taken to be negative. (-) .
L = Horizontal distance between BVC and EVC in ft.

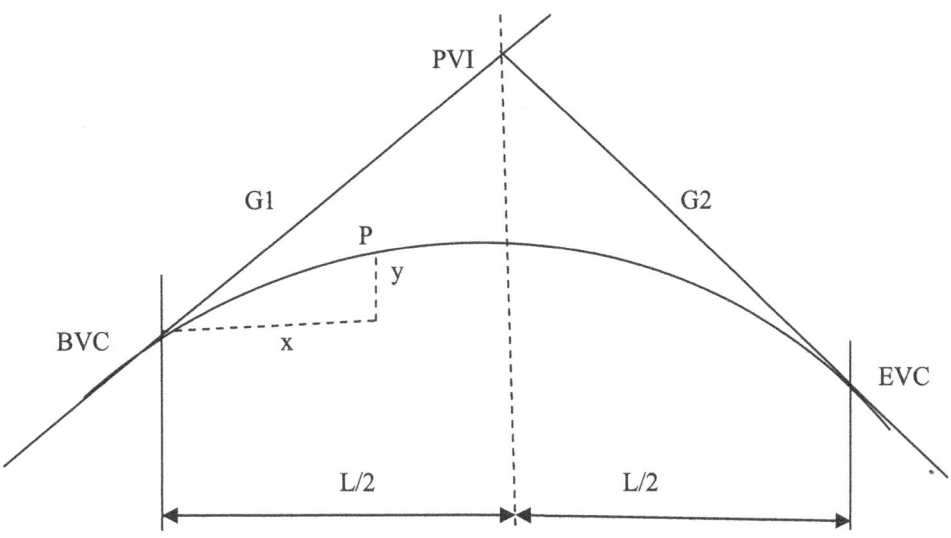

Assume point "P" located at "x" and "y" distance from the BVC point as shown. If "x" is known, y can be found using the vertical curve equation.

It is important to remember that the highest point on the vertical curve may or may not occur at the center. However, PVI is always at the center of the curve.

Practice Problem: Following parameters are given for a vertical crest curve.
G1 = 2.3%,   G2 = 5.4%,   L = 200 ft
What is the height of a point that lies 75 ft away from BVC?

A) 0.6422   B) 1.2901   C) 0.9871   D) 2.9812

Solution:

STEP 1:   Convert G1 and G2 into decimals.
G1 = +0.023;   G2 = -0.054;   L = 200;   x = 75
G1 is positive since it is an upward grade.   G2 is negative since it is a downward grave.

STEP 2:   Apply the equation;

$$y = \frac{(G2 - G1)}{2\,L}. x^2 +   G1. x$$

$$y = \frac{(-0.054 - 0.023)}{2 \times 200}. x^2 +   0.023 \times 75$$

$y = -0.077/400 \times 75^2 + 1.725$
$y = -1.08 + 1.725 = 0.6422$ ft   Ans A

Practice Problem: Following information given for a crest vertical curve.
G1 = 1.3%,   G2 = 2.4%,   L = 300 ft
BVC station = 12+00;   EVC station = 20+00
Elevation at BVC = 102.89
What is the elevation of a point 600 ft from the BVC?

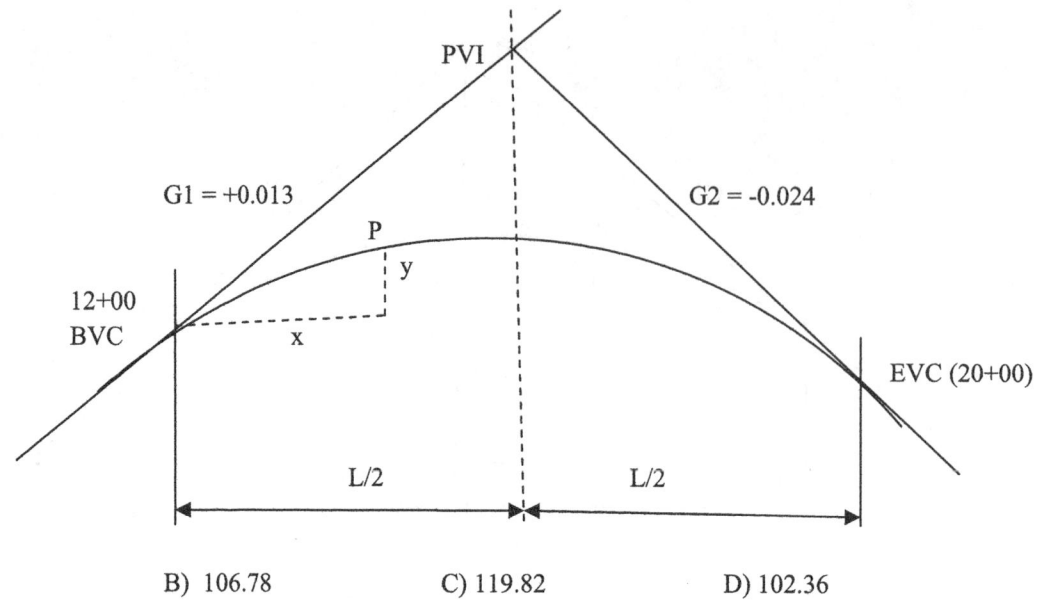

A) 113.67   B) 106.78   C) 119.82   D) 102.36

Solution:

STEP 1:   Find "L"

L = 20 + 00 − 12 + 00 = 2,000 − 1,200 = 800 ft

STEP 2:   Apply the equation;

$$y = \frac{(G2 - G1)}{2\,L}.\, x^2 + \quad G1.\, x$$

Find "y" when "x" = 600
G1 = +0.013 (upward);   G2 = -0.024 (downward);   L = 800 ft

$$y = \frac{(-0.024 - 0.013)}{2 \times 800}.\, 600^2 + \quad 0.013 \times 600$$

y = -8.325 + 7.80 = -0.525

Elevation at BVC = 102.89 (given)

Elevation at a point 600 ft away from BVC = 102.89 − 0.525 = 102.365
Ans D

6.5.1 Sag Curve:

Sag curves are needed for valleys. Typical sag curve is shown below.

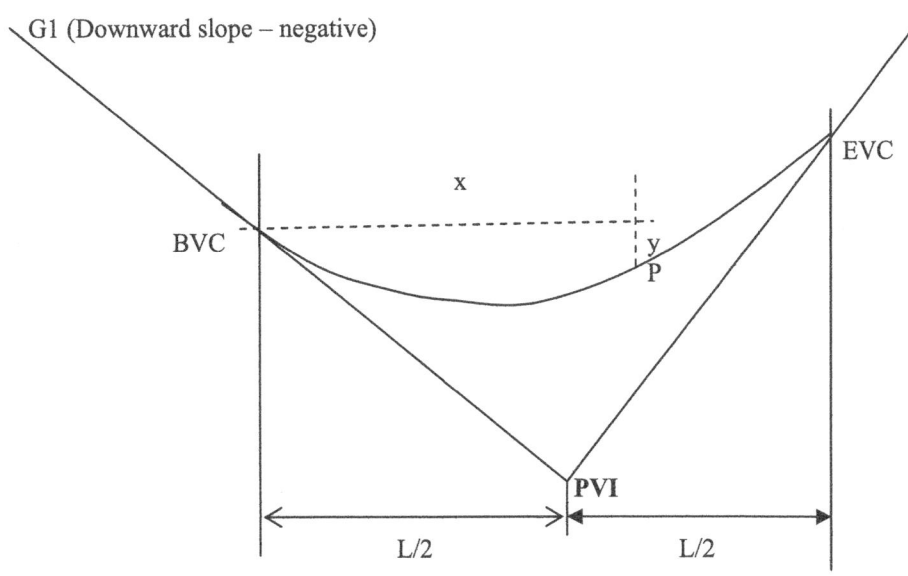

BVC = Beginning of vertical curve
EVC = End of vertical curve
PVI = Point of vertical intersection   (Occurs at the midpoint of the vertical curve).

Practice Problem: Following information available for a sag curve.
BVC = Beginning of vertical curve = Station 12+90; Elevation 302.8 ft
EVC = End of vertical curve   = Station 15+30
G1 = 3.8%;   G2 = 2.2%

a) What is the station of PVI
b) What is the elevation of EVC
c) What is the elevation of station 15+00

Solution:
a)   Find the station of PVI;

Station of PVI is in between BVC and EVC.

The distance between BVC and EVC = (15+30) – (12+90) = 240 ft
PVI is 240/2 distance from BVC.   The stations in vertical curves are measured horizontally.
PVI station = 12+90 + 120 ft = 14+10

b) What is the elevation of EVC;

STEP 1: Apply the vertical curve equation;

$$y = \frac{(G2 - G1)}{2\,L}\, x^2 + \quad G1 . x$$

G1 = 3.8% = -0.038   (Note that in sag curves, G1 is downward. Hence it is negative).
G2 = 2.2% = +0.022   (Note that in sag curves, G2 is upward. Hence it is positive).
x = 240 ft
(EVC is 240 ft from BVC).
Also L = 240 ft.

$$y = \frac{(0.022 - -0.038).\ 240^2}{2 \times 240} + \ -0.038 \times 240$$

$$y = \frac{(0.022 + 0.038).\ 240^2}{2 \times 240} - \ 0.038 \times 240$$

$$y = 7.2 - 9.12 = -1.92$$

BVC elevation = 302.8
EVC elevation = 302.8 – 1.92 = 300.88 ft

c) What is the elevation of station 15+00;

Apply the vertical curve equation;

$$y = \frac{(G2 - G1).}{2\,L}\ x^2 + \ G1.\ x$$

G1 = 3.8% = -0.038
G2 = 2.2% = +0.022

x = 15+00 – 12+90 = 210 ft
L = 240 ft.

$$y = \frac{(0.022 - -0.038).\ 210^2}{2 \times 240} + \ -0.038 \times 210$$

$$y = \frac{(0.022 + 0.038).\ 210^2}{2 \times 240} - \ 0.038 \times 210$$

$$y = 5.51 - 7.98 = -2.47$$

BVC elevation = 302.8
EVC elevation = 302.8 – 2.47 = 300.33 ft

Practice Problem: A Highway Engineer is designing a vertical sag curve. BVC station is 7+00 and BVC elevation is 155.8 ft. EVC station is 11+30. There is an bridge overpass at station 10+00 going above the sag curve. The elevation of the overpass bridge is 164.3. Trucks need a clearance of 13 ft below the bridge. Gradient of the downward curve is 3.8%. What is the required gradient for the upward slope of the sag curve?

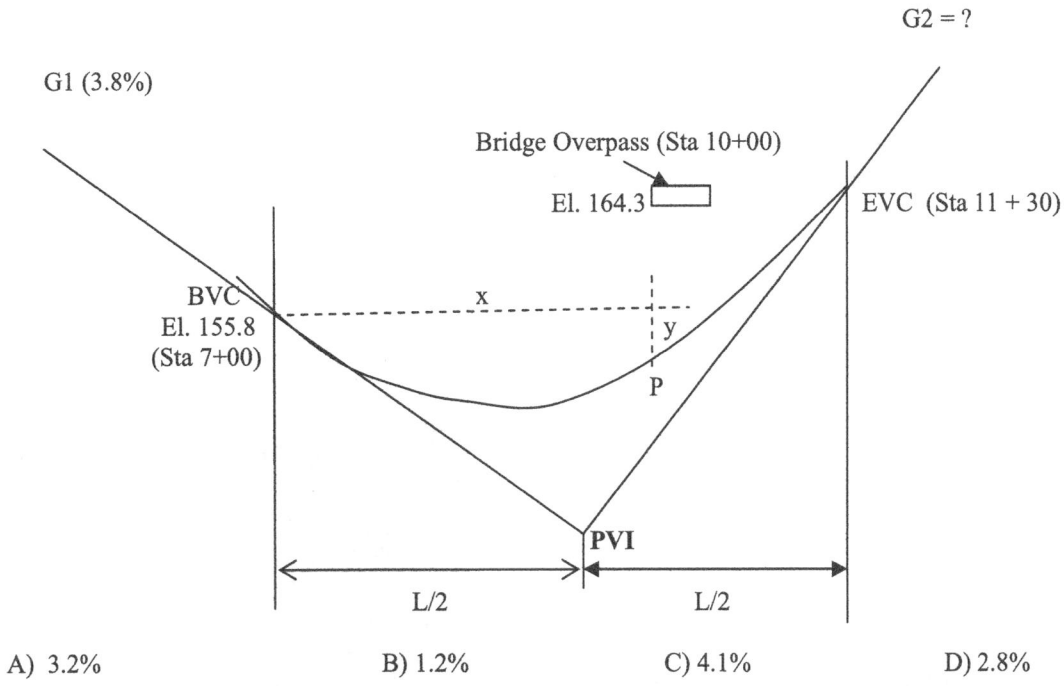

A) 3.2%   B) 1.2%   C) 4.1%   D) 2.8%

Solution:
STEP 1:   Let us gather all the information for the sag curve;

G1 = -0.038   (Downward slope is negative)
G2 = ?
L = EVC station – BVC station = [11+30] – [7+00] = 1,130 – 700 = 430 ft
Station at overpass = 10+00
Horizontal distance between BVC and overpass = (10+00) - 7+00 = 300 ft

Elevation of the overpass = 164.3
Trucks need a clearance of 13 ft between overpass and ground of the sag curve.
Elevation of ground under the overpass (Point P) = 164.3 – 13 = 151.3
Elevation difference between BVC and P = y = 151.3 - 155.8 = -4.5
Note that "y" is negative since point P is lower than BVC.

Horizontal distance between point P and BVC = x = [10+00] – [7+00] = 300 ft

STEP 2: Apply the vertical curve equation;

$$y = \frac{(G2 - G1)}{2\,L}.\, x^2 +\quad G1.\, x$$

G1 = 3.8% = -0.038   (Note that in a sag curve G1 is downward. Hence it is negative).
G2 = ?
L = 430 ft
x = 300
y = -4.5.

$$y = \frac{(G2 - G1)}{2L} . x^2 + G1 . x$$

$$-4.5 = \frac{(G2 - -0.038)}{2 \times 430} . 300^2 + -0.038 \times 300$$

$$-4.5 = (G2 + 0.038) \times 104.65 \; - 11.4$$

$$6.9 = (G2 + 0.038) \times 104.65$$
$$G2 = 0.0279 = 2.79\%$$
Ans D

Sag curve meeting an overpass bridge is shown.

6.6   Stopping Sight Distance (Equations):

Stopping Sight Distance on Flat Ground: In the previous chapter we discussed the stopping sight distance. AASHTO provided a table to find the stopping sight distance. AASHTO also provides an equation to find the stopping sight distance. Some problems can be solved easily using equations. Hence I recommend learning both equations and tables.

AASHTO defines stopping sight distance as the length of the roadway ahead that is visible to the driver. Stopping sight distance is the sum of:

1. The distance traversed by the vehicle from the instant the driver sights an object necessitating a stop to the instant the brakes are applied; and

2. The distance needed to stop the vehicle after applying brakes

AASHTO assumes that the height of Driver's eye = 3.5 feet from ground
Height of Object = 2.0 feet

The equation for stopping sight distance is:

$$d = 1.47 \, V.t + 1.075. \, V^2/a$$

d = Stopping sight distance in ft;
t = brake reaction time in seconds
(AASHTO recommends 2.5 sec. for break reaction time).
V = design speed in mph
a = deceleration rate, ft/s^2
(AASHTO recommends 11.2 ft/s^2 for deceleration rate).

Practice Problem;   Find the stopping sight distance of a vehicle travelling at 45 mph in flat ground. Assume brake reaction time to be 2.5 seconds and vehicle deceleration to be 11.2 ft/sec^2.

A) 359.8   B) 198.3 ft   C) 291.6 ft   D) 76.2 ft

Solution:

Apply the equation;

d = 1.47 V.t + 1.075. V^2/a
d = 1.47 x 45 x 2.5 + 1.075 x 45^2/11.2
d = 165.375 + 194.36 = 359.8 ft   Ans A

4.1.3.1 <u>Stopping Sight Distance on Sloped Ground</u>:   If a vehicle is travelling in a downward grade, it would take more time to stop. In addition, it would need more distance to stop. On the other hand, if a vehicle is travelling in an upward slope, it would take less time and distance to come to a stop after applying brakes.

$$d = 1.47V.t + \frac{V^2}{30[a/32.2 +/- G]}$$

d = Stopping sight distance in ft;
V = design speed in mph
t = brake reaction time in seconds
a = deceleration rate in ft/s
G = Grade in decimals   (Upward slope is positive and downward slope is negative).

Practice Problem: Find the stopping sight distance of a vehicle travelling at 45 mph in an upward slope of 3%. Assume brake reaction time to be 2.5 seconds and vehicle deceleration to be 11.2 ft/sec^2.

A) 344.0   B) 198.3 ft   C) 291.6 ft   D) 76.2 ft

Solution:

Write down the equation;

d = 1.47V.t + $\dfrac{V^2}{30(a/32.2 +/- G)}$

V = design speed, mph
t = brake reaction time, 2.5 sec.
a = deceleration rate, ft/s = 11.2 ft/s$_2$
G = Grade in decimals.

V = 45 mph;   G = +0.03   (G is positive since it is an upward slope).

$$d = 1.47V.t + \frac{V^2}{30(a/32.2 +/- G)}$$

$$d = 1.47 \times 45 \times 2.5 + \frac{45^2}{30 (11.2/32.2 + 0.03)}$$

$$d = 165.37 + 178.65 = 344.02 \text{ ft}$$

6.7 Superelevation:

In horizontal curves, there is a radial acceleration. Think about a merry go round. In a merry go round, if the rope breaks, the person would be thrown away from the circle.

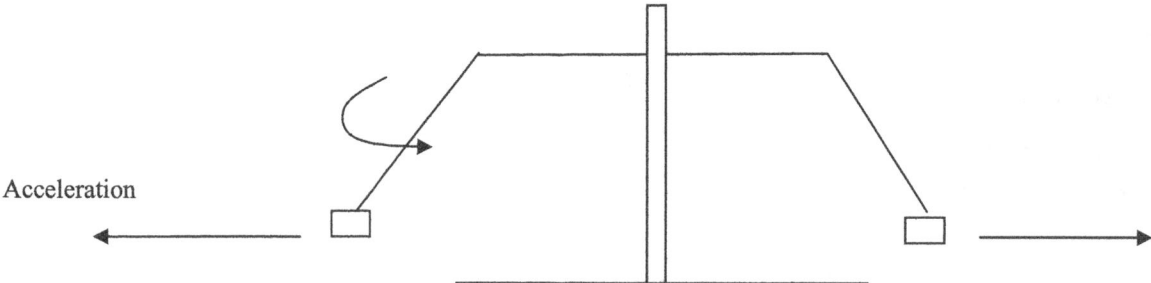

Acceleration

There is an acceleration in a merry go round that acts outwards from the circle. This is known as centrifugal acceleration. Centrifugal acceleration gives rise to centrifugal force.

The curve is sloped to counter centrifugal forces acting on the train:

Forces on a Superlevation;

Following figure shows forces acting on a vehicle travelling on a superelevated curve.

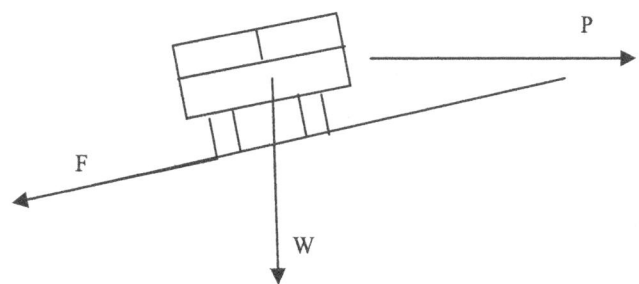

There are three forces acting on a vehicle travelling in a superelevation. They are;
1)   Weight (W)
2)   Centrifugal force (P). This force acts outwards from the center of the circle.
3)   Friction on tires (F)

Centrifugal force is balanced by the tire friction. Centrifugal force is dependent upon the speed of the vehicle. Higher speeds would generate higher centrifugal force. If the friction of the tires cannot balance the centrifugal force, the vehicle would go out of the road.

Superelevation Equation;

AASHTO Geometric Design of Highways known as green book gives the following equation;

$$f = V^2/15.R \ - \ e/100$$

f = Side friction coefficient (decimals)
V = Velocity in mph
R = Radius of the horizontal curve (ft)
e = Superelevation in percent

Practice Problem: A road has a superelevation of 2.6%. Radius of the horizontal curve is 300 ft. The friction coefficient between tires and road is 0.42. What is the maximum speed vehicles can achieve safely.

A) 44.8 mph;   B) 35.7 mph   C) 65.0 mph   D) 42.5 mph

Solution:

STEP 1: Write down all the known parameters;

R = 300 ft;   f = 0.42;   e = 2.6

STEP 2:   Apply the AASHTO superelevation equation;

$$f = V^2/15.R - e/100$$
$$0.42 = V^2/15 \times 300 - 2.6/100$$
$$V = 44.8 \text{ mph}$$

6.8 Vertical and horizontal clearances:

<u>Horizontal Clearance</u>:   Lane width should be 10 ft to 12 ft. Roads with curbs should have a clearance of 1.5 ft with any object. Clearance of 3ft is required to objects near curves.

<u>Vertical Clearances</u>: New structures should provide 16 ft vertical clearance over the entire roadway width. Existing structures may be 14 ft unless prohibited by local statues. Highly urbanized areas, vertical clearance of 14 ft is acceptable as long as there is an alternate route with 16 ft clearance.

6.9 Acceleration and Deceleration:

There are three dynamic equations that combines distance travelled, time elapsed, velocity and acceleration . They are

$$v = u + a.t \quad (1)$$
$$s = ut + 1/2at^2 \quad (2)$$
$$v^2 = u^2 + 2a.s \quad (3)$$

u = Initial velocity; (ft/sec)
v = Final velocity;   (ft/sec)
s = Distance;   (ft)
t = time;   (sec)
a = acceleration   (ft/sec^2)

If a vehicle is travelling at the speed of "u" at an acceleration of "a" then its velocity after "t" time period would be equal to v = u + at. If you need to find the distance travelled during a given time period, then equation 2 can be used.

Practice Problem: Initial speed of a vehicle is 30 mph. It is accelerating at a rate of 1.3 ft/sec^2. What is the velocity of the vehicle after 1 minute?
A) 35.7 mph   B) 83.1 mph   C) 65.9 mph   D) 29.2 mph

Solution:

STEP 1: Convert mph to ft/sec.

u = 30 mph = 30 x 5280/3600 ft/sec = 44 ft/sec
t = 1 minute = 60 seconds
a = Acceleration = 1.3 ft/sec^2.

STEP 2: Apply the equation (1).

v = u + a.t
v = 44 + 1.3 x 60 = 122 ft/sec
Convert this value to mph
v = 122 ft/sec = 122 x 3600/5,280 mph = 83.1 mph

Practice Problem: A vehicle is travelling at a speed of 60 mph. The driver sees a stopped car in the middle of the road and applies brakes. Brake reaction time is 2.5 seconds. The stopped car is 350 ft away when the driver saw it for the first time. What is the brake deceleration required to avoid a collision?

A) -43.5 ft/sec^2 B) -12.9 ft/sec^2.   C) -29.8 ft/sec^2.   D) 12.9 ft/sec^2.

Solution:

The stopped car is 350 ft away when the driver happened to see it. After seeing the car, driver would travel at the same speed for 2.5 seconds. The vehicle will travel at the same speed during the brake reaction time.

STEP 1:   Find the distance travelled during brake reaction time.

Speed of the vehicle = 60 mph = 88 ft/sec
Brake reaction time = 2.5 seconds.
Distance travelled during brake reaction time = 88 x 2.5 = 220 ft.

STEP 2:   Find the braking Distance;

Braking distance = 350 – 220 = 130 ft

STEP 3:   Apply the equation;

u = 88 ft/sec;   v = 0;   s = 130 ft   a = ?

Apply equation 3;

$$v^2 = u^2 + 2a.s$$

0 = 88^2 + 2 .a x 130
a = -29.8 ft/sec^2.   Ans C
(Negative value indicates a deceleration.)

Newton's Second Law;   Newton's second law states that when a force is applied to a mass, the force is equal to mass multiplied by the acceleration.

$$P = m . a$$

P = Acceleration force (lbf);

a = acceleration (ft/sec^2)
m = mass (lbs)

Practice Problem: 10 lb object is accelerated to 3 ft/sec^2. What is the force required to achieve this acceleration?

Solution: Apply the Newton's equation;

          P = m . a
m = 10 lbs (given);   a = 3 ft/sec^2.   (given)
P = m. a = 10 x 3 = 30 lbf.

Practice Problem: A vehicle was parked on a sloped road. Due to brake inactivation, the vehicle start to roll down the slope. The slope is at an angle of 30 degrees to the horizontal. What is the acceleration of the vehicle? (Assume the friction of the road to be zero).
A) 16.1 ft/sec^2 B) 26.7 ft/sec^2 C) 7.1 ft/sec^2 D) 19.8 ft/sec^2

Solution:

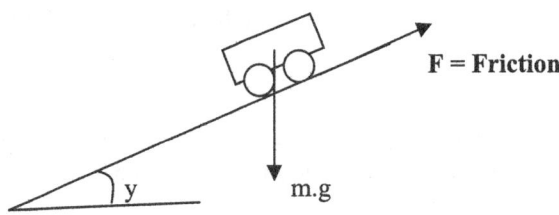

The weight acts vertically downward. Friction acts against the movement of the vehicle. In this case, friction is assumed to be zero.

Resolve forces, along the slope and perpendicular to slope.

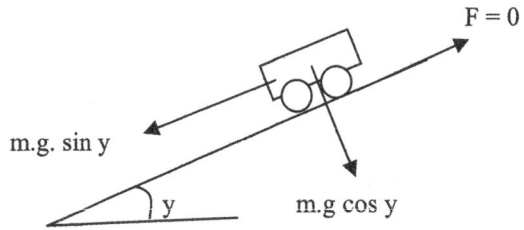

y = Angle of the slope = 30 degrees.

STEP 1:   Find the force parallel to the slope.

P = m.g. sin y – Friction = m.g. sin y – 0 = m.g. sin y

STEP 2:   Apply Newton's equation;

P = m. a
m.g. sin y = m. a   (Divide by m)
a = g. sin y

g = 32.2 ft/sec^2.
a = 32.2 sin 30 = 16.1 ft/sec^2.   Ans A

Practice Problem: A vehicle was parked on a sloped road. Due to brake inactivation, the vehicle start to roll down the slope. The slope is at an angle of 30 degrees to the horizontal. The coefficient of friction between road and tires is 0.35. What is the acceleration of the vehicle?

A) 4.9 ft/sec^2 B) 5.9 ft/sec^2 C) 12.1 ft/sec^2 D) 6.4 ft/sec^2

Solution:

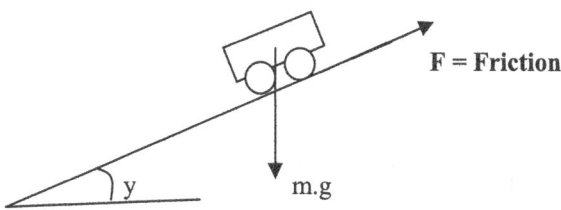

The weight acts vertically downward. Friction acts against the movement of the vehicle. Resolve forces, along the slope and perpendicular to slope.

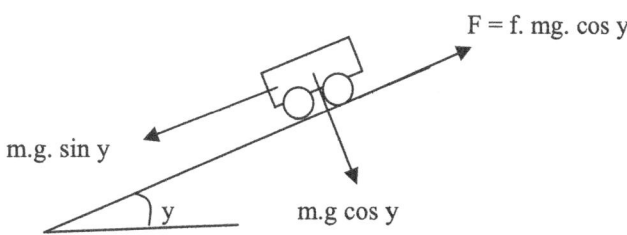

f = Friction coefficient

y = Angle of the slope = 30 degrees.

Friction coefficient = f = 0.35
F = f. m.g cos y = 0.35 x m.g cos y

STEP 1:   Find the force parallel to the slope.

P = m.g. sin y – Friction = m.g. sin y – 0.35 x m.g cos y = m.g. (sin y - 0.35 x cos y)
y = 30 degrees;

P = m. g (sin 30 – 0.35 x cos 30) = 0.197 mg

STEP 2:   Apply Newton's equation;

P = m. a
0.197 mg = m. a   (Divide by m)
a = 0.197.g

g = 32.2 ft/sec^2.
a = 0.197 x 32.2 = 6.43 ft/sec^2.   Ans D

Practice Problem: A car is travelling downhill of a slope at a velocity of 50 mph. The slope gradient is 5H to 1V. The friction coefficient is 0.40. The driver sees a stopped truck 350 ft away in the middle of the road and applies brakes. Brake reaction time is 2.5 seconds. Will there be a collision?

Solution: STEP 1:   <u>Draw the force diagram;</u>

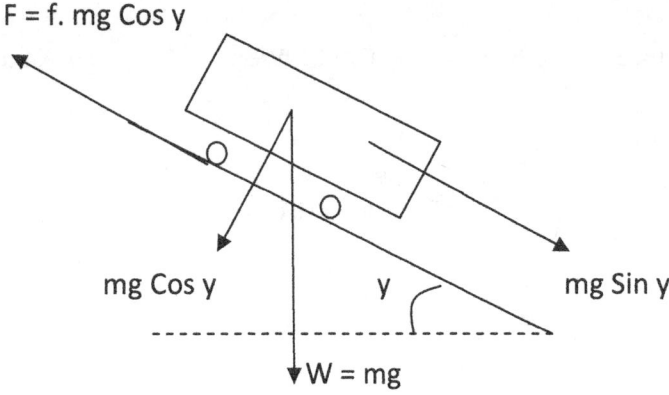

STEP 2:   Find the angle "y"

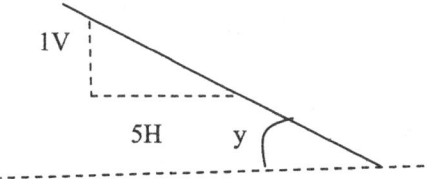

tan y = 1/5
y = tan^{-1} 1/5
y = 11.31 degrees

STEP 3:   Resolve the forces along the slope;

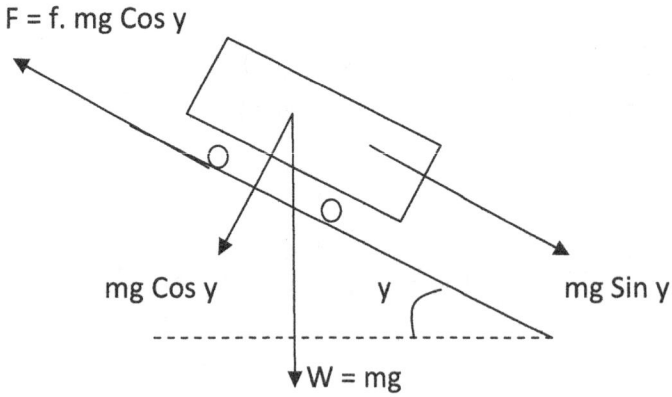

Downward force (P) = mg. sin y – f. mg. cos y

STEP 4:   Apply Newton's equation;
Apply Newton's second law (P = m. a) to find the acceleration of the vehicle.

$$P = m \times a$$

P = Acceleration force (lbf);
a = acceleration (ft/sec^2)

$mg. \sin y - f. mg. \cos y = m. a$
$g. \sin 11.31^0 - f. g \cos 11.31^0 = a$ (Divide by m)
g = 32.2 and f = 0.40
Hence;
$32.2 \times \sin 11.31^0 - 0.40 \times 32.2 \times \cos 11.31^0 = -6.31$ ft/sec^2.

STEP 5:   Initial speed of the vehicle is 50 mph.
50 mph = 50 x 5280/3600 = 73.33 ft/sec

Brake reaction time = 2.5 sec

STEP 6: Find the distance travelled after applying brakes;

Initial speed at application of brakes = 73.33 ft/sec
Distance travelled during brake reaction time = 2.5 x 73.33 = 183.33 ft
Distance between car and stopped truck = 350 – 183.33 = 166.68 ft

When the car reaches the truck, if the velocity is a positive value, then there will be a crash.
Find the velocity of the vehicle at a distance of 166.68 ft;
s = 166.68 ft;   u = 73.33;   a = -6.31 ft/sec^2.

Apply the following dynamic equation;

$$v^2 = u^2 + 2a.s \text{ -------------------------------------(3)}$$

u = Initial velocity;
v = Final velocity;
s = Distance;
a = acceleration
v = ?;   u = 73.33 ft/sec;   s = 166.68 ft;   a = -6.31 ft/sec^2.

$v^2 = u^2 + 2.a.s$
$v^2 = 73.33^2 + 2 \times -6.31 \times 166.68$
$v^2 = 3,273.8$
$v = 57.21$ ft/sec

When the car reached the stopped truck it would have a velocity of 57.21 ft/sec.
There will be an accident.

6.10 Practice Problems from Four Sample Exams for the Civil Engineering PE Exam:
(Detailed step by step solutions are given in Four Sample Exams for the Civil PE Exam, Second Edition):

Problem 1.14)   Intersection angle of a horizontal curve is 144^0 12' 14". Radius of the curve is 531.3 ft. Find the area of the hatched section.

A) 1.544 acres   B) 3.549 acres   C) 8.155 acres   D) 9.534 acres

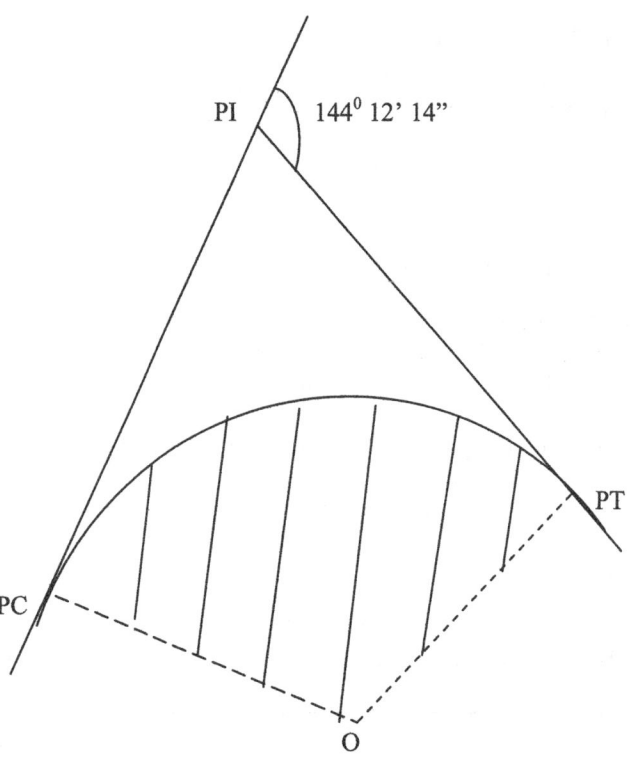

Ans C

Problem 1.17)   Station at point of curvature (PC) is given to be 4+10 and the radius of the curve is 140 ft. Find the station at PI. (Angle of intersection is 50 degrees).

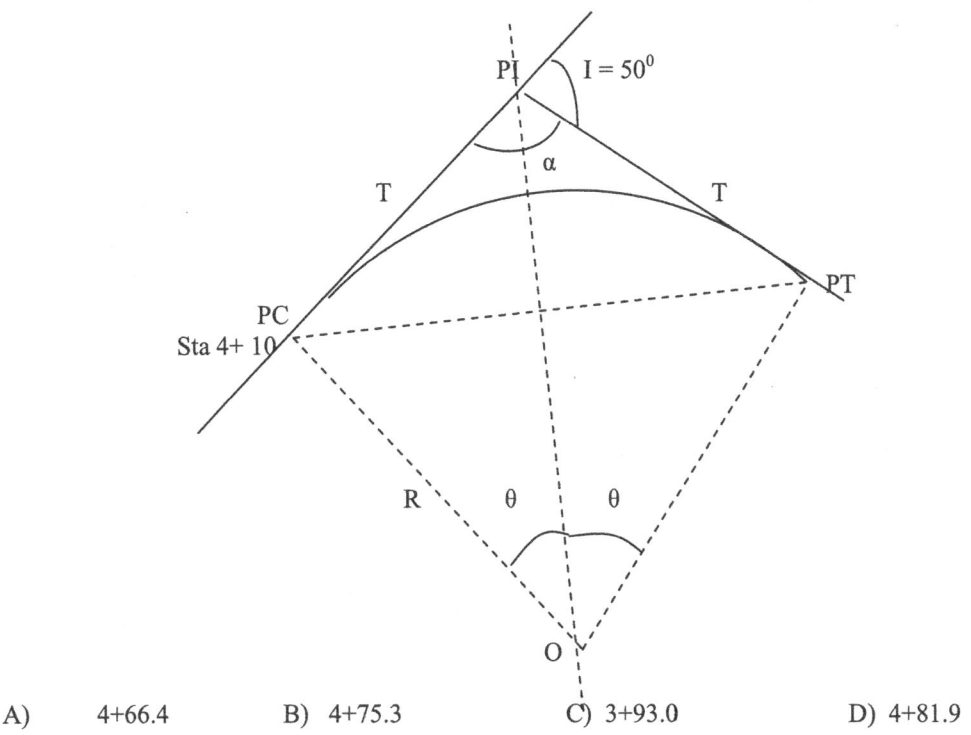

A)   4+66.4   B) 4+75.3   C) 3+93.0   D) 4+81.9

Problem 1.18):   Engineers have designed a horizontal curve so that traffic at PC station will be able to see the traffic at PT station. Property owner would like to build a structure as shown. What's the closest distance to the road that the structure can be built?.   (Distance AB)

Radius of the curve = 900 ft

Angle of intersection is 18^0.

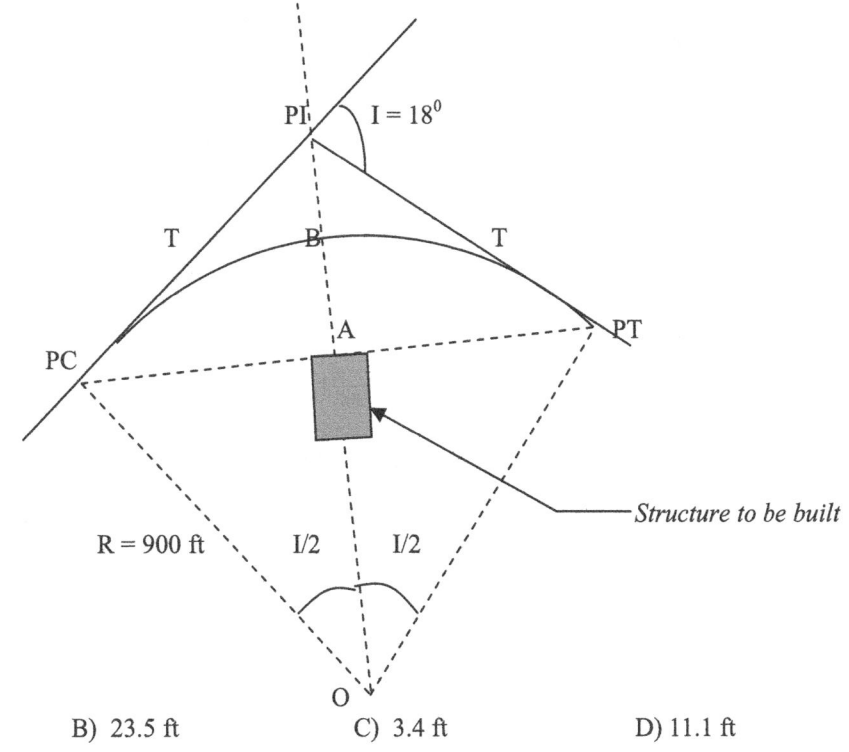

A) 13.4 ft   B) 23.5 ft   C) 3.4 ft   D) 11.1 ft

Problem 1.19):   Degree of curve of a horizontal curve (arc method) is given to be 4.3 degrees. What's the radius of the curve?

    A) 1,332.4 ft   B) 291.8 ft   C) 1,621.1 ft   D) 871.8 ft

Problem 2.11) A car was travelling on an asphalt road and ran out of the road and collided with a car parked away from the road in a sandy area. Impact speed was found to be 30 mph. Friction coefficient (μ) of asphalt is 0.6 and friction coefficient (μ) of sand is 0.5. Length of travel in asphalt was 200 ft and length of travel in sand was 120 ft. What was the speed of the vehicle just before applying brakes?

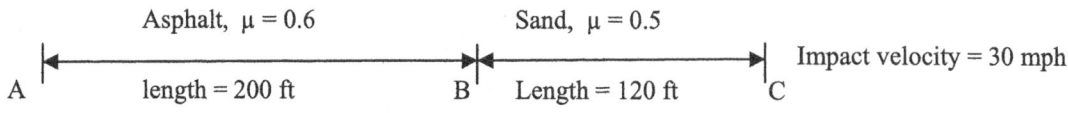

A) 116.3 ft/sec   B) 231.2 ft/sec   C) 78.9 ft/sec   D) 43.9 ft/sec

Problem 2.12)   Stopping sight distance of a road is 460 ft. Radius of the horizontal curve is 600 ft. Find the horizontal sight offset required for the horizontal curve.

    A)   123.56 ft   B) 25.89 ft   C) 91.00 ft   D) 43.55 ft

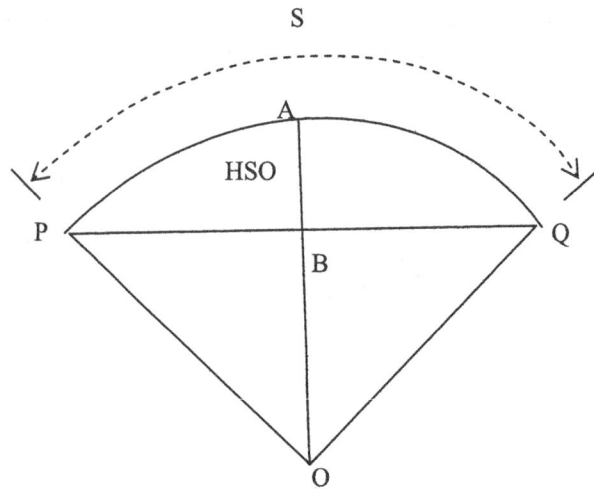

AB = HSO (Horizontal sight offset)
Ans D

Problem 2.13):   A traverse is done as shown. Internal angles of the traverse is as given. What is the total error of internal angles?

A) 4.662^0 B) 3.632^0 C) 4.169^0 D) 6.167^0

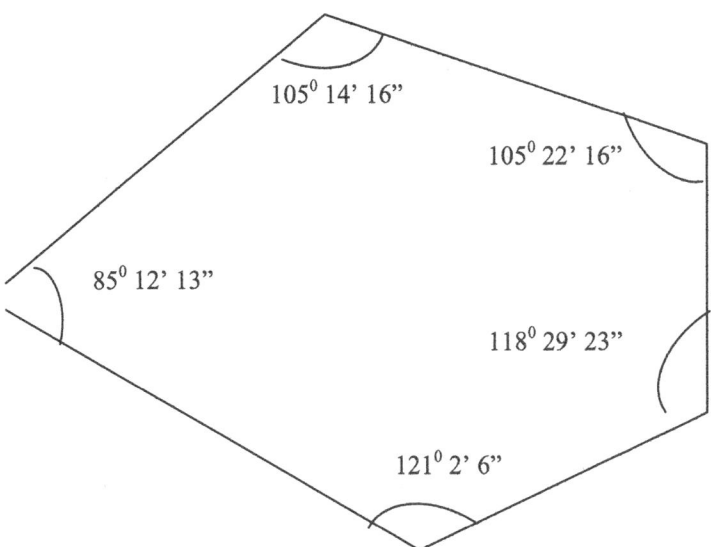

Problem 2.14)   Two vehicles are travelling in the same direction in a highway next to each other. First vehicle is travelling at a speed of 50 mph and the second vehicle is travelling at a speed of 40 mph. Both drivers apply brakes at the same time. First vehicle decelerates at 12 ft/sec^2 and the second vehicle decelerates at 8 ft/sec^2. When both vehicles come to a complete stop, what's the distance between the two vehicles?

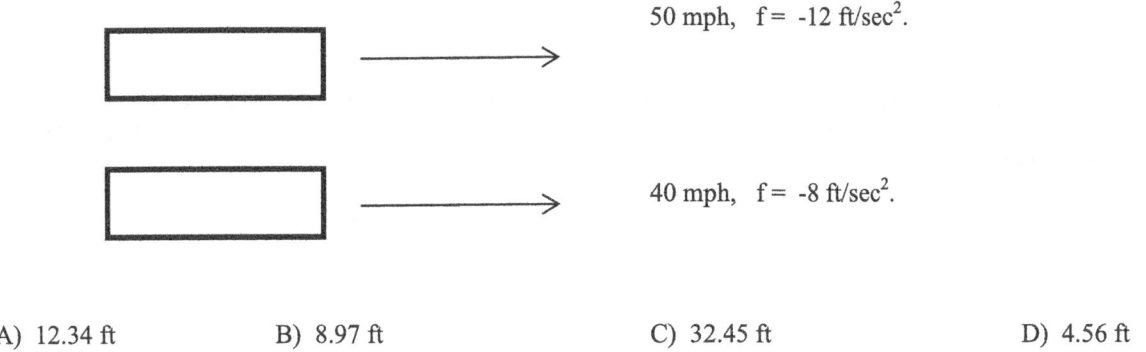

A) 12.34 ft   B) 8.97 ft   C) 32.45 ft   D) 4.56 ft

Problem 2.15):   Station at point of intersection (PI) is given to be 14+70 and the radius of the curve is 1,400 ft. Find the station at PT. (Angle of intersection is 53 degrees).

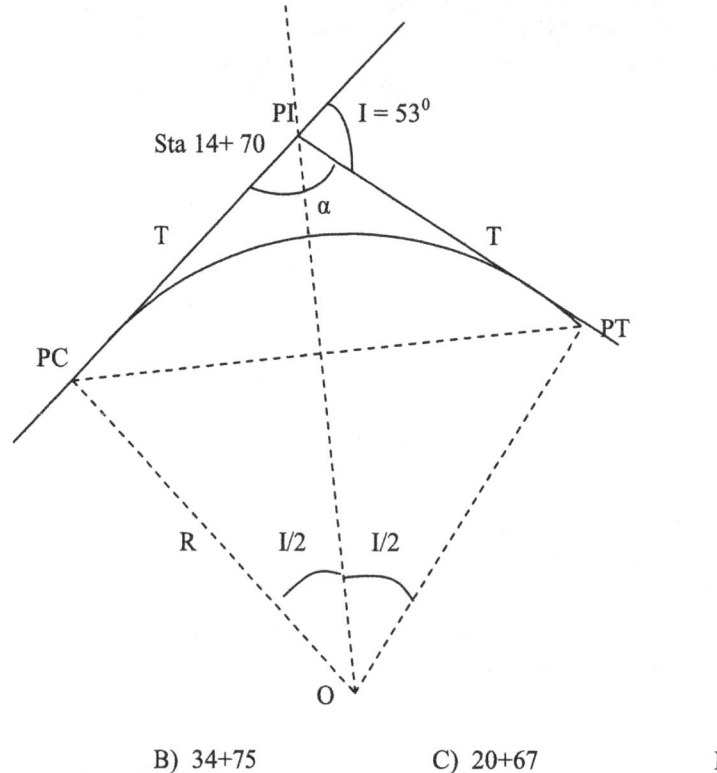

A)   14+66   B) 34+75   C) 20+67   D) 24+81

Problem 2.16):   Vertical curve is shown in the figure. Upward gradient is 1.8% and downward gradient is 3.2% as shown. BVC elevation is 132.71 and BVC station is 11+53. Station of point P is given to be 20 +00. EVC station is 22+48. Find the elevation at point P.

A) 105.94   B) 113.87   C) 106.79   D) 121.97

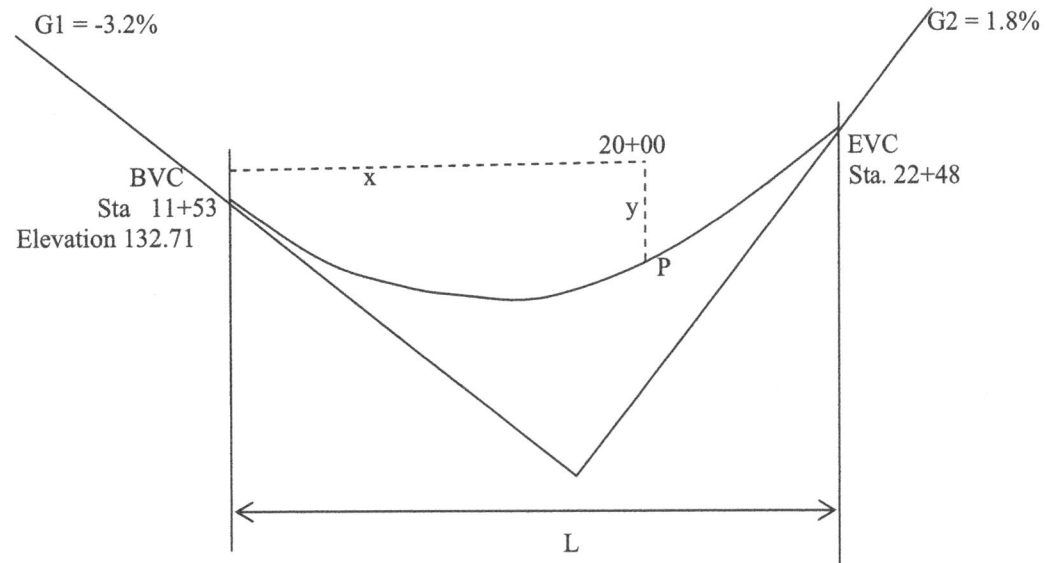

BVC = Beginning of vertical curve
EVC = End of vertical curve

Problem 2.17)   Engineers are designing a roadway horizontal curve. Side friction factor of the roadway is 0.2 and the speed of vehicles is 60 mph. Radius of the horizontal curve is 700 ft. What is the superelevation required at the curve to maintain the speed.

      A) 12.3%   B) 14.3%   C) 3.2%   D) 7.9%

Problem 2.18)   Horizontal curve is designed with a superelevation for a highway. The road has a normal crown and after the superelevated horizontal curve it switches back to the normal crown. What is the correct process of changing a normal crown to superelevation and then back to a normal crown?

A)   Normal crown, Tangent runout, superelevation runoff, full superelevation, superelevation runoff, Tangent runout, Normal crown

B)   Normal crown, superelevation runoff , Tangent runout, full superelevation, superelevation runoff, Tangent runout, Normal crown

C)   Normal crown, superelevation runoff , full superelevation, superelevation runoff, Tangent runout, Normal crown

D)   superelevation runoff, superelevation runout, Tangent runout, Normal crown, superelevation, superelevation runoff, Tangent runout, Normal crown

Problem 3.17)   Station at point of intersection (PI) is given to be 112 +70 and the degree of curve (arc method) is given to be 2.3^0. Find the station at PT. (Angle of intersection is 43^0 13' 12").

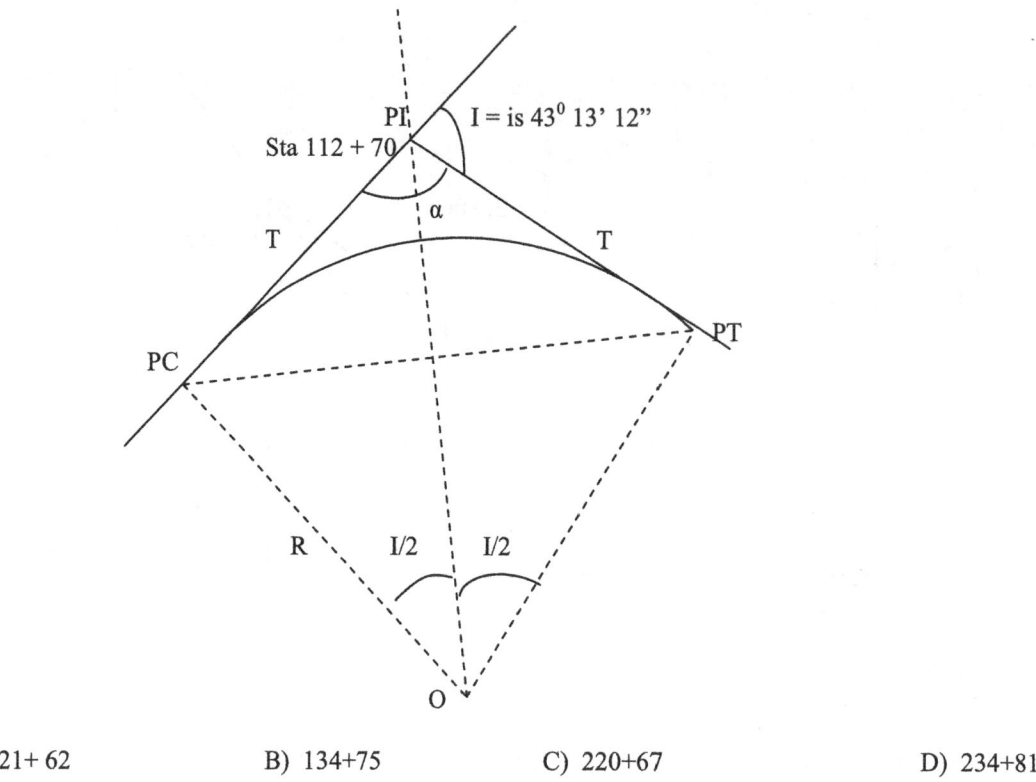

A)   121+ 62   B) 134+75   C) 220+67   D) 234+81

Problem 3.18):   Sag vertical curve of a roadway is shown in the figure. Find the elevation at point P.

Station of BVC = 105 + 45,
Elevation of BVC = 124.56 ft
Station of PVI = 109 + 34
Station at point P = 113 + 08

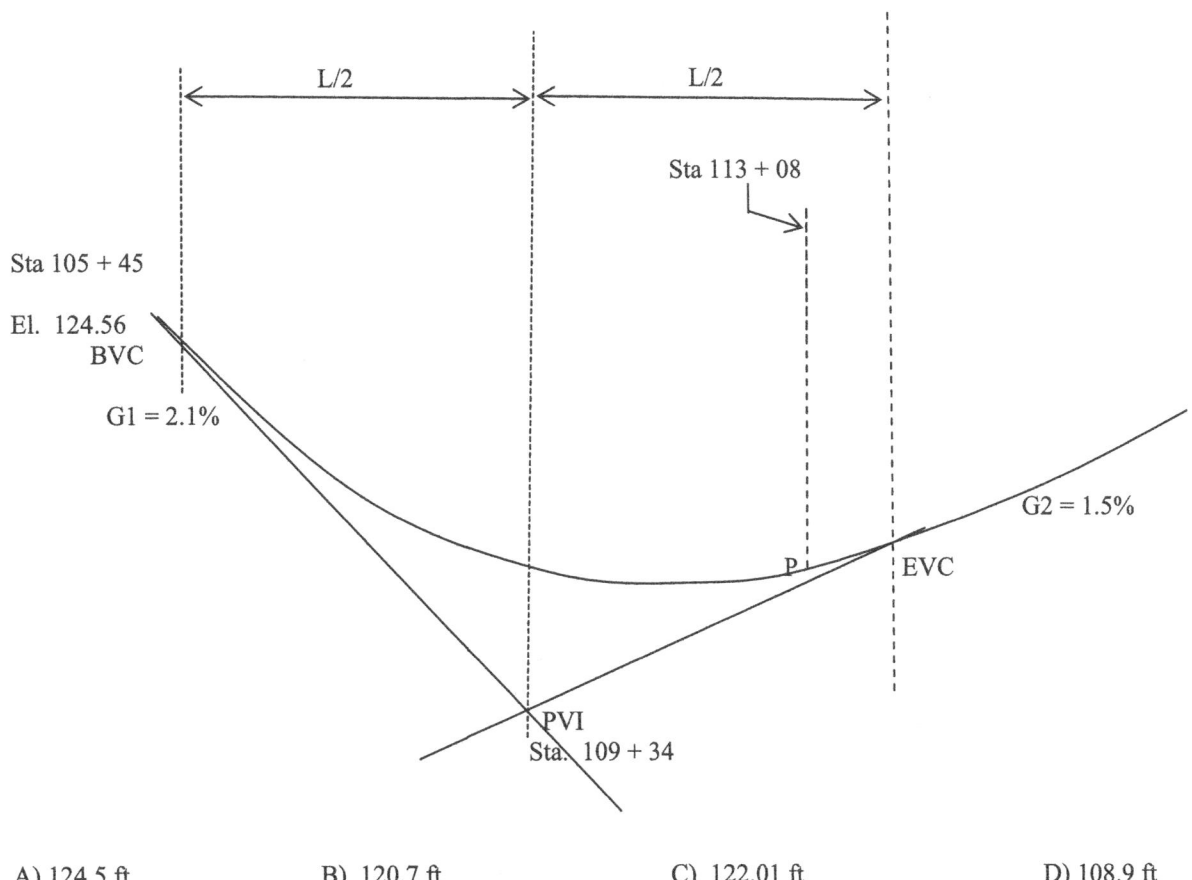

A) 124.5 ft   B) 120.7 ft   C) 122.01 ft   D) 108.9 ft

Use the data given in 3.19 to solve 3.20.

Problem 3.19)   Engineers are designing a horizontal curve with a superelevation. Following information is provided regarding the highway.

Number of lanes = 3,

Width of one lane = 12 ft

Lane adjustment factor = 0.67

Superelevation = 4%

Relative gradient = 0.5%

Find the runoff length required prior to achieving the full superelevation.

A) 123 ft   B) 231 ft   C) 193 ft   D) 351 ft

Problem 3.20):   Find the tangent runout for the horizontal curve given in the previous problem.

Normal cross slope = 2%

Design superelevation = 4%.

A) 102.3 ft   B) 96.5 ft   C) 234.8 ft   D) 78.9 ft

Problem 4.19):   Station at point of intersection (PI) is given to be 112 +70 and the degree of curve (arc method) is given to be 2.3^0. Find the station at PT. (Angle of intersection is 28^0 15' 08").

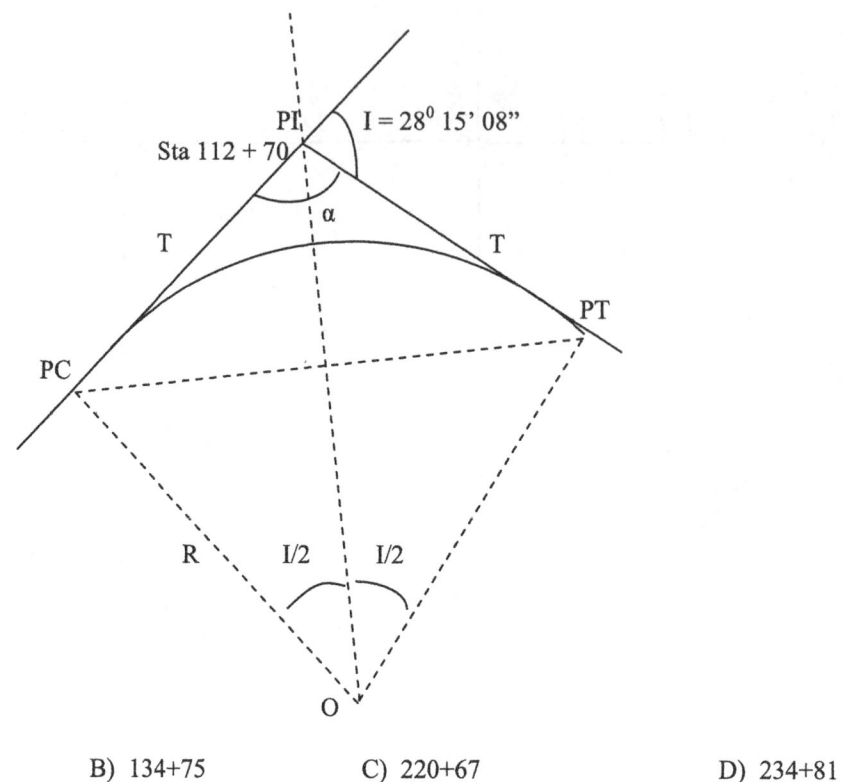

A) 118 + 71   B) 134+75   C) 220+67   D) 234+81

Problem 4.20)   Radius of a horizontal curve is given to be 1,300 ft. Station at PC is given to be 187+34 and angle of intersection is 73°.   Find the deflection angle to station 190+00.

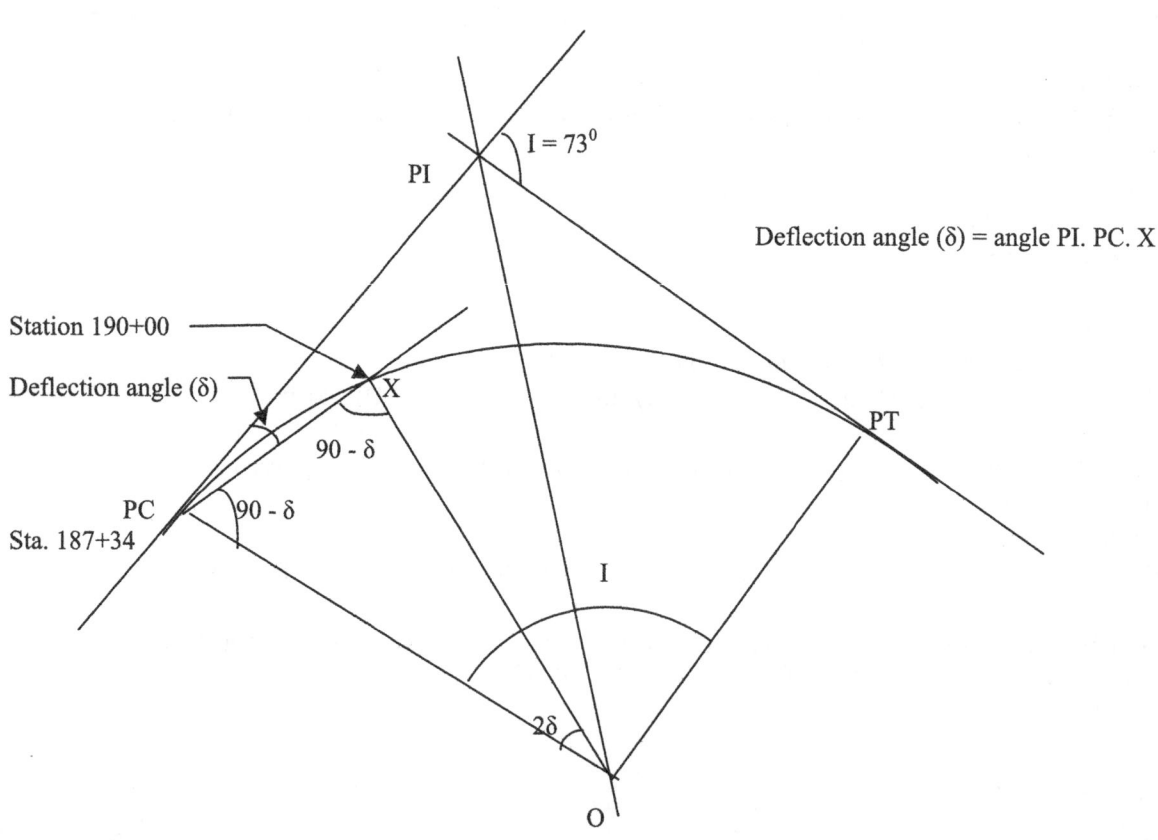

Deflection angle (δ) = angle PI. PC. X

A) 3.95^0 B) 9.20^0 C) 5.86^0 D) 3.67^0

Problem 4.23):   Sag vertical curve of a roadway is shown in the figure. The roadway goes under an overpass. Station of BVC = 115+45,
Elevation of BVC = 134.56 ft

Station of PVI = 120 + 34
Station of overpass = 123 + 13
Elevation of overpass = 143.25 ft

What is the maximum height of trucks that can be allowed in the roadway assuming 1 ft clearance between roof of trucks and the overpass.
Note that vertical curves are constructed in a manner so that PVI station is at the center of BVC and EVC.

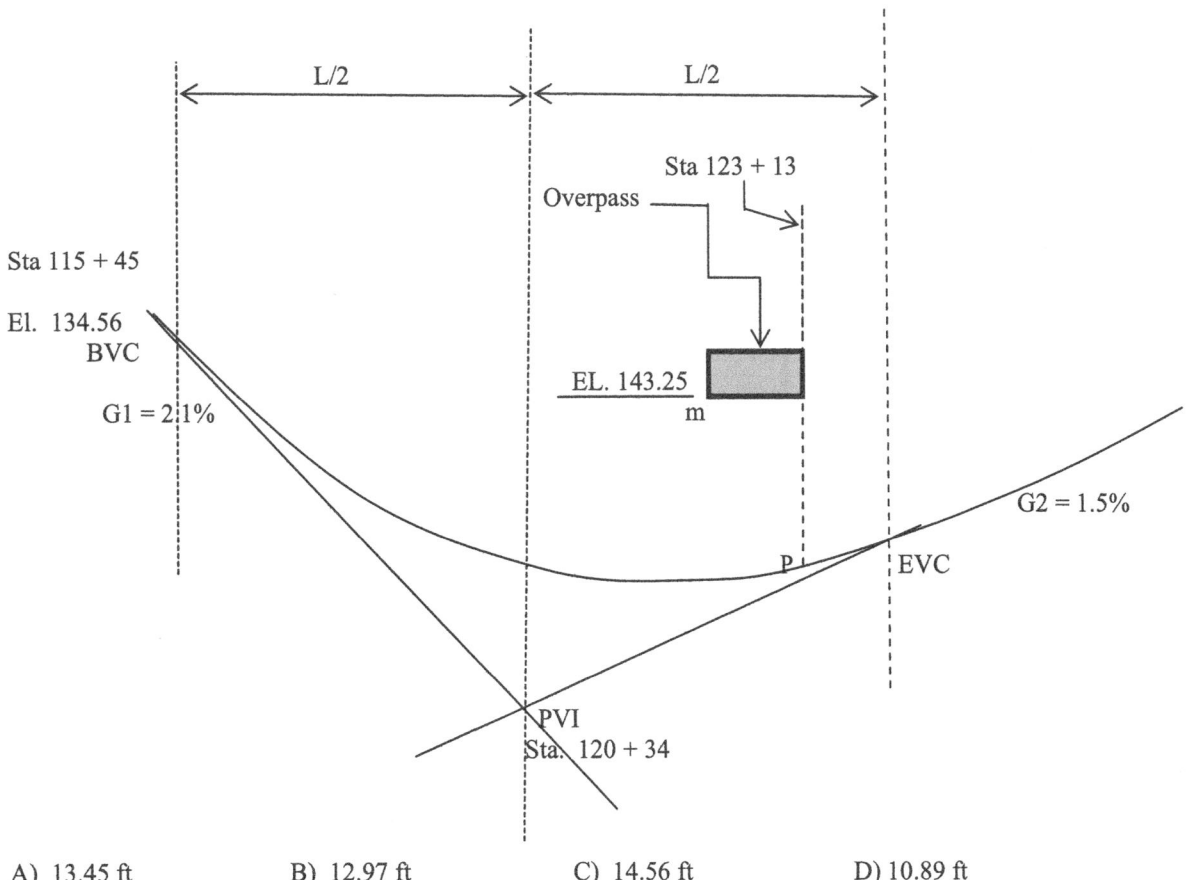

A) 13.45 ft   B) 12.97 ft   C) 14.56 ft   D) 10.89 ft

Answers;

1.14)   Ans C
1.17)   Ans B
1.18)   Ans D
1.19)   Ans A
2.11)   Ans A
2.12)   Ans D
2.13)   Ans A
2.14)   Ans B
2.15)   Ans C
2.16)   Ans D

2.17)   Ans B
2.18)   Ans A
3.17)   Ans A
3.18)   Ans C
3.19)   Ans C
3.20)   Ans B
4.19)   Ans A
4.20)   Ans C
4.23)   Ans B

Please see "**Four Sample Exams for the Civil PE Exam, Second Edition**" for detailed step by step solutions.

7.0 Materials   -   6 Questions

7.1   Soil Classification:

Soils are classified into different categories. Major categories of soils are:
- Gravels
- Sands
- Silts
- Organic Clays and peat
- Inorganic clays

Silts and Sands: Sand particles can be seen with the naked eye. Silt and clay particles cannot be seen with the naked eye. The difference between silts and clays is that clay particles have cohesion. Pure silt particles do not possess cohesive properties. Cohesion is an electro chemical process while friction is a physical process.

In many cases silt particles and clay particles are mixed together. Such soils are known as silty clay or clayey silt. When the predominant constituent is clay, and silt is the secondary component, such soils are known as silty clay. If the predominant constituent is silt, such soils are known as clayey silt.

Gradation: When a sand sample contains particles of small and large sizes, it is called a well graded sand. When sand particles are of more or less similar in size, the sand is considered to be poorly graded. Well graded sands can be compacted better than poorly graded sands.

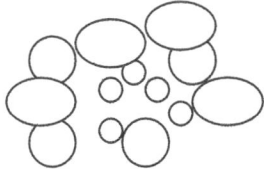   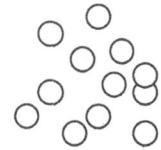

Well Graded Sands (various sizes)   Poorly Graded Sands (similar sizes)

7.1.1 Unified Soil Classification:

Sands: SP - Poorly graded sand
          SW - Well graded sand

Since symbol "S" was used for sands, another symbol was needed to represent silts. Symbol "M" is used to represent silts.

Figure C.1   Poorly graded gravel (on left) and well graded gravel (on right).

Gravels: (Particles greater than No. 4 sieve (4.75 mm) considered to be gravel)
          GP - Poorly graded gravel   **GW** - Well graded gravel

Silts: Silts could be high plastic or low plastic. Plasticity of a clay or silt is determined by the plasticity test, which would be described later.
          ML - Low plastic silts   **MH** - High plastic silts
Silts get the letter "M" since "S" is used for sands.

Clays: Some clays contain plenty of organic matter. Organic matter is mainly decomposed trees and roots. Clays with large quantity of organic matter is known as organic clays. When the major constituent is organic matter then such soils are known as peat. Clays could be high plastic to low plastic. Highly cohesive clays are known as high plastic clays and vice versa. High plastic clays are known as fat clays also.

        CL - Low plastic inorganic clays
        CH - High plastic inorganic clays
        OL - Low plastic organic clays or silts
        OH - High plastic organic clays or silts
        PT - Predominantly organic soils, peat, muck, marsh soils.
Before we go further it is important to discuss the sieve analysis procedure.

Sieve Analysis: Sieve analysis is conducted to classify soil into sands, gravels, silts and clays. Sieves are used to separate soil particles and group them based on their size. This test is used for the purpose of classification of soil.

Standard sieve sizes are shown below.

Gravel: Particles greater than #4 sieve is considered to be gravel.
(Mesh size of #4 sieve is 4.75 mm. Mesh size of #200 sieve is 0.075 mm).
Sands: Particles in the range #4 to #200 sieves are considered to be sands.
Silts and Clays: Particles smaller than #200 is considered to be silts and clays.
Differentiation of silts and clays cannot be done using sieve analysis. Clays are bound together due to chemical and electromagnetic forces. Silt particles are not bound together due to chemical forces.

Mesh size of No. 4 sieve is 4.75 mm and mesh size of No. 200 sieve is 0.075 mm.

Set of sieves

	Mesh size (mm)
Sieve No:	
No. 4	4.75
No. 6	3.35
No. 8	2.36
No. 10	2.00
No. 12	1.68
No. 16	1.18
No. 20	0.85
No. 30	0.60
No. 40	0.425
No. 50	0.30
No. 60	0.25
No. 80	0.18
No. 100	0.15
No. 200	0.075
No. 270	0.053

US sieve No. and mesh size

Practice Problem: Draw the sieve analysis curve for the sieve information provided.

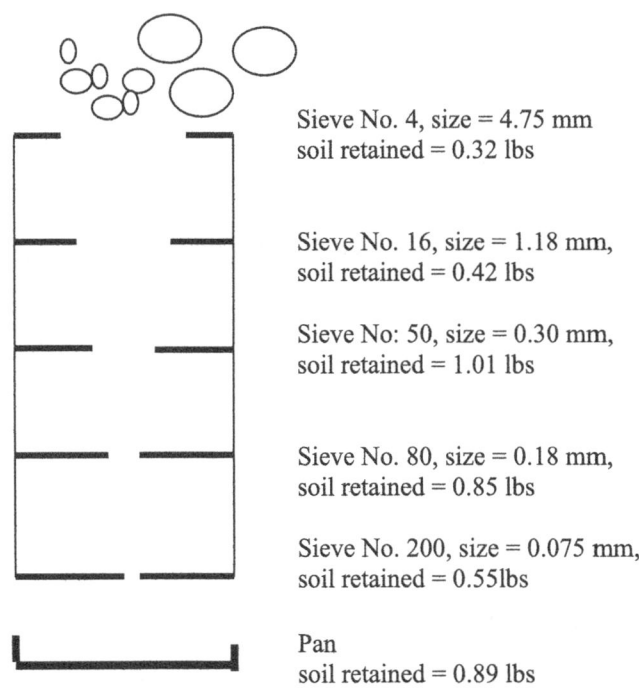

Sieve No. 4, size = 4.75 mm
soil retained = 0.32 lbs

Sieve No. 16, size = 1.18 mm,
soil retained = 0.42 lbs

Sieve No: 50, size = 0.30 mm,
soil retained = 1.01 lbs

Sieve No. 80, size = 0.18 mm,
soil retained = 0.85 lbs

Sieve No. 200, size = 0.075 mm,
soil retained = 0.55lbs

Pan
soil retained = 0.89 lbs

Find the fine content percentage?

A) 18%   B) 22%   C) 32%   D) 88%

Solution:

Fines are defined as particles smaller than No. 200 sieve. Particles larger than No. 200 sieve is known as coarse material. In order to solve this problem you need to find the percentage of soil passing the No. 200 sieve.
Total soil = 0.32 + 0.42 + 1.01 + 0.85 + 0.55 + 0.89 = 4.04 lbs

Weight of soil passing No. 200 sieve = 0.89 lbs
% passing the No. 200 sieve = 0.89/4.04 = 22.0%
Ans B

Practice Problem: Find the % passing No. 50 sieve   (Use the data given in the previous problem)

A) 57%   B) 22%   C) 42%   D) 43%

Solution:
We found total weight of soil to be 4.04 lbs.
We need to find the weight of soil passing No. 50 sieve.

Weight of soil passing No. 50 sieve = 0.85 + 0.55 + 0.89 = 2.29 lbs

1.01 lbs was retained on the No. 50 sieve and did not go thru. Hence, we do not use 1.01 lbs for the calculation.

% passing No. 50 sieve = 2.29/4.04 lbs = 56.7 %
Ans A

If we know the percentage of soil retained in a given sieve, we can find the percentage of soil passed that sieve.

- Sieve No 4 (size 4.75 mm):   All soil went passed sieve No. 4.   (4.75 mm)

Percent retained at this sieve is 0%.
Percent passed = 100%.

- Sieve No 16 (1.18 mm):   20% of soil was retained in sieve No. 16. (1.18 mm)
Percent retained at sieve No. 16 = 20%
Percent passed = 100 – 20 = 80%.

- Sieve No 50 (0.30 mm):   25% of soil was retained in sieve No. 50. (0.30 mm)
Total retained so far = 20 + 25 = 45%
Percent passed = 100 – 45 = 55%

- Sieve No 80 (0.18 mm):   20% retained in sieve No. 80. (0.18 mm)
 Total retained so far = 20 + 25 + 20 = 65%
Percent passed = 100 – 65 = 35%.

- Sieve No 200 (0.075 mm):   30% retained in sieve No. 200. (0.075 mm)
Total retained so far = 20 + 25 + 20 + 30 = 95%
Percent passed through sieve No. 200 = 100 – 95 = 5%.

Now it is possible to draw a graph indicating % passing at each sieve.

D_{60}: D_{60} is defined as the size of the sieve that allows 60% of the soil to pass. (See figure below). This value is used for soil classification purposes and frequently appears in geotechnical engineering correlations. To find the D_{60} value, draw a line at 60% passing point. Then drop it down to obtain the D_{60} value.

Finding D_{60}:   As shown in figure below, D_{60} is closer to 0.5 mm for this particular soil.

Find D_{30}:   As before, draw a line at 30% passing line. In this case, D_{30} happens to be approximately 0.1 mm.

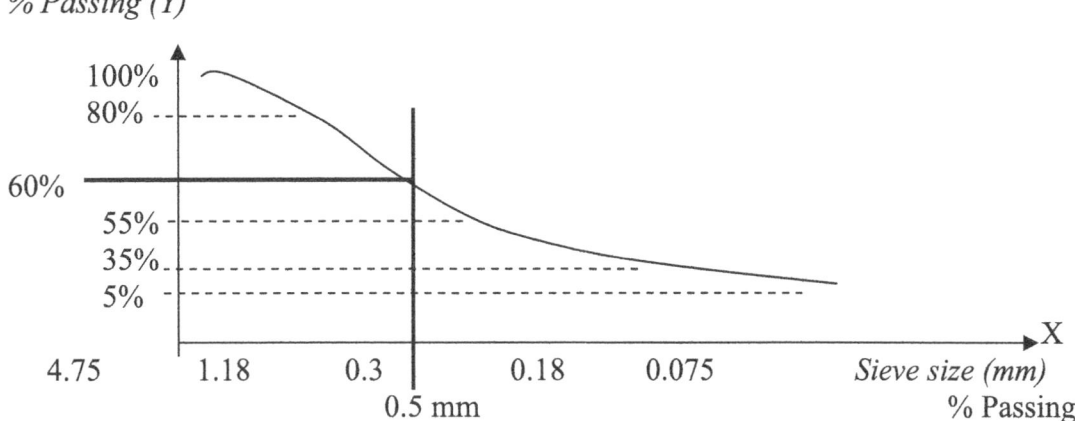

X – axis - sieve size.
Y - axis - Percent passing.

Sieve analysis curve

Soil Classification Procedure:

Now we are ready to discuss the soil classification procedure.

Coarse Grained Soils and Fine Grained Soils: Sands and silts are known as coarse-grained soils and silts and clays are known as fine-grained soils.

- Particles larger than 0.075 mm (No. 200 sieve size) are classified as coarse-grained soils or **coarse fraction.** (sands and gravel)
- Particles smaller than 0.075 mm (No. 200 sieve size) are classified as fine-grained soils or **fine fraction**. (silts and clays)

Classification of Gravels:

If 50% or more of the **coarse fraction** is larger than 4.75 mm (No. 4 sieve) such soils are classified as gravels.
If 50% or more of the **coarse fraction** is smaller than 4.75 mm (No. 4 sieve) such soils are classified as sands.
How to differentiate between poorly graded gravel and well graded gravel?

Conditions for well graded gravels (GW):   Following two conditions have to be met to be classified as well graded gravel.

Condition 1: $D_{60}/D_{10} > 4$
Condition 2: $1 < D_{30}^2/(D_{10} \times D_{60}) < 3$

We discussed before how to find D_{60} and D_{30}.
If any of the conditions given above are violated such soils are classified as GP.
If gravel contains silts, then it would be classified as GM. Note that "M" is used for silts since "S" is reserved for sands.
Similarly, if the gravel contains clay, it would be classified as GC.

Classification of Sands: If more than 50% of the soil sample is larger than 0.075 mm (No. 200 sieve) such soils are classified as gravels and sands. If 50% or more of the **coarse fraction** is smaller than 4.75 mm (No. 4 sieve) such soils are classified as sands.

Conditions for well graded sands (SW):
Condition 1: $D_{60}/D_{10} > 6$
Condition 2: $1 < D_{30}^2/(D_{10} \times D_{60}) < 3$

If any of these conditions are violated such soils are classified as poorly graded sands (SP). If sand contains silts, then it would be classified as SM. Similarly, if the sand contains clays, it would be classified as SC.

Graph for Silts and Clays: Graph shown in fig C5 is used to determine CL, CH, MH, OH, ML and CL. To use this graph, one may need to find the liquid limit and plasticity index. Liquid limit and plastic limit are known as Atterberg limits and explained below. The line at the center of the graph is known as the A- line.

Index Properties: Liquid limit (LL), Plastic limit (PL) and Shrinkage limits (SL) are known as soil indexes.

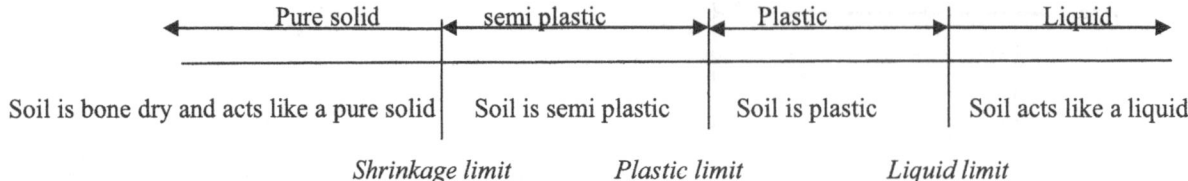

Plasticity Index (PI) = LL – PL   (LL = Liquid limit;   PL = Plastic limit

Liquid Limit: Liquid limit is the water content where soil starts to behave like a liquid. Liquid limit is measur by placing a clay sample in a standard cup and making a separation (groove) using a spatula. The cup is dropped until t separation vanishes. Water content of the soil is obtained at this sample. The test is performed again by increasing the wat content. Soil with low water content would yield more blows and soil with high water content would yield less blows. A gra is drawn between number of blows and the water content. (See fig. C2). LL is defined as the water content that corresponds 25 blows.

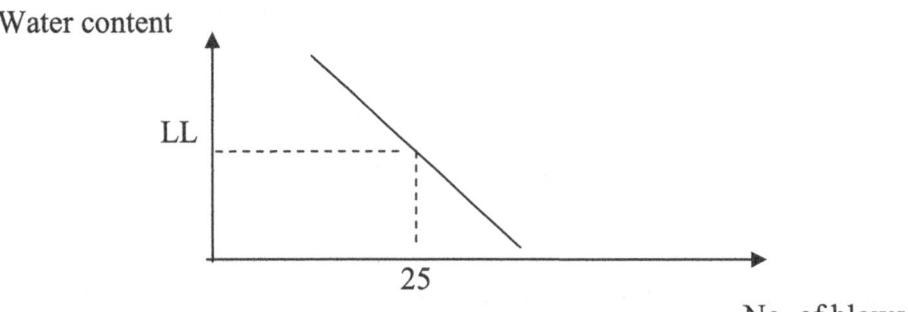

Figure C.2   Graph for liquid limit test
For liquid limit test

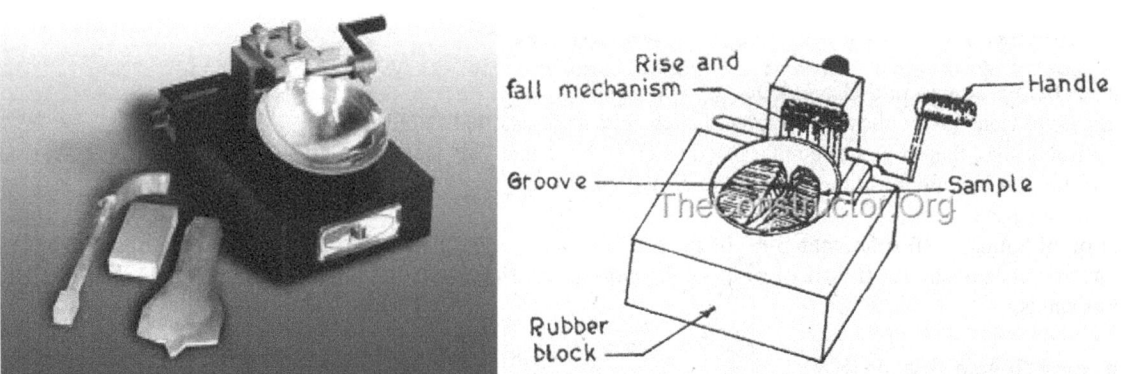

Liquid limit test apparatus is shown above. The handle is rotated until the groove is closed. The number of blows required to close the groove is noted.

Plastic limit: Plastic limit is the water content of the soil when cracks start to appear. Plastic limit is measured by rolling a clay sample to a 3 mm diameter cylindrical shape. During continuous rolling at this size, the clay sample tends to lose moisture and cracks start to appear. Water content where cracks start to appear is defined as the plastic limit.

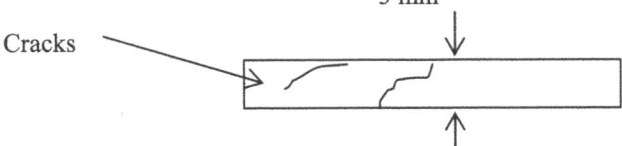

Fig C.3 Plastic limit test (Clay threads)

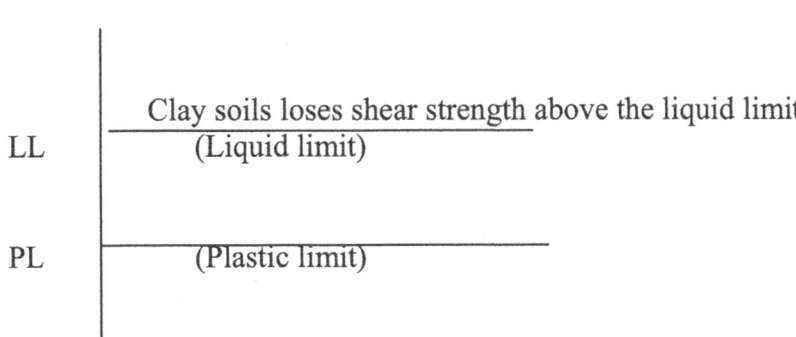

Figure C.4 Liquid limit and plastic limit

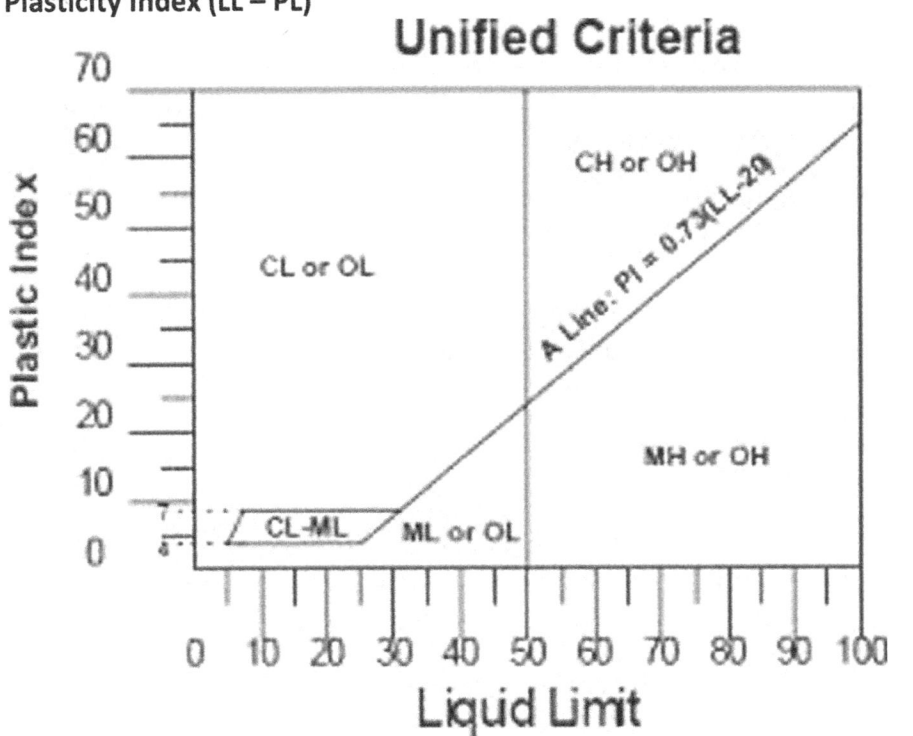

Figure C.5   A-Line (Soil classification for fine grained soils)

Practical considerations of liquid limit and plastic limit: Water content where a soil converts to a liquid like state is known as liquid limit. Consider two slopes as shown below. Assuming all other factors to be equal which slope would fail first?

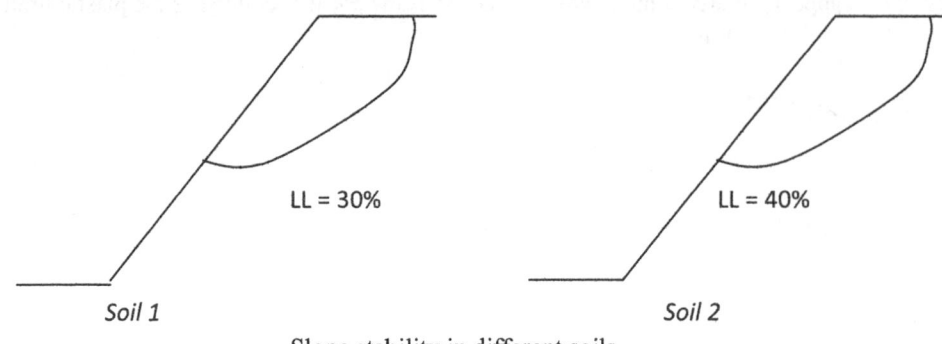

LL = 30%   LL = 40%

Soil 1   *Soil 2*

Slope stability in different soils

During a rain event, soil 1 would reach the liquid limit prior to soil 2. Hence, soil 1 would fail before soil 2. During earthquakes, water tends to rise. If soils with low liquid limit were to be present, those soils would lose its strength and fail. Plastic limit indicates the limit of plasticity. When the water content goes below the plastic limit of a soil, then cracks would start to appear in that soil. Soils lose its cohesion below the plastic limit.

Practice Problem: Find the classification of soils given below.

Soil (visual observation)	Liquid Limit	Plastic Limit	Plasticity Index (LL - PL)
Soil 1 - Silty clay	40	30	10
Soil 2 - Clay	60	50	10
Soil 3 - Organic clay (visual observation)	80	40	40
Soil 4 - Clay	40	10	30

Table D.1

From the graph given in fig C.5, locate the segment using liquid limit and plasticity Index.
Soil 1   (LL= 40, PI = 10)   = ML or OL (from fig C.5).
Visual observation does not indicate organic matter. Hence correct answer is ML.

Soil – 2   Clay   = MH or OH
(Visual observation does not indicate organic matter. Hence correct answer is MH.)

Soil – 3   Organic clay = MH or OH
(Since visual observation indicates organic soil it cannot be MH. Correct answer is OH)

Soil – 4   Clay   = CL or OL
(Visual observation does not indicate organic matter. Hence correct answer is CL.)

Particle Sizes and General Classification:-

Soil	Size in inches	Size in mm.	Comments
Boulders	6 in. or larger	150 mm or larger	
Cobbles	3 to 6 inches	75 mm to 150 mm	
Gravel	0.187 in. to 3 in.	4.76 mm to 75 mm	Greater than #4 sieve size
Sand	0.003 in. to 0.187 in.	0.074 mm to 4.76 mm	Sieve #200 to sieve #4
Silt	0.00024 in. to 0.003 in.	0.006 mm to 0.074 mm	Smaller than sieve #200
Clay	0.00004 in to 0.00008in	0.001 mm to 0.002 mm	Smaller than sieve #200
Colloids	Less than 0.00004 in	Less than 0.001 mm	

Table D.2　　　　Size Ranges for Soils and Gravels

Soil	Specific Gravity
Gravel	2.65 – 2.68
Sand	2.65 – 2.68
Silt (inorganic)	2.62 – 2.68
Organic Clay	2.58 – 2.65
Inorganic Clay	2.68 – 2.75

Table D.3　Specific Gravity of Soils

Laboratory Test Program:

Laboratory tests on soil samples are conducted to obtain soil parameters. Laboratory test program is dependent on the project requirements. Some of the common laboratory tests required for geotechnical work is given below.

1) 　Sieve analysis
2) 　Hydrometer
3) 　Water content
4) 　Atterberg limit tests (Liquid limit and plastic limit)
5) 　Permeability test
6) 　UU tests (Undrained unconfined tests)
7) 　Density of soil
8) 　Consolidation test
9) 　Tri-axial tests
10) 　Direct shear test
11) 　Consolidation test

7.2   Boring Log Interpretation (Soil Profile):

After conducting a boring program, a soil profile is developed. Development of a soil profile can be illustrated using an example.

Practice Problem   Consider a site as shown in figure below. Proposed building is shown in dotted lines and the borings are shown in dark circles.

Elevations of top of borings were found to be B1 (107 ft),   B2 (112 ft),   B3 (113 ft),   B4 (110 ft),   B5 (109 ft).

Depths of borings are B1 (40 ft),   B2 (42 ft),   B3 (50 ft),   B4 (35 ft),   B5 (40 ft).

Now it is possible to draw a soil profile along line X-Y. (See below).

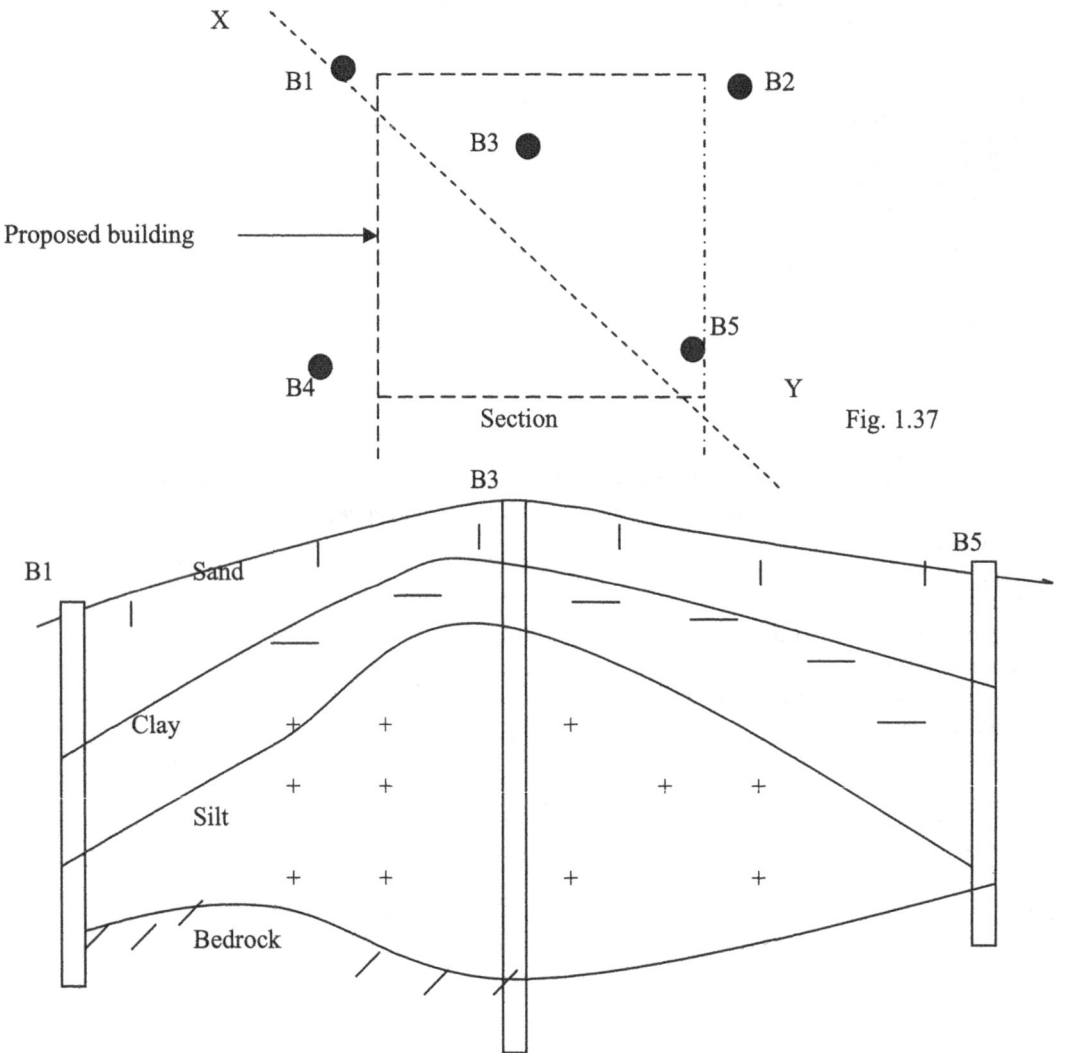

Figure C.6   Borings along line X-Y, Soil stratification

Based on boring log information, one can draw the soil profile. Soil profiles can be used to determine type of footing needed. If piles are needed, the engineer could decide the depth of piles by looking at the soil profile. For an instance if the engineer were to decide piles to be extended to bedrock, approximate length of piles at each location can be determined by looking at the soil profile. For example, piles at B3 will be longer than B5 and B1. Another important engineering aspect is depth to groundwater. One can assume that groundwater to be below the clay layer. If the engineers were to design shallow foundations, consolidation settlement at B5 would be larger than B3 since the thickness of the clay layer is much larger at B5.

7.3 Permeability:

Water travels thru soil due to pressure head, velocity head and the potential head due to elevation. Potential head due to elevation is the most important parameter. Water does not flow upwards against gravity unless a force is applied. Permeability of sandy soils is higher than that of clayey soils.

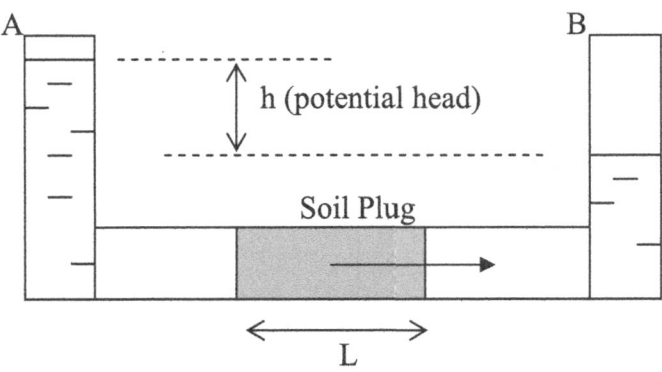

Water flowing through soil

In the above figure, water would travel from point A to B due to high potential head. Velocity of traveling water is given by the Darcy equation.

$v = k \cdot i$ (Darcy's Equation)

v = Velocity;
k = Coefficient of Permeability (cm/sec or in/sec)
i = Hydraulic gradient = h/L
L = Length of soil
Volume of water flow = Q = A x v
  A = Area;   v = Velocity

Practice Problem: Find the volume of water flowing in the pipe shown. Soil permeability (k) is 10^{-5} cm/sec. Area of the pipe is 5 cm^2. Length of soil plug is 50 cm.

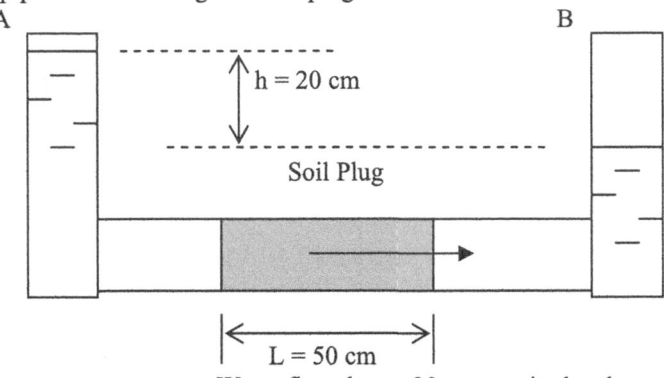

Water flow due to 20 cm gravity head

Solution: Apply the Darcy equation;
  $v = k \cdot i$ (Darcy's Equation)

v = Velocity;
k = Coefficient of Permeability (cm/sec or in/sec) = 10^{-5} cm/sec.
i = Hydraulic gradient = h/L   (h = 20 cm and L = 50 cm)
L = Length of soil

$v = k \times (h/L)$
$v = 10^{-5} \times (20/50) = 4 \times 10^{-6}$ cm/sec

A = Area = 5 sq. cm
Volume of water flow = $A \times v = 5 \times 4 \times 10^{-6}$ cm^3/sec = 2×10^{-5} cm^3/sec

7.4   Soil Phase Relationships:

Roads are constructed over uneven ground. During construction of a road, some locations required to be cut and other locations have to be filled. Consider the terrain shown below. Point A to B has to be **cut** and point B to C has to be **filled**.

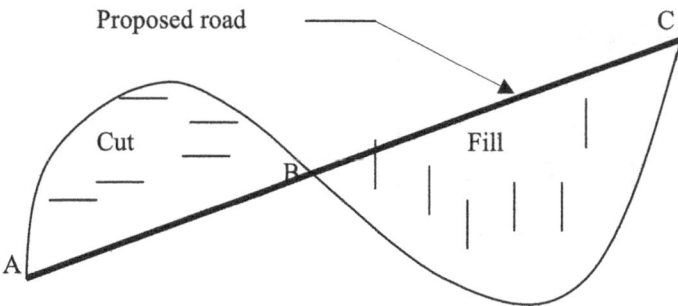

Cut and Fill: In some situations, the soil removed due to cut can be used for fill. If the soil removed is not suitable, then suitable soil has to be imported to the site for fill purposes.

Borrow Pit Volume Problems: To solve borrow pit problems, the student needs a good knowledge of phase relationships in soil.

Soil Phase Relationships: Soil consists of solids, air and water. Solids are soil particles. Soil matrix can be schematically represented as shown below.

Volume		*Mass*
V_a	Air	$M_a = 0$ (Mass of air is taken to be zero).
V_w	Water	M_w (Mass of water)
V_s	Solid	M_s (Mass of solids)

Soil phase diagram

M_a = Mass of air = 0 (Usually taken to be zero)
V_a = Volume of air (Volume of air is **not** zero)
M_w = Mass of water
V_w = Volume of water

M_s = Mass of solids
V_s = Volume of solids

M = Total mass of soil   = $M_s + M_w$ (mass of air ignored).
V = Total volume of soil = $V_s + V_w + V_a$
V_v = Volume of voids = $V_a + V_w$

Density of Water (γ_w) = Density of water can be expressed in many units.

$$\gamma_w = M_w/V_w$$

SI Units: $\gamma_w = 1 \text{ g/cm}^3 = 1{,}000$ g per liter $= 1{,}000 \text{ kg/m}^3 = 9.81 \text{ kN/m}^3$.
fps Units: $\gamma_w = 62.42$ pounds per cu. feet (pcf)

Total Density of Soil γ_t, γ_{wet} or γ: Some books use γ_{wet} and some other books uses γ_t or simply γ to denote total density of soil. Total density (also known as wet density) is simply mass of soil (including water) divided by the volume of soil.

$$\gamma_{wet} = M/V$$
$$M = M_w + M_s \text{ and } V = V_w + V_a + V_s$$
{M = Total mass of soil including water. Mass of air ignored}
{V = Total volume of soil including soil, water and air. Volume of air is **not** ignored.}

Dry Density of Soil (γ_d): $\gamma_d = M_s/V$
          M_s = Mass of solid only.
          V = Total volume of soil including soil, water and air
          $V = V_w + V_a + V_s$

Density of Solids = M_s/V_s

Specific Gravity (G_s):   Specific gravity is defined as density of solids divided by the density of water. Density of solids represented by G_s or by simply G.
Specific Gravity (G_s) = $M_s/(V_s \cdot \gamma_w)$

Void Ratio (e):   Void ratio (e) is defined as the ratio of volume of voids to volume of solids.
          $e = V_v/V_s$
V_v = Volume of voids (Volume of water + Volume of air) = $V_a + V_w$
V_s = Volume of solids

Moisture Content (w):   Moisture content (w) = M_w/M_s
M_w = Mass of water; M_s = Mass of solids

Porosity (n):
          $n = V_v/V$
V_v = Volume of voids   V = Total volume = $V_s + V_w + V_a$

What does porosity means? If you look at the top term V_v, basically tells us how much voids are there in the soil. The ratio between voids and total volume is given by porosity. In other words, porosity gives us an indication of pores in a soil. Soil with high porosity would have more pores than soil with low porosity. It is reasonable to assume that soils with high porosity would have a higher permeability.

Degree of saturation (S): $S = V_w/V_v$

          V_w = Volume of water
          V_v = Total of volume of voids

When total volume of voids is filled with water S = 100%.
Degree of saturation tells us how much water is in the voids.

Some Relationships to Remember:
Relationship 1:

$$\gamma_d = \gamma_{wet}/(1 + w)$$

This relationship appears in soil compaction section as well.
It can be proven as follows.

$\gamma_{wet} = M/V$, hence $V = M/\gamma_{wet}$
$\gamma_d = M_s/V$;
Replace V with M/γ_{wet}
$\gamma_d = M_s/(M/\gamma_{wet}) = M_s \times \gamma_{wet}/M$
$M = M_s + M_w$ (mass of air is ignored)
$\gamma_d = M_s \times \gamma_{wet}/(M_s + M_w)$

Divide both top and bottom by M_s.
$\gamma_d = \gamma_{wet}/(1 + w)$

Relationship 2:

$$S.e = G_s.w$$

See geotechnical section in this book for the proof.

Relationship 3:

$$n = e/(1 + e)$$

See geotechnical section in this book for the proof.

Relationship 4:

$$e = n/(1 - n)$$

Relationship 5:

$$\gamma_d = \gamma_w. G_s/[1 + (w/S)G_s]$$

Relationship 6:

$$\gamma_{wet} = \frac{\gamma_w. G_s \times (1 + w)}{[1 + e]}$$

Relationship 7:

$$\gamma_d = \frac{\gamma_w \cdot G_s}{[1 + e]}$$

Practice Problem: Specific gravity of a soil sample is given to be 2.65. Moisture content and degree of saturation are 0.6 and 0.7 respectively. Find the void ratio.

Solution: S.e = G.w

0.7 x e = 2.65 x 0.6

e = 2.27

Practice Problem: Total density of a soil sample was found to be 110 pcf and moisture content to be 60%. What is the dry density of the soil sample.

Solution: $\gamma_d = \gamma_{wet}/(1 + w) = 110/(1 + 0.6) = 68.75$ pcf

Practice Problem: Soil sample obtained from the ground and measured the weight to be 1 lb and total soil volume to be 0.01 cu. ft. The soil sample is then put in the oven and dried. Dried soil sample was weighed to be 0.7 lbs. Specific gravity of the soil was known to be 2.6.

Find the following;
   a) Total density or wet density
   b) Dry density
   c) Porosity
   d) Void ratio
   e) degree of saturation

Solution:
   a) Total Density $= M/V = 1/0.01 = 100$ lbs/ft^3.
   b) Dry density $= \gamma_d = \gamma/(1 + w)$
Weight of dry soil $(M_s) = 0.7$ lbs
Weight of water in the soil sample $(M_w) = 1 - 0.7 = 0.3$ lbs

Water content $(w) = M_w/M_s = 0.3/0.7 = 0.428$
Dry density $= \gamma_d = \gamma/(1 + w) = 100/(1 + 0.428) = 69.9$ lbs/ft^3.

   c) Porosity $(n) = V_v/V$
$V = 0.01$ cu. ft. This is the total volume of the soil sample
Specific gravity (G) of the soil is given to be 2.6.

Find V_s:
          $G = 2.6 = M_s/(V_s \cdot \gamma_w) = 0.7/(V_s \times 62.4)$
          Since $\gamma_w = 62.4$ lbs/ft^3.
          Hence $V_s = 0.0043$ ft^3.
          $V = V_v + V_s$ (Total volume = Volume of voids + Volume of solids)
          $0.01 = V_v + 0.0043$
          $V_v = 0.01 - 0.0043 = 0.0057$
          Porosity $(n) = V_v/V = 0.0057/0.01 = 0.57$

   d) Void ratio (e) can be found using the following equation

$$e = n/(1 - n)$$
$$e = 0.57/(1 - 0.57) = 1.326$$

e) $S.e = G.w$
   $S = 2.6 \times 0.428/(1.326) = 0.839$

Practice Problem: A soil sample obtained from the ground was measured and weighed. The soil sample has a diameter of 4 in and a height 6 in. The weight of the soil sample was measured to be 4.8 lbs. The soil sample was oven dried and weighed again. Dry weight of the soil sample was found to be 3.9 lbs. Specific gravity of the soil sample is known to be 2.65.

Find the following;

   a) Total density
   b) Water content
   c) Dry density
   d) Porosity
   e) Void ratio
   f) Degree of saturation

Solution:

Total Density:

   Volume of the soil sample (V) $= \pi \times d^2/4 \times h = \pi \times (4/12)^2/4 \times (6/12) = 0.044$ cu. ft
   Wet weight of the soil sample $=$ 4.8 lbs
   Total density (or wet density) $= M/V = 4.8/0.044 = 109.1$ lbs/cu. ft

Moisture Content (w) $=$ M_w/M_s
$M_s = 3.9$ lbs; $M_w = 4.8 - 3.9$ lbs $= 0.9$ lbs
$w = 0.9/3.9 = 0.23$
$\gamma_d = \gamma/(1 + w) = 109.1/(1 + 0.23) = 88.7$ lbs/cu. ft

Find V_s:
$G = 2.65 = M_s/(V_s . \gamma_w) = 3.9/(V_s \times 62.4)$,   Since $\gamma_w = 62.4$ lbs/ft^3.
Hence $V_s = 0.024$ ft^3.
$V = V_v + V_s$ (Total volume = Volume of voids + Volume of solids)
$0.044 = V_v + 0.024$

$V_v = 0.02$

Porosity (n) $= V_v/V = 0.02/0.044 = 0.45$
Void Ratio (e) $= n/(1 - n) = 0.45/(1 - 0.45) = 0.82$

Degree of saturation (S): $S.e = G.w$
                $S = 2.65 \times 0.23/0.82 = 0.74$

Practice Problem: Degree of saturation, water content and specific gravity are respectively 75%, 42% and 2.68.
Find

   1) Total density (γ_{wet}) 2) Void ratio (e)   3) Porosity (n)

Solution:
STEP 1: Use relationship 5;

$$\gamma_d = \frac{\gamma_w . G_s}{[1 + (w/S)G_s]}$$

$$\gamma_d = \frac{62.4 \times 2.68}{[1 + (0.42/0.75) \times 2.68]} = 66.9 \text{ pcf}$$

From relationship 1, $\gamma_d = \gamma_t/(1 + w)$

$\gamma_{wet} = \gamma_d \times (1 + w) = 66.9 \times (1 + 0.42) = 95 \text{ pcf}$

From relationship 2, $S.e = G.w$

$e = 2.68 \times 0.42/0.75 = 1.5$

From relationship 3, $n = e/(1 + e) = 1.5/(1.5 + 1) = 0.6$

7.4.1 Borrow Pit Problems:

Note: The student should master the previous chapter on soil relationships thoroughly in order to understand borrow pit problem.

Fill material for civil engineering work is obtained from borrow pits. The question is how much soil should be removed from the borrow pit for a given project?

Usually, final product is the controlled fill or the compacted soil. Total density, optimum moisture content and dry density of the compacted soil will be available. This information can be used to obtain the mass of solids required from the borrow pit. If the soil in borrow pit is too dry, water can always be added in the site. If the water content is too high, then soil can be dried prior to use. This could take some time in the field since one has to wait for few sunny days to get rid of water.

Water can be added or removed from soil.
What cannot be changed is the mass of **solids**. Mass of solids is the link between borrow pit soil and soil that has been transported.

Procedure:
Find the mass of solids required for the compacted fill
Excavate and transport the same mass of solids from the borrow pit.

Practice Problem: Road construction project needs compacted soil to construct a road 10 ft wide, 500 ft long. The road project needs 2 ft layer of soil. Soil density after compaction was found to be 112.1 pcf at optimum moisture content of 10.5%.

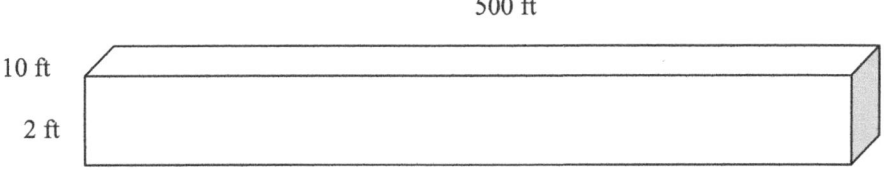

500 ft

10 ft

2 ft

Compacted Soil

The soil in the borrow pit has following properties:
Total density of the borrow pit soil = 105 pcf
Moisture content of borrow pit soil = 8.5%.
Find the total volume of soil that need to be hauled from the borrow pit.
Total Density = $(M_w + M_s)/V$
Dry density = M_s/V
(See below for definitions of all terms).

Solution:

STEP 1:   Find the mass of solids (M_s) required for the controlled fill:
Volume of compacted soil required = 500 x 10 x 2 = 10,000 cu. ft

Soil density after compaction (dry density) = 112.1 pcf
Moisture content required = 10.5%
Draw the phase diagram for the controlled fill:

Volume   **Mass**

V_a | Air | $M_a = 0$ (Usually mass of air is taken to be zero).

V_w | Water | M_w (Mass of water)

V_s | Solid | M_s (Mass of solids)

Soil phase diagram

V = Total Volume = $V_s + V_w + V_a$
M = Total mass = $M_w + M_s$ (Note that mass of air is taken to be zero)
V_a = Volume of air (Volume of air is not zero)

M_w = Mass of water
V_w = Volume of water

M_s = Mass of solids
V_s = Volume of solids

M = Total mass of soil   = $M_s + M_w$
V = Total volume of soil   = $V_s + V_w + V_a$
Total Density = $(M_w + M_s)/V$
Dry density = M_s/V

Soil in the site after compaction has a dry density of 112.1 pcf and moisture content of 10.5%.

Dry density = M_s/V = 112.1 pcf
Note that total density is $(M_w + M_s)/V$
Moisture content = M_w/M_s = 10.5% = 0.105
The road needed 2 ft layer of soil at a width of 10 ft and length of 500 ft.
Hence the total volume of soil = 2 x 10 x 500 = 10,000 cu. ft.
V = Total volume = 10,000 cu. ft
Since M_s/V = Dry density
$M_s/10,000 = 112.1$
$M_s = 1,121,000$ lbs.
M_s is the mass of solids. This mass of solids should be hauled in from the borrow pit.

STEP 3:   Find the mass of water in compacted soil:
Moisture content in the compacted soil = M_w/M_s = 10.5% = 0.105
M_w = 0.105 x 1,121,000 lbs = 117,705 lbs

STEP 4:   Find the total volume of soil that needs to be hauled from the borrow pit:
The contractor needs to obtain 1,121,000 lbs of solids from the borrow pit.
Contractor can add water to the soil in the field if needed.
Mass of solids needed (M_s) = 1,121,000 lbs.
Density and moisture content of borrow pit soil is known.
Total density of borrow pit soil = M/V = 105 pcf

Moisture content of borrow pit soil = $M_w/M_s = 8.5\% = 0.085$
Since $M_s = 1,121,000$ lbs,   (M_s is the mass of solids required).
$M_w/M_s = 0.085$
$M_w = 0.085 \times 1,121,000$ lbs $= 95,285$ lbs.

Solid mass of 1,121,000 lbs of soil in the borrow pit contains 95,285 lbs of water.
Total mass of borrow pit soil = 1,121,000 + 95,285 = 1,216,285 lbs
Total density of borrow pit soil is known to be 105 pcf.
Total density of borrow pit soil = $M/V = (M_w + M_s)/V = 105$ pcf
Insert known values for M_s and M_w.

$M/V = (M_w + M_s)/V = (95,285 + 1,121,000)/V = 105$ pcf
Hence   V = 11,583.7 cu. ft

The contractor needs to extract 11,583.7 cu. ft of soil from the borrow pit.
The borrow pit soil comes with 95,285 lbs of water.
Compacted soil should have 117,705 lbs of water. (see above step 3).
Hence water needs to be added to the borrow pit soil
Amount of water needs to be added to the borrow pit soil =   117,705 – 95,285 = 22,420 lbs.
Weight of water is usually converted to gallons. One gallon is equal to 8.34 lbs.
Amount of water needs to be added = 2,688 gallons.

Summary:

STEP 1: Obtain all the requirements for compacted soil.
STEP 2: Find M_s or the mass of solids in the compacted soil.
   This is the mass of solids that needs to be obtained from the borrow pit.
STEP 3: Find the information about the borrow pit. Usually the moisture content in the borrow pit and total density of the
   borrow pit can be easily obtained.
STEP 4: The contractor needs to obtain M_s of soil from the borrow pit.
STEP 5: Find total volume of soil that needs to be removed in order to obtain M_s mass of solids.
STEP 6: Find M_w of the borrow pit. (mass of water that comes along with soil).
STEP 7: Find M_w (mass of water in compacted soil).
STEP 8: The difference in above two masses is the amount of water needs to be added.

Soil Borrow Pit

Note: Refer to my "Civil PE Construction Module, Fifth Edition" for more problems"

7.5 Soil Compaction:

Shallow foundations can be rested on controlled fill, also known as engineered fill or structural fill. Typically such fill material are carefully selected and compacted to 95% of the modified Proctor density.
Modified Proctor test is conducted by placing soil in a standard mould and compacted with a standard ram.

7.5.1   Modified Proctor Test Procedure:

STEP 1:   Soil that needs to be compacted is placed in a standard mould and compacted.

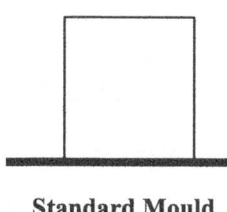

Standard Mould

STEP 2:   Compaction of soil is done by dropping a standard ram 25 times for each layer of soil from a standard distance. Typically soil is placed in five layers and compacted.

STEP 3: After compaction of all five layers, the weight of the soil is obtained. The soil contains solids and water. Solid is basically soil particles.

$$M = M_s + M_w$$

M = Total mass of soil including water
M_s = Mass of solid portion of soil
M_w = Mass of water

Proctor test

STEP 4: Find the moisture content of the soil.
Moisture content is defined as M_w/M_s
Small sample of soil is taken and placed in the oven and measured.

STEP 5: Find the dry density of soil.
Dry density of soil is given by M_s/V

M_s is the dry weight of soil and "V" is the total volume.

STEP 6: Repeat the test few times with different moisture contents and plot a graph between dry density and moisture content.

STEP 7: Obtain the maximum dry density and the optimum moisture content.

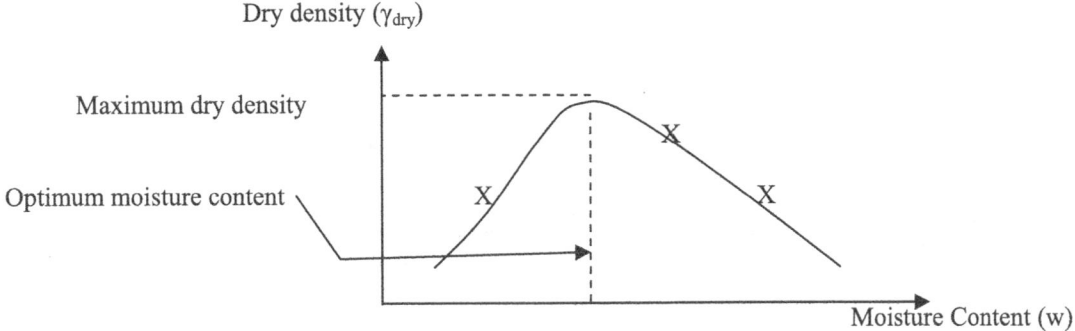

Dry density vs. Moisture Content

For a given soil, there is an optimum moisture content that would provide the maximum dry density.
It is not easy to attain the optimum moisture content in the field. Usually soil that is too wet is not properly compacted. If the soil is too dry, water is added to increase the moisture content.

Compacted sample ejected

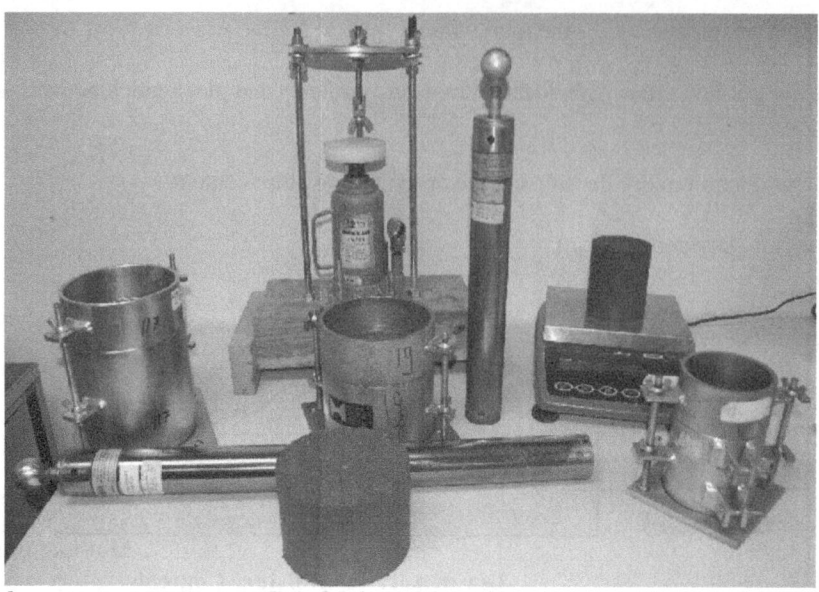

Standard and modified proctor test apparatus. Standard moulds and hammers are smaller. Modified proctor moulds and hammers are larger. (Source: Timely Engineering Soil tests LLC).

Controlled Fill Applications: Controlled fill can be used to build shallow foundations, roadways, parking lots, building slabs and equipment pads.

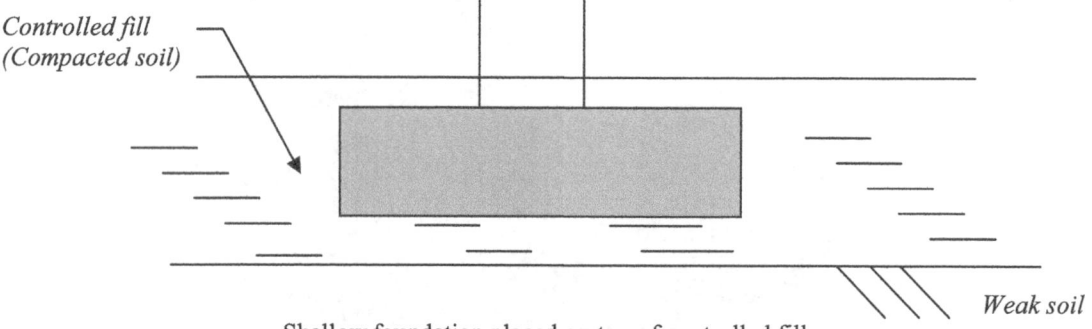

Controlled fill
(Compacted soil)

Weak soil

Shallow foundation placed on top of controlled fill

Checking the compaction with a nuclear gauge

7.6 Concrete:

Concrete is a product made of cement, sand, stones and water. Sand is known as fine aggregates and stone is known as coarse aggregates. Chemical compounds known as admixtures are also added to concrete to obtain special properties.

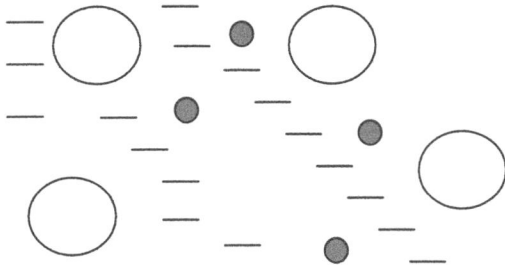
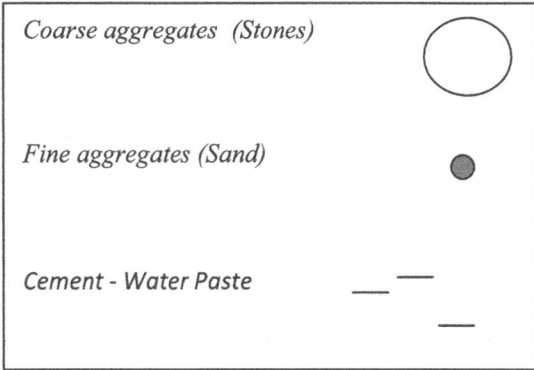

Coarse aggregates (Stones)

Fine aggregates (Sand)

Cement - Water Paste

Cement, fine aggregates and coarse aggregates:
Cement water mixture acts as the binder between coarse and fine aggregates. Generally higher the cement content, higher the strength. It has been found that high water content would lead to lesser strength. Hence, to achieve high strength, one should minimize the water content. If the water content is reduced, the concrete may not be workable. Also in most cases, concrete needs to be pumped. Certain amount of flowability is needed to pump concrete. When one needs to maintain high workability and also needs to have high strength then chemical admixtures can be added to increase the workability without reducing the strength.

High water content   ⟶   *Low strength + High workability*

Low water content   ⟶   *High strength + Low workability*

Low water content + Admixtures   ⟶   *High strength + High workability*

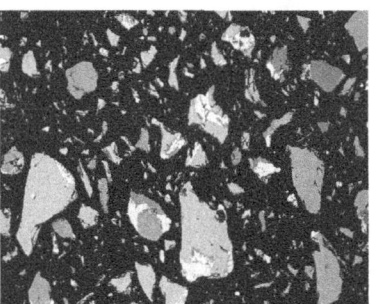

Cement particles seen with a high-resolution microscope

Cement Types:
Five major types of cements are available in the US market.

Type I Cement - Type I cement is known as general-purpose cement and widely used. This is the cheapest type of cement.

Type II Cement - Type II cement generates less heat than type I cement. This property can be useful for mass concretes. When large mass of concrete is poured (ex. dams, large footings, retaining walls) heat generated inside the core may not be able to escape. High temperatures give rise to low strength. In such situations, type II cement can be used. Another property of type II cement is its resistance to sulphate attack. Sulphates are present in some soils and groundwater.

Type III Cement: Type III cement is known as high early strength cement. High early strength is required to remove forms and move forward. Type III cement is expensive than type I cement. Hence, one may have to consider cost vs. schedule benefits when recommending type III cement. Typically, type III cement will achieve the 28-day strength of type I cement in 7 days. Eventually they both will have the same strength assuming other ingredients are the same.

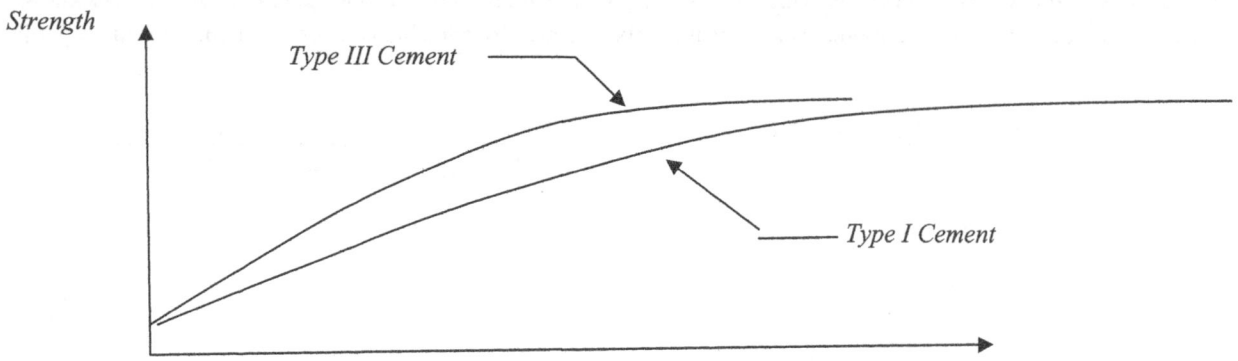

Type IV Cement: Type IV cement generates much less heat during hydration. Hence, this cement is used for very large concrete structures. These cements have the lowest heat generation. Type IV cements are not readily available in the market.

Type V Cement: Type V cements are known as sulphate resistance cements. These cements are used when the groundwater or the soil contain large concentrations of sulphates.

Pozzalans: Pozzalans are known as supplementary cementitious material. Pozzalans can be mixed with cement to reduce cost without reducing the strength. Pozzalans react with byproducts of cement hydration. One of the main byproduct of cement hydration is Calcium Hydroxide. Pozzalans react with this byproduct to generate strength. Hence, Pozzalans do not contribute to initial strength. However, later Pozzalans react with byproducts of cement hydration and generate additional strength. Other than the cost benefit, there is another benefit of pozzalans. Since most pozzalans are materials that are wastes from other processes, owners can gain LEEDs environmental points by using them in the mixture. Three widely used pozzolans are;

Fly Ash
Blast furnace slag
Micro silica or silica fume

Fly Ash:

Left: Coal Plant
Right: Fly Ash is a byproduct of coal burning

Coal is burnt to generate electricity. It has been reported that 50% of electricity generated in USA comes from coal plants. Fly ash is a byproduct of coal burning plants. In the past, fly ash was sent to landfills. Recently it has been found that fly ash could be used as a supplement to cement without affecting the strength. Fly ash is the most commonly used supplementary cementitios material.

Since fly ash particles are more spherical in shape than cement particles, workability and pumpability can be improved by adding fly ash. Some fly ashes cause low early strength. This can be a problem when strength needs to be attained sooner. Adding fly ash usually improve the resistance against sulphate attack. Another property of fly ash is to reduce the air content in concrete. In freezing and thawing conditions, air entrained concrete is preferred. In such situations, fly ash should be avoided.

Blast Furnace Slag: Slag is produced in blast furnaces that produce iron and steel. Slag also can be used as a supplementary cementitious material. Slag tends to improve resistance for sulphate attack. In addition, it has been reported that slag develop higher long-term strength.

Silica Fume: Silica fume is a byproduct of silica alloy industry. Silica fume is also known as micro silica. Micro silica particles are 100 times smaller than cement particles. Main advantage of Silica fume is high durability. Since micro silica particles are extremely small, permeability of concrete is reduced. This is an important quality since rebars would be better protected from water permeation.

Concrete Admixtures:
Chemical admixtures are widely used to improve required properties of concrete.

Air Entraining Admixtures: Concrete that is subjected to repeated freezing and thawing would develop cracks. This can be avoided by increasing entrapped air. Wide array of chemicals are used for air entrainment. Vinsol resin is the most popular air entrainment admixture.

Water Reducing Admixtures: Less water in the concrete would generate higher strengths. On the other hand, there need to be enough water for workability. Water reducing agents can maintain a low water content while maintaining workability.

Accelerating Admixtures: Accelerating agents are used to accelerate the setting of concrete. Some accelerators are capable of increasing the early strength of concrete as well

Superplasticizers (Commonly known as Super Ps):   Superplastizers are used to maintain high workability at the same time maintaining strength. When concreting highly reinforced structures, concrete has to be able to flow freely. Superplastizers can be used in such situations to increase the flowability without compromising strength.

Concrete Retarders: Concrete retarders are added to delay the setting of concrete. Delaying of concrete setting is required in following situations.
Concrete has to be transported longer distances
   1) Provide more time for the workers to carve grooves, curves and architectural features.
   2) To avoid cold joints

Concrete Slump Test:
Concrete slump test should be done as described in ASTM C 143. Slump test is widely used to check the workability and consistency of concrete.

<u>Left</u>:   Above left photo shows the slump cone, rod and hand trowel.
<u>Middle</u>: After concrete is placed inside the slump cone, the slump cone is removed.
<u>Right</u>:   Next, the slump is measured. Higher the slump, higher the water content.

Slump Test Procedure:

STEP 1: Obtain fresh concrete from the truck.
STEP 2: Fill 1/3rd of the slump cone and tamp 25 times with a rod.   (This is commonly called rodding)
STEP 3: Fill another 1/3rd and tamp 25 times. Fill the last one third and tamp 25 times.
STEP 4: Lift the cone
STEP 5: Measure the slump.

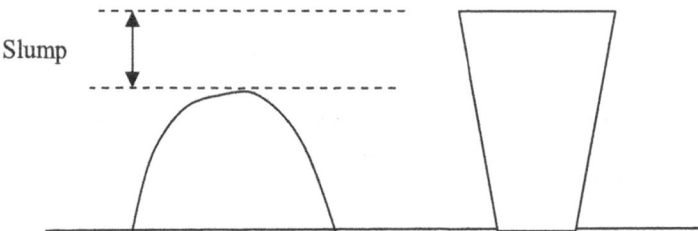

Concrete Cylinders:

Concrete cylinders are taken to conduct compressive strength tests. Concrete compressive strength tests need to be done as per ASTM C 31 and ASTM C 39. Concrete cylinders are 6 in diameter and 12 in. high. Small size cylinders (4 in. diameter and 8 in. tall) are also used.

Procedure to obtain concrete cylinders;

STEP 1: Fill 1/3rd of the cylinder and tamp 25 times with a rod.
STEP 2: Fill another 1/3rd of the cylinder and tamp 25 times. Finally, fill the last 1/3rd of the cylinder and tamp 25 times.
Concrete cylinders should be placed in the job site in a controlled environment at temperature 60F to 80F. Cylinders should be transported to the lab within 48 hours.

Acceptance Criteria:

Typically, three cylinders are taken. One cylinder is broken after 7 days. The strength of the 7-day test should be approximately 65 to 70% of the 28-day strength. 7-day break is for informational purposes only.
Other two cylinders are broken after 28 days. Average of these two cylinders should be equal or more than the required strength. In addition, none of the 28 days tests should fall below required strength by more than 500 psi.
It is a good practice to obtain 4 cylinders. If the 28-day breaks do not reach the required strength, 4th cylinder can be tested at a later time.

Left picture shows concrete cylinders taken in the field. The cylinders are taken to the lab and covers are removed. Picture at the right shows compressive strength test.

Concrete cylinders should be placed in concrete curing boxes when on site. During wintertime, these boxes need to be powered and heat should be provided to attain proper temperatures.

Left: Temperature controlled concrete curing box
Right: Concrete cylinders placed inside the box

7.7   Structural Steel:

Ancient civilizations used rocks, bricks and wood to construct structures. Iron was known to man for thousands of years. Iron was used to make swords, arrows and various other equipment. Very first example of iron being used as a construction material comes to us from Konark Sun Temple in Orissa India. Iron beams were used at openings to hold rocks above.

Konak Sun Temple – *Iron beams were used at entrances to carry the load from above. This is the very first use of metal in construction.*

Konark temple do not have metal to metal connections. Ditherington Flax Mill in England is the very first iron framed building.

Ditherington Flax Mill in England – World's first iron framed building. Actually the walls are taking large portion of horizontal loads. Iron posts and beams take the vertical loads.

Steel Mills: Steel mills manufacture standard shapes. Standard shapes are given in AISC steel contraction manual.

Left: Standard steel beams   **Right:** Standard steel channels

Steel Fabricators: Steel fabricators obtain standard shapes from steel mills and make necessary changes so that it can be used. Some of the work of fabricators are;

- Cutting beams, columns and plates to correct lengths
- Drilling holes for bolts
- Provide curvature to beams.
- Welding pieces to obtain necessary shapes

Left: Fabricating a arched bridge using rolled sections
Right: Fabricating stiffener plates

8.0   Site Development   -   5 Questions

8.1   Excavation and Embankment (Cut and Fill)

Earthwork Mass Diagrams:

Earthwork and mass diagrams are used to compute the cut and fill quantity required for road construction projects.

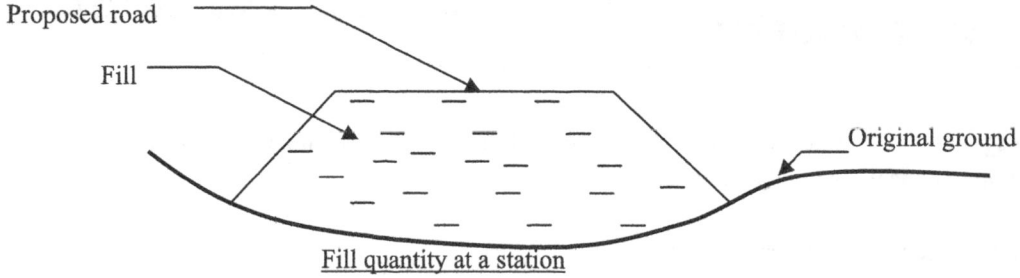

During road projects, the road is divided into stations. Typically, stations are given as 0+00, 1+00, 2+00 etc. 1+00 means 100 ft from the reference point and 2+00 means 200 ft from the reference point. Similarly, 2+30 means 230 ft from the reference point. Fill quantity required at each station varies since the ground surface tends to vary.

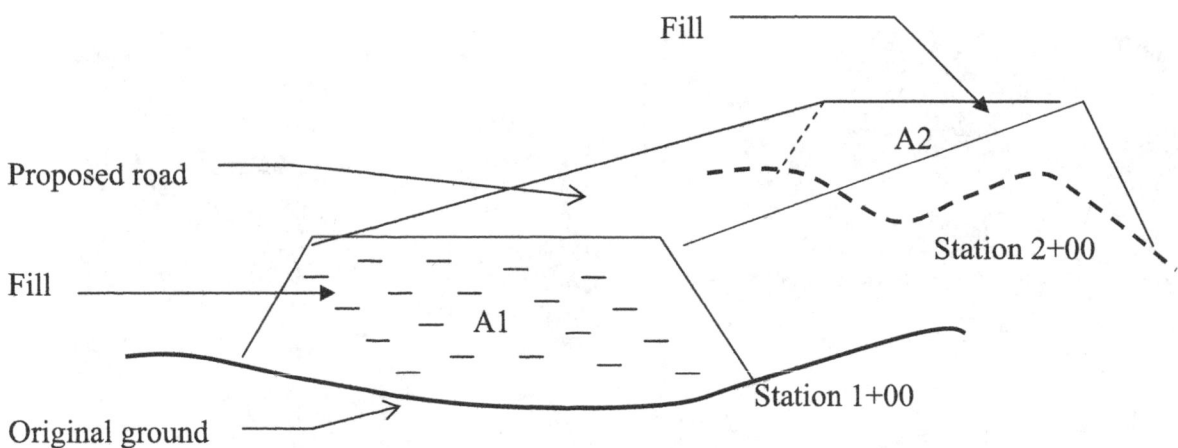

Placement of fill in road construction

Fill quantity required at station 1+00 is different from the quantity required at station 2+00.
Volume of fill required is obtained by multiplying the average area of two stations by the distance between stations.

Volume of fill required = (A1 + A2)/2 x d
d = distance between stations

How to obtain the areas A1 and A2?
In the real world, computer programs and planimeters are used to obtain A1 and A2 areas since they cannot be computed due to their irregular shapes. In the exam, these areas will be provided.

Above figure shows a highway been constructed thru a mountain range. Such projects involve huge quantity of soil been cut, transported and filled.

Practice Problem: Find the volume of fill required from station 1+00 to 1+50. Fill area at station 1+00 is found to be 60 ft^2 and station 1+50 found to be 80 ft^2.

Solution: Volume of fill required = (A1 + A2)/2 x distance between stations
                                = (60 + 80)/2 x 50 ft^3.
                                = 3,500 ft^3 = 129.6 yd^3.

Cut: During road construction, some locations may have to be cut. This happens when the proposed road is at a lower elevation than the existing ground.

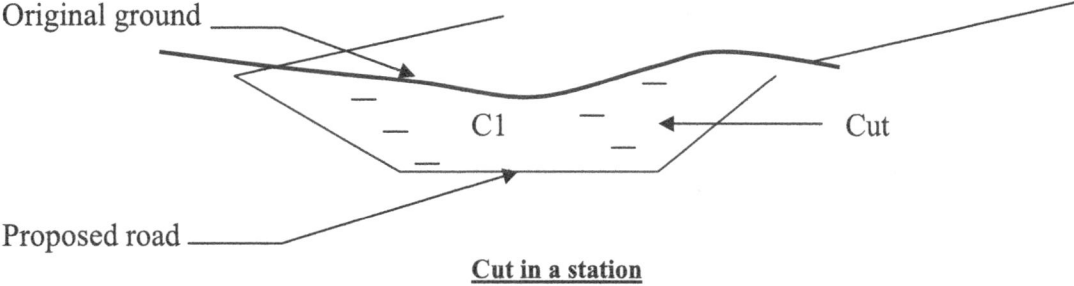

Cut in a station

In some stations both cut and fill can occur.

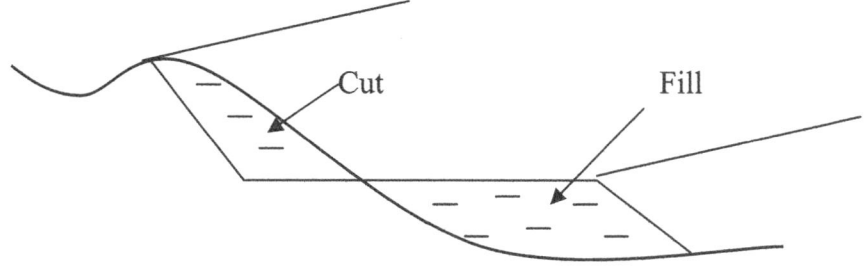

Cut and fill at the same station

When both cut and fill occurs at the same station, net value is obtained. The process is easily explained using an example.

Practice Problem: Find the net cut or fill between station 1+50 and 2+00.

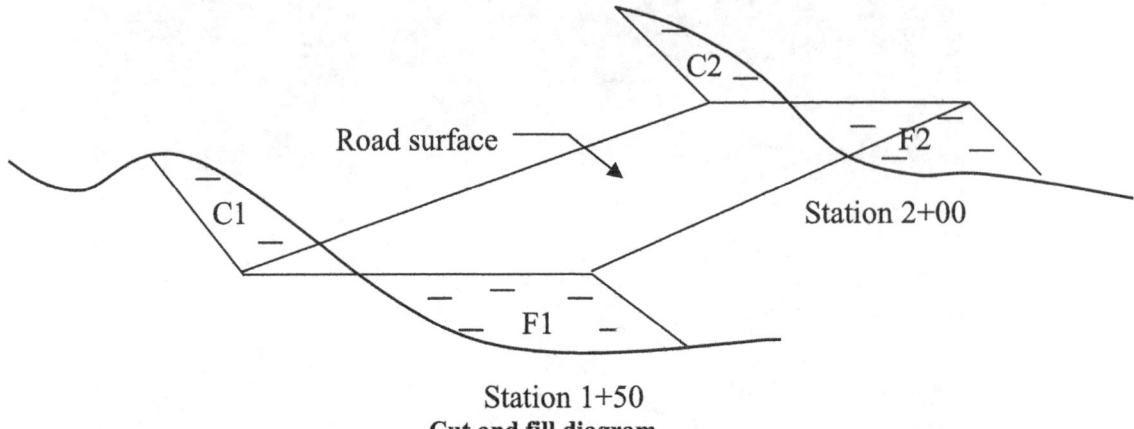

Station 1+50
Cut and fill diagram

Following information provided.
Station 1+50 $C1 = 120 \ ft^2$ $F1 = 170 \ ft^2$
Station 2+00 $C2 = 200 \ ft^2$ $F2 = 60 \ ft^2$
Usually this information is tabulated.

Station No.	Cut Area	Fill Area	Cut Vol	Fill Vol	Net Cut (yd^3)
1+50	120	170			
2+00	200	60			

Solution:
Total cut between two stations = (C1 + C2)/2 x 50
= (120 + 200)/2 x 50 = 8,000 ft^3 = 296.3 yd^3.

Total fill between two stations = (F1 + F2)/2 x 50
= (170 + 60)/2 x 50 = 5,750 ft^3 = 213 yd^3.

Net amount = 296.3 – 213 = 83.3 yd^3. (cut)
Cut is represented with positive while fill is represented with negative.

Station No.	Cut Area	Fill Area	Cut Vol (ft^3)	Fill Vol (ft^3)	Net Cut (ft^3)
1+50	120	170			
2+00	200	60	296.3	213	83.3

Mass Diagrams: Mass diagram is drawn using the net cut values. Mass diagrams are easily explained using an example.

Practice Problem: Mass diagram example:

Cut areas and fill areas are tabulated as shown. Complete the blank columns and draw a mass diagram.

Station No.	Cut Area	Fill Area	Cut Vol (ft^3)	Fill Vol (ft^3)	Net Cut (ft^3)	Cumulative Cut (ft^3)
0+00	175	125				
0+50	117	123				
1+00	238	250				
1+50	211	240				
2+00	198	180				
2+50	140	141				
3+00	258	200				

Solution:

STEP 1: Complete the "Cut Vol" column.

Cut volume (1st entry) = (175+117)/2 x 50   = 7,300
Cut volume (2nd entry) = (117+238)/2 x 50   = 8,875
Cut volume (3rd entry) = (238+211)/2 x 50   = 11,225
Cut volume (4th entry) = (211+198)/2 x 50   = 10,225
Cut volume (5th entry) = (198+140)/2 x 50   = 8,450
Cut volume (6th entry) = (140+258)/2 x 50   = 9,950

Station No.	Cut Area	Fill Area	Cut Vol (ft^3)	Fill Vol (ft^3)	Net Cut (ft^3)	Cumulative Cut (ft^3)
0+00	175	125	0	0	0	
0+50	117	123	7,300			
1+00	238	250	8,875			
1+50	211	240	11,225			
2+00	198	180	10,225			
2+50	140	141	8,450			
3+00	258	200	9,950			

STEP 2: Complete the "Fill Vol" column.

Fill volume (1st entry) = (125+123)/2 x 50   = 6,200
Fill volume (2nd entry) = (123+250)/2 x 50   = 9,325
Fill volume (3rd entry) = (250+240)/2 x 50   = 12,250
Fill volume (4th entry) = (240+180)/2 x 50   = 10,500
Fill volume (5th entry) = (180+141)/2 x 50   = 8,025
Fill volume (6th entry) = (141+200)/2 x 50   = 8,525

Station No.	Cut Area	Fill Area	Cut Vol	Fill Vol	Net Cut (ft^3)	Cumulative Cut
0+00	175	125	0	0	0	
0+50	117	123	7,300	6,200		
1+00	238	250	8,875	9,325		
1+50	211	240	11,225	12,250		
2+00	198	180	10,225	10,500		
2+50	140	141	8,450	8,025		
3+00	258	200	9,950	8,525		

STEP 3: Complete the "Net Cut" column
Net cut (1st entry) = 7,300 – 6,200 = 1,100

Net cut (2nd entry) = $8{,}875 - 9{,}25 = -450$
Net cut (3rd entry) = $11{,}225 - 6{,}200 = -1{,}025$
Net cut (4th entry) = $7{,}300 - 6{,}200 = -275$
Net cut (5th entry) = $7{,}300 - 6{,}200 = 425$
Net cut (6th entry) = $7{,}300 - 6{,}200 = 1{,}425$

Station No.	Cut Area	Fill Area	Cut Vol	Fill Vol	Net Cut (ft³)	Cumulative Cut
0+00	175	125	0	0	0	0
0+50	117	123	7,300	6,200	1,100	
1+00	238	250	8,875	9,325	-450	
1+50	211	240	11,225	12,250	-1,025	
2+00	198	180	10,225	10,500	-275	
2+50	140	141	8,450	8,025	425	
3+00	258	200	9,950	8,525	1,425	

STEP 4: Complete the "Cumulative Cut" column:
Cumulative cut is obtained by adding the present net cut to the previous net total.

Cumulative cut (1st entry) = $1{,}100 + 0 = 1{,}100$
Cumulative cut (2nd entry) = $1{,}100 + (-450) = 650$
Cumulative cut (3rd entry) = $650 + (-1{,}025) = -375$
Cumulative cut (4th entry) = $-375 + (-275) = -650$
Cumulative cut (5th entry) = $-650 + 425 = -225$
Cumulative cut (6th entry) = $-225 + 1{,}425 = 1{,}200$

Station No.	Cut Area	Fill Area	Cut Vol	Fill Vol	Net Cut (ft³)	Cumulative Cut
0+00	175	125	0	0	0	0
0+50	117	123	7,300	6,200	1,100	1,100
1+00	238	250	8,875	9,325	-450	650
1+50	211	240	11,225	12,250	-1,025	-375
2+00	198	180	10,225	10,500	-275	-650
2+50	140	141	8,450	8,025	425	-225
3+00	258	200	9,950	8,525	1,425	1,200

STEP 5: Draw the mass diagram:
Mass diagram is drawn between the "cumulative cut and the stations.

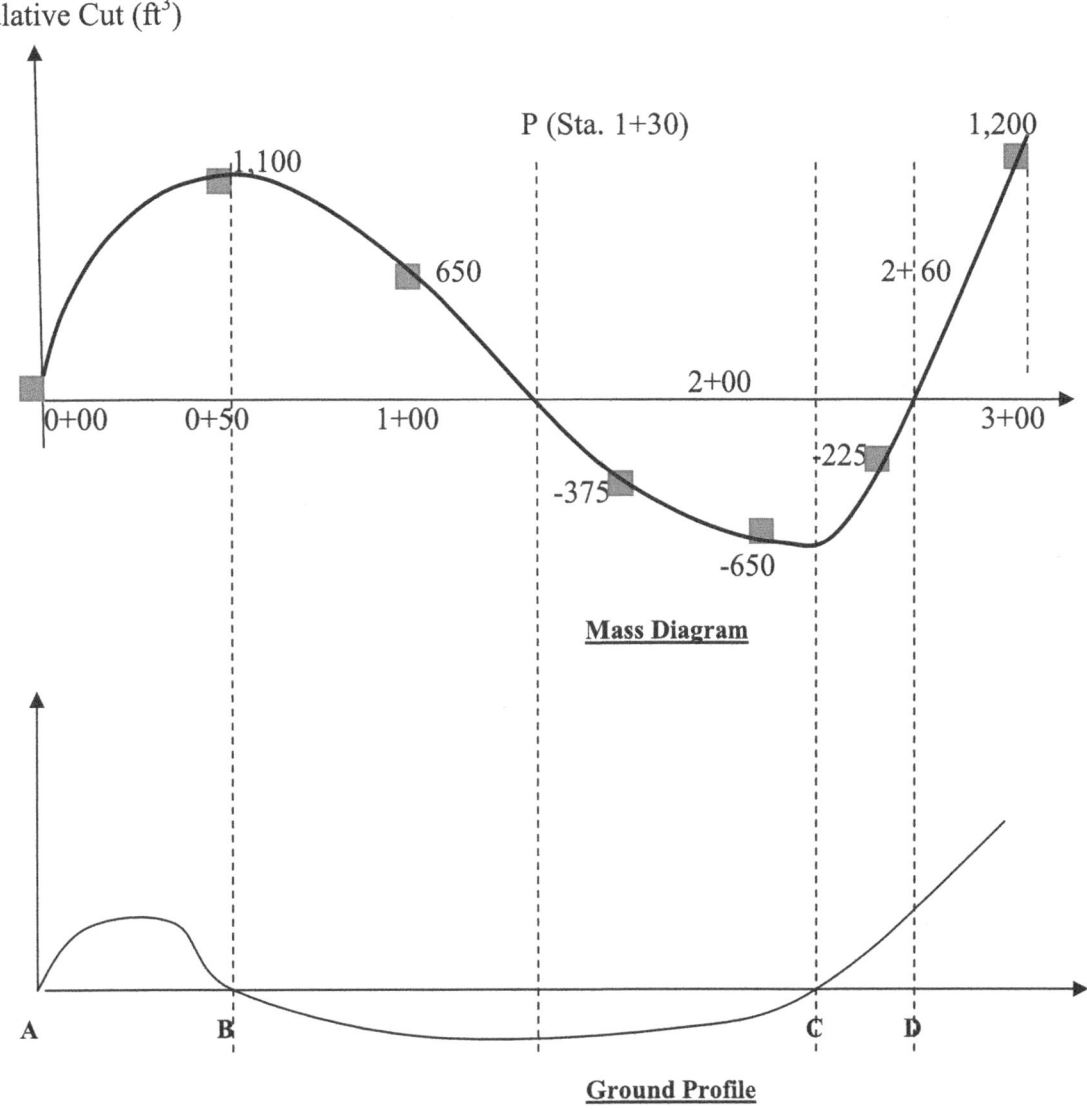

Above top figure shows the mass diagram. Above lower figure shows the ground profile. Look at the ground profile.

Point A to B: This section is all cut. Hence, mass diagram keeps going up.
Point B to C: This section is all fill. Mass diagram changes direction at point B and start to go down.
Point C to D: Point C to D is all cut. Mass diagram changes direction and start going up.
Beyond Point D: Beyond point D is cut. Hence cumulative cut keep going up.

Mass Diagram Implications: Consider point "P" shown with a vertical line. At point "P", the curve crosses the "X" axis. Point "P" lies between station 1+00 and 2 + 00. Let us assume point "P" to be at station 1+30.
At station 1+30, the mass curve crosses the "X" axis. At point P cumulative cut is zero.
What does this mean?
It means that between station 0+00 and station 1+30, total required fill is zero. In other words, all the cut material has been used for fill purposes.

Balanced points: Net cut between balanced points is zero. Sta. 0+00 to 1+30, net cut is zero. Hence, station 0+00 and 1+30 are balanced points.   In other words, all the cut material in this region is utilized for fill.   Similarly, station 1+30 to 2+60 also are two balanced points.

8.2 Construction Site Layout and Control:

<u>Horizontal Control</u>:   Design documents typically specify that the coordinates should be obtained using a monument nearby. Surveyors need to use the given monuments provided by the client and establish control points near the site. In many instances these control points get runover by machines and new control points need to be installed. It is important to make sure that the control points are protected. If not the building would be constructed at wrong cordinates.

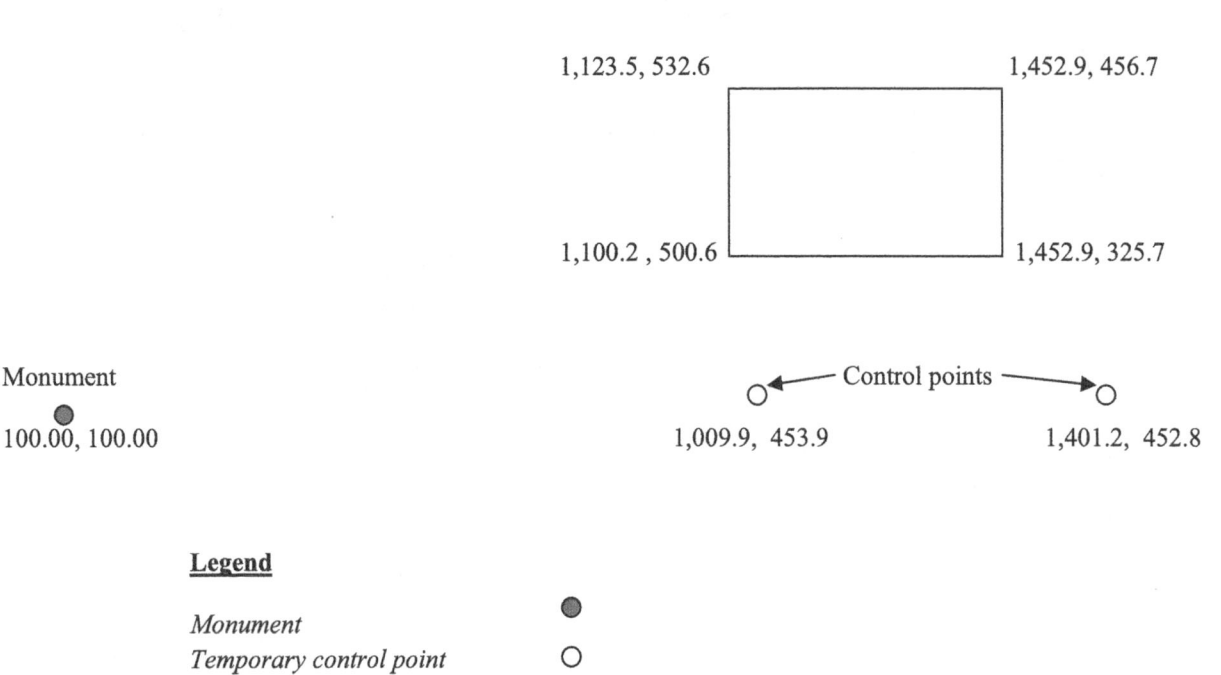

Above figure shows a proposed building, monument and two control points. Control points need to be closer to the building. However when too close, they get damaged due to construction activities.

<u>Vertical Control</u>:   Similar to horizontal control, design documents should indicate the elevation and the datum used. Surveyors need to establish temporary bench marks near and around the site to be used for grading, establish footing and slab elevations and elevations of utility pipes.

<u>Manhole Construction</u>:   Manholes are required to clean out pipes. Manholes can be temporary or permanent. Site vertical control is important in construction of pipelines and manholes.

<u>Surveying</u>:   Knowledge of surveying is important for construction work. Hence, you will be tested in the exam on surveying topics.

Magnetic North and Geographic North:

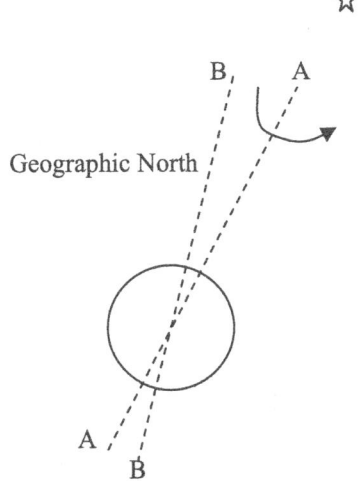

A-A = Axis of rotation or geographic rotation
B- B = Magnetic north
Geographic north is the direction of the axis of rotation of earth and pointed to polar star. Polaris is a very bright star 430 light years away. It is 2,500 times brighter than our sun and 5 times heavier. Polaris is getting brighter. It is 2.5 times brighter today than Greek astronomer Ptolemy observed it 2,000 years ago. However, earth's axis of rotation is not a constant. The axis of rotation of the earth is changing every year by a very small amount. It has been calculated that 25,000 years from now axis of rotation would point to a different star known as Vega.

Polaris is at the end of the little dipper. It is in line with the two stars in big dipper, Dubne and Merak.

<u>Magnetic North</u>: Geographic north depends on the rotation of the earth. Magnetic north depends on the magnetic field of the earth. It is a coincidence that earth's magnetic north is very close to the earth rotational axis. For an instance, the difference between Neptune's rotational axis and its magnetic north is 40^0.

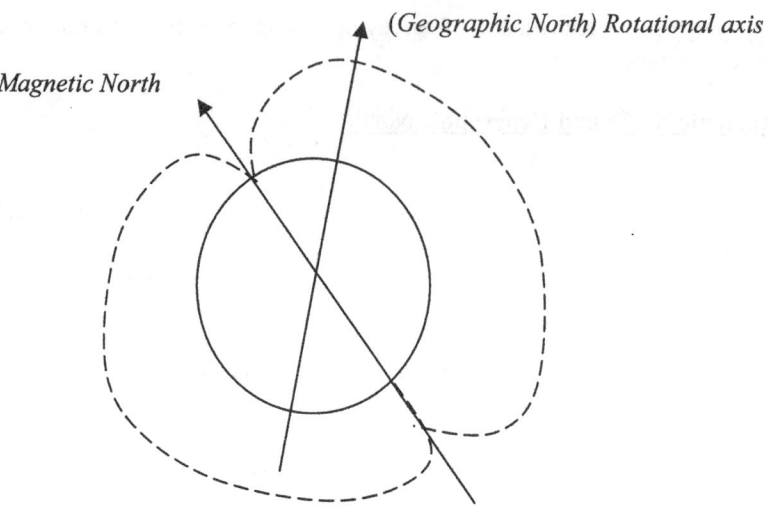

Planet Neptune's rotational axis and magnetic north are 40^0 apart

In the case of earth, the difference between geographic north and magnetic north is very small.

Meridian: Meridian is any longitude. Meridian at a location is the longitude of that location.

Azimuth: Azimuth is the horizontal angle made with respect to the **geographic north**. (Axis of rotation of earth)

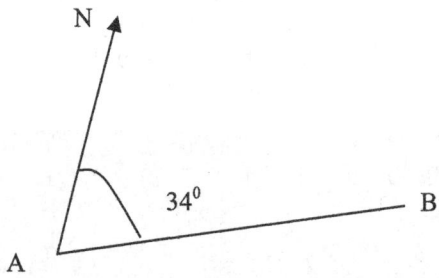

Azimuth of line AB is 34^0 in this case.

Zenith and Nadir: Line drawn vertically to the sky is known as zenith. Line drawn directly to the center of the earth is known as Nadir.

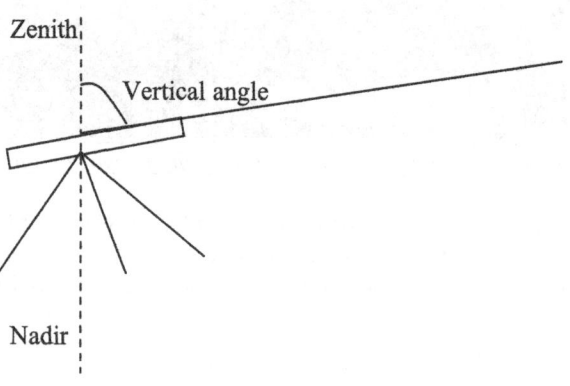

Surveying Instruments:

Level: levels are used to measure the vertical heights. One person would hold a rod and the other person would take a reading.

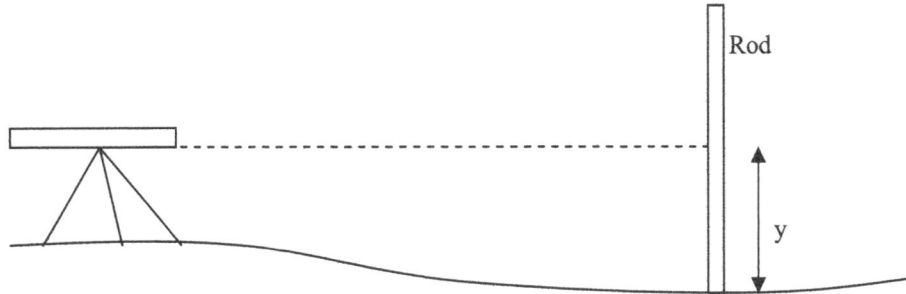

Some levels are equipped with a cross hair so that they can be used to measure horizontal distances. Cross hairs are located in such a manner, the length is computed by multiplying the distance between cross hairs by 100.

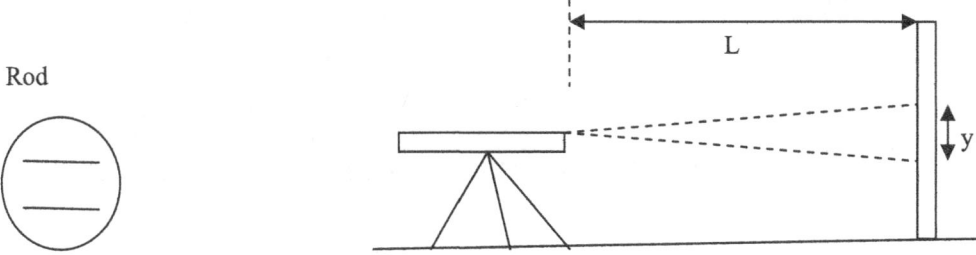

The instruments are designed in a such a manner so that $L = 100 \times y$
If the reading y is 1.2 ft, then the distance (L) would be 120 ft.
Levels cannot be used for measurement of angles.

Theodolites: Theodolites are designed to measure horizontal and vertical angles.

EDM: Electronic distance measurement or EDM can be used to measure horizontal and vertical distances.

Total Stations: Total stations are electronic instruments equipped with computers that can be used to measure angles and distances.

Total stations can measure horizontal and vertical angles and also can measure distances

Bearing: Bearing is the angle measured from North or South.

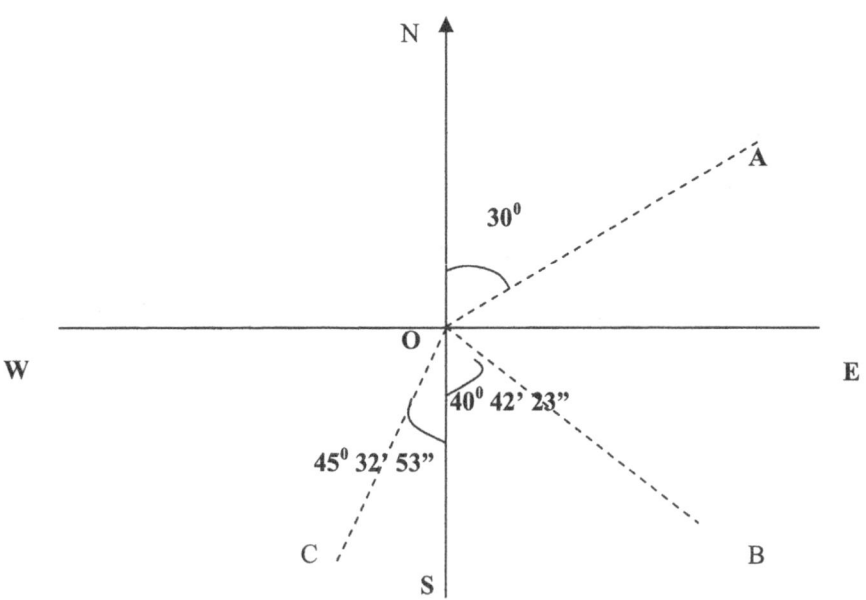

Bearing of OA = N 30⁰ E
This means OA line is 30 degrees measured from N towards east.
Similarly, bearings of other lines are given below.
          OB = S 40⁰ 42' 23" E
          OC = S 45⁰ 32' 53" W

Practice Problem: Provide the bearing for above line OC from North.

Solution: Angle between OC and North is $180 + 45^0$ 32' 53" $= 225^0$ 32' 53"

Traverse: Traverse is conducted by starting from a known point. Angles and distance measurements are taken to new points from the initial point.

Practice Problem: Bunch of new houses were built in a remote area. Surveyors were called upon to locate the new houses with better accuracy relative to a known benchmark in the vicinity.

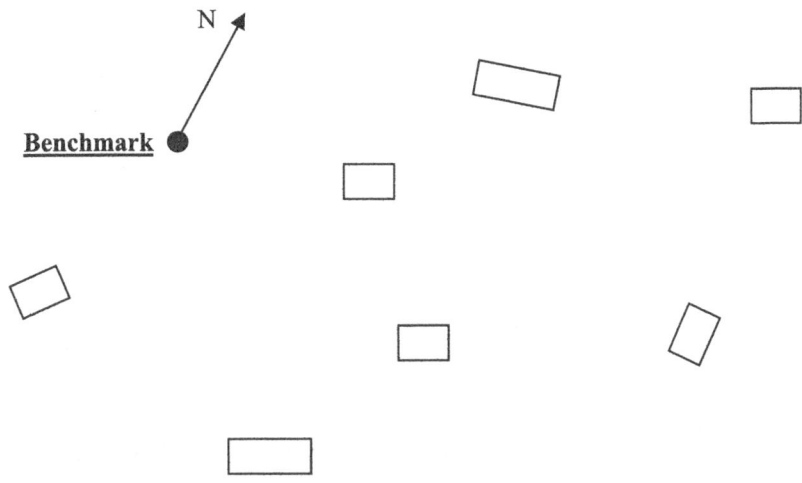

Solution:
STEP 1:   Locate the benchmark and call it point A. Select a point B, which is closer to existing houses. Find the angle of AB relative to geographic north.

STEP 2:   Measure the distance AB. Now point B can be located in a map.

STEP 3:   Measure distance Ap.

STEP 4:   Measure pp' perpendicular to AB. Now point p' can be located in a map. If the house has to be exactly located in a map, more than one measurement is needed to its edges.

STEP 5:   Select another convenient point C, closer to houses that need to be measured. Find the interior angle ABC and the distance BC. Now point C can be established in a map.

STEP 6:   Measure Br and rr' distances.
This way all houses can be located in a map.

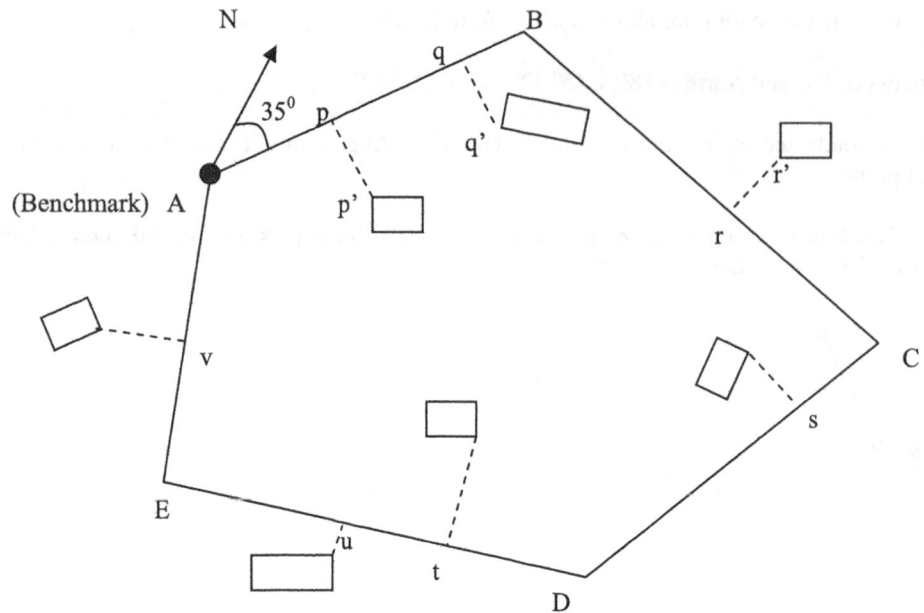

Accuracy of such a procedure heavily depends upon the interior angle measurements at points A, B, C, D and E.

Construction workers need to know the building footprint, column locations and wall lines. Locations of these structural elements are provided by surveyors. Nevertheless, construction engineers need to have a good understanding of the process.

Building Line: Surveyors usually provide a string line to indicate the building footprint.

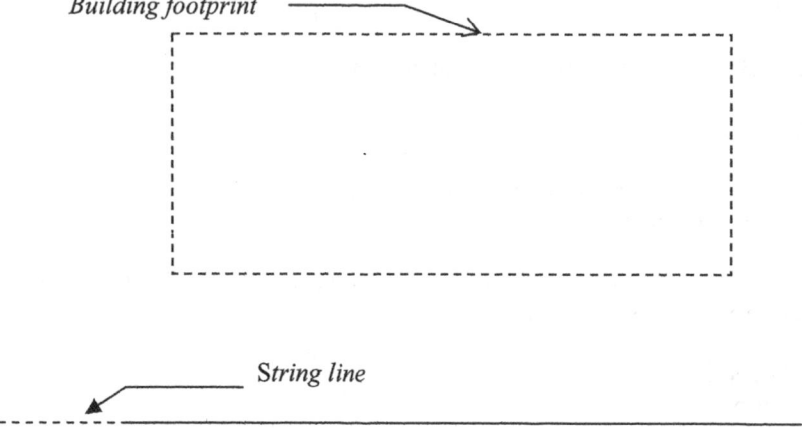

String line is used to take measurements to the building footprint. String line should not be too close to the building since construction equipment could damage it. In addition, it should not be too far since that would make taking measurements difficult.

Elevations: It is important to obtain elevations for construction. Elevations are obtained with reference to a given benchmark.

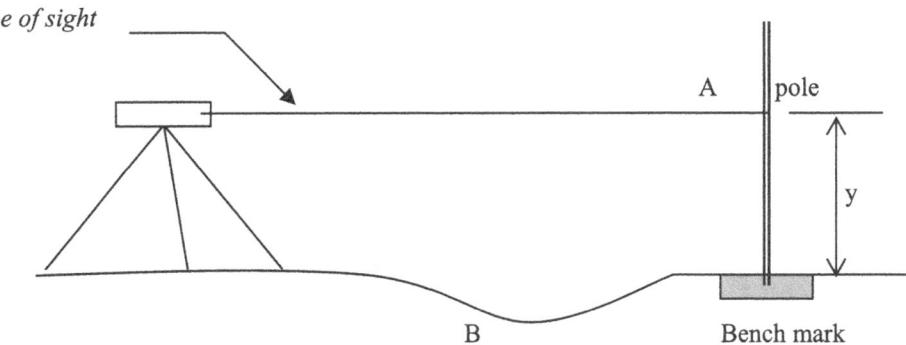

Assume the benchmark elevation is 100 ft.
Assume pole reading is 5.5 ft.
Elevation of line of sight = 100 + 5.5 = 105.5
Now place the pole on point B.

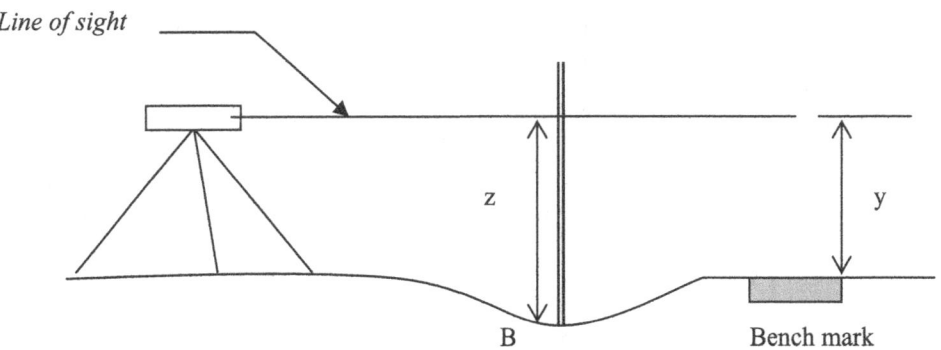

Obtain the new reading "z'.
Assume that new reading "z" is 7 ft.
Elevation of point B = Line of sight elevation - z = 105.5 - 7 = 98.5 ft

Practice Problem: Surveyor finds a US geological benchmark in the site. The benchmark elevation was found to be 98.7 ft above mean sea level. (MSL). The surveyor places the pole on top of the benchmark and obtained a reading of 5.7 ft. Then surveyor obtained a reading of 3.2 for point A, 6.7 for point B and 5.0 for point C. Find elevations of point A, B and C.

Solution:
Benchmark elevation = 98.7
Pole reading on top of the benchmark = 5.7 ft
Elevation of line of sight = 98.7 + 5.7 = 104.4 ft
Reading of point A = 3.2
Elevation of point A = 104.4 - 3.2 = 101.2 ft
Reading of point B = 6.7
Elevation of point B = 104.4 - 6.7 = 97.7 ft
Reading of point C = 5.0
Elevation of point C = 104.4 - 5.0 = 99.4 ft

Practice Problem: Portion of surveyor's logbook is shown below. Elevation of known benchmark is given to be 101.23 ft. Find the elevation of point C.

Location	Back sight	Fore sight	Elevation of line of sight	Elevation
BM	5.23			101.23
Point A		6.12		
Point B		7.23		
Point C		2.45		

Solution:

It is advisable to draw a level and line of sights until you are familiar with back sight and fore sight readings.

Elevation of line of sight = 101.23 + 5.23 = 106.46 ft

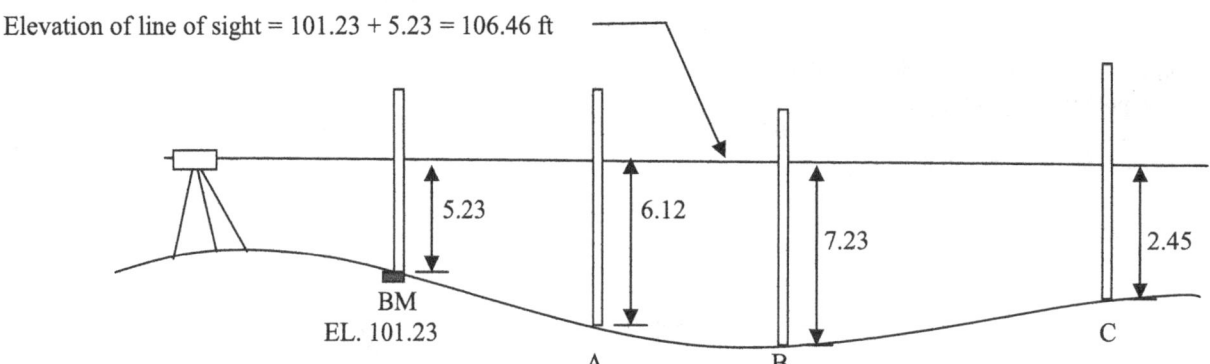

Elevation of line of sight can be computed by adding the reading at benchmark to the elevation of benchmark.

Elevation of line of sight = 101.23 + 5.23 = 106.46 ft

Elevation of point A   = Elevation of line of sight - Reading at point A
                            = 106.46 - 6.12 = 100.34

Elevation of point B   = Elevation of line of sight - Reading at point B
                            = 106.46 - 7.23 = 99.23

Elevation of point C   = Elevation of line of sight - Reading at point C
                            = 106.46 - 2.45 = 104.01

Hence, elevation of point C is 104.01.

You do not need to find elevations of points A and B to find the elevation of point C.
Though it is not necessary to fill the table in the exam, it may be useful to learn how to fill the table.

Location	Back sight	Fore sight	Elevation of line of sight	Elevation
BM	5.23		106.46	101.23
Point A		6.12		100.34
Point B		7.23		99.23
Point C		2.45		104.01

Distance Measurement: Distances are measured with tapes, Theodalites and EDM (Electronic distance measurement).

Practice Problem: A surveyor had to measure the distance between points A and B. Surveyor locates a third point C and obtains angle measurements as shown. Find the distance AB.

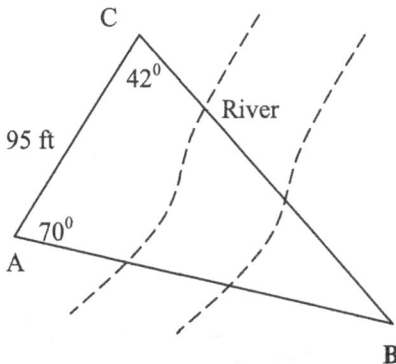

Solution: This problem can be easily solved using the Sine law.

Sine Law: AC/Sin B = AB/Sin C = BC/Sin A

Angle B = 180 − (42 + 70) = 68

AC/Sin B = AB/Sin C

95/Sin 68 = AB/Sin 42
AB = 68.6 ft

Problems:

Problem 1:   An engineer decided that a certain earthwork project has less cutting. He also found out that cut and fill locations are far apart. What is the best equipment for this project?

Problem 2:   Borrow pits are used for what purpose?

Problem 3:   What is the difference between geographic north and magnetic north?

Problem 4:   What is the difference between a level and a Theodolite?

Solutions:

Solution 1:   Dozers can be used to cut. Backhoes and dump trucks to load and transport soil.
Solution 2:   In some projects there is more fill volume compared to cut volume. In these cases, soil has to be brought in from a borrow pit.
Solution 3:   The earth rotates around the geographic north. Magnetic north is the direction of the magnetic field.
Solution 4:   Levels can measure only vertical heights. Theodolites can be used to measure horizontal angles, vertical angles and horizontal distances. On the other hand, total station does everything with electronics.

8.3   Erosion and Sediment Control:

Many states require an erosion and sediment control plan be submitted prior to start of any construction work. Silt fences are used to stop soil eroding away. Near river beds, rip rap is provided. Hay bales and geo-fabrics also can be used to stop erosion.

Left: Riprap to stop erosion near a stream

Silt Fence: Silt fence is a geo-fabric installed using posts. Silt fence should be buried few inches deep into soil. Silt fences stop silt and sand from eroding away due to water flow.

Left: Protecting a manhole using hay bales

Right: Hay bales used to stop erosion of soil

8.3.1 Detention and Retention Basins:

To understand the function of detention ponds let us look at an example. Let us say there is a train station. Train platform has room for 200 people. However, at peak times, train platform has a gathering of 500 people. This is considered to be an unsafe condition. To solve the problem we can build a waiting area for people to stay. When their train arrive the passengers would go to the platform and get on the train. Number of people in the platform at a given time is minimal. This does not mean number of people using the train station went down.

Due to the construction of the waiting area, we have reduced the congestion in the train platform.

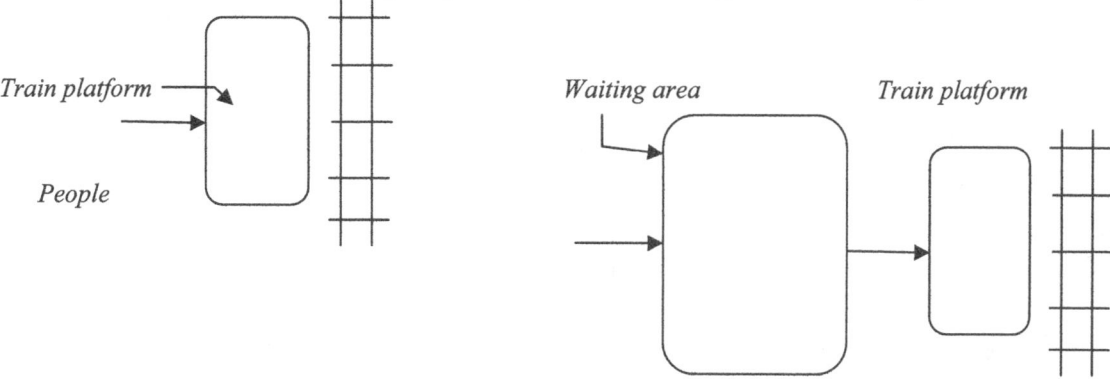

Due to the waiting area, number of people accumulated in the train platform will be diminished.

Detention ponds perform the same function. During a rain event some storm water pipes and culverts would not be enough. In such situations highways and land areas would get flooded. In such situations, detention ponds are built. Water temporarily stored in detention ponds. This way flooding due to under capacity of storm water pipes can be avoided.

Difference Between Detention Basins and Retention basins:

There is a difference between detention basins and retention basins. Main difference is duration of water accumulation. Typically retention basins store water for a long period of time. On the other hand, detention basins are emptied just after a rainfall event. Detention basins are known as dry basins while retention basins are known as wet basins.

Retention Basin *– (Also known as wet basins. Retention basins are constructed to store water for a long period of time)*

Retention basins are designed to attract birds and fish. Many plant species are maintained in retention basins. Water is reintroduced into soil and atmosphere.

Detention Basin *– Detention basins are also known as dry basins. Water is accumulated during a rain event. After rain stops, water is discharged. the basin is kept dry so that it can be used for the next rain event.*

Let us consider a strip mall been built on the side of a road. Presently the land is covered with vegetation. The runoff in vegetated covered land is minimal. Once the strip mall is built, roofs and parking lots would have zero percent infiltration and large quantity of water would flow to the street storm water system. Let us assume that street storm water pipes are not capable of handling the extra flow. There are two solutions.

- Build a detention basin or
- Build a retention basin.

In wet basins, water is not drained after a rain event. For an example let us assume a 4 inch rain fall occurred. Water was transferred to the wet basin. Water is not discharged from wet basins. What would happen if there is another similar rain event to occur in few days? In that case wet basin would not have enough room to absorb any additional water. In this situation, water has to be sent to storm water pipes. This would overload the storm water pipes and flooding would occur. On the other hand, it is possible to design wet basins to have certain amount of capacity in the basin for a second or third rainfall event in a short period of time. In this case larger wet basin is required. Certain volume of water is maintained in the wet basin to accommodate plants and animals. In addition, certain amount of empty volume is maintained for a second rainfall in quick succession. To achieve this, larger retention basin is required. In addition, people need to be trained or hired to operate a wet basin.

On the other hand, dry basins (detention basins) are simpler. After every rain event, dry basins are fully discharged. Hence the dry basin is always available to absorb the flow from next rainfall. Main disadvantage of dry basins is that they do not contribute to the welfare of the environment. On the other hand, wet basins can be made a sanctuary for plants, geese, birds and fish.

Detention Ponds (Dry Basins) Illustration:

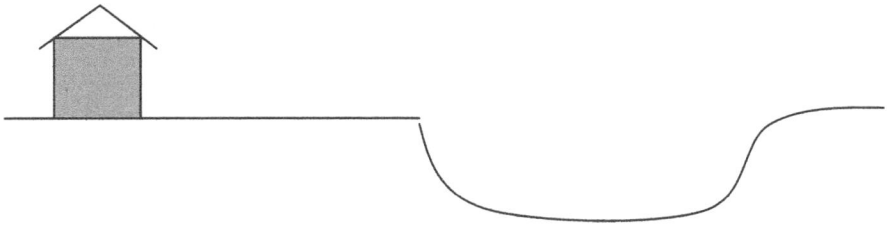

Dry basin prior to a rain event

First Rain Event:

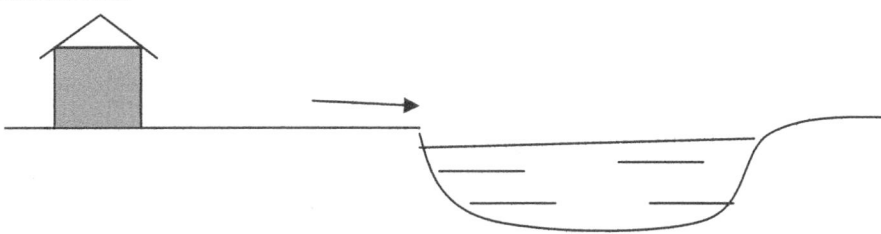

Dry basin after a rain event

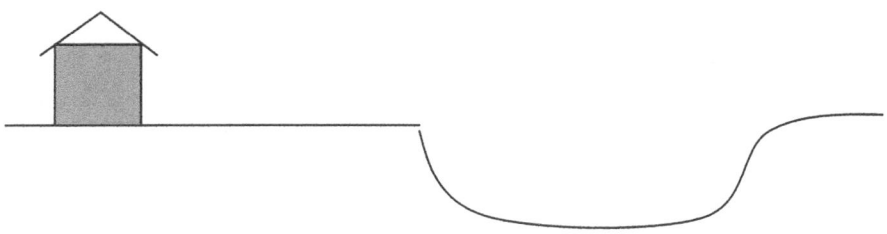

Dry basin is emptied immediately after the rain event.

Second Rain Event:

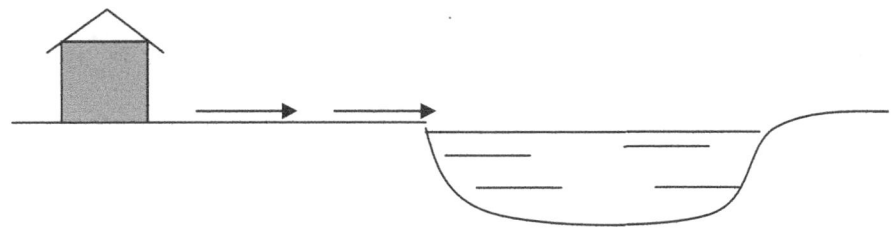

Dry basin can absorb rain water from a second rain event in quick succession

Retention Ponds (Wet Basin) Illustration:

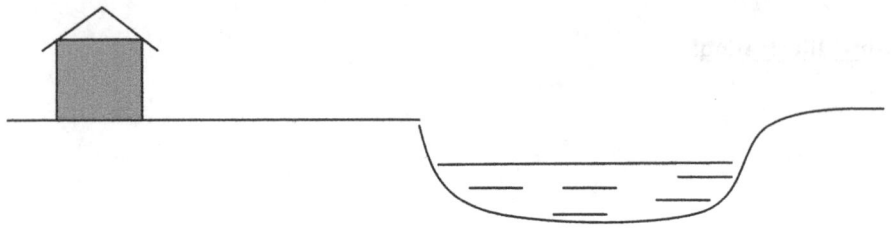

Wet basin prior to a rain event

First Rain Event:

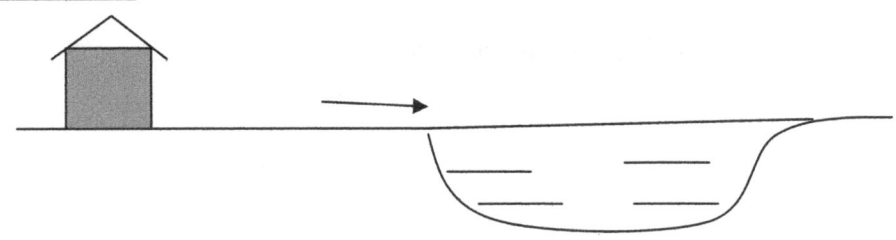

Wet basin after a rain event   (water not discharged)

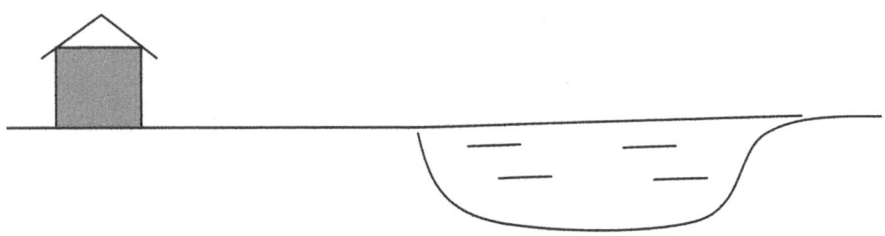

Wet basins are not emptied after the rain event.

Second Rain Event:

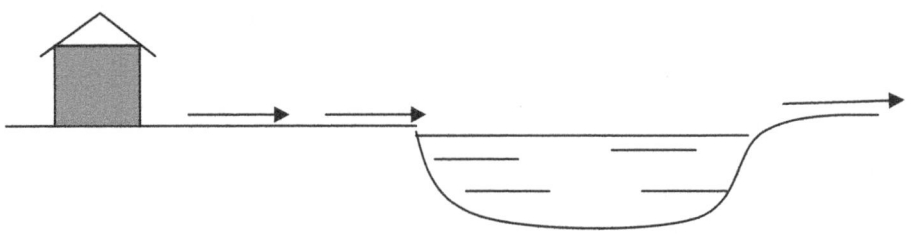

Overflow to storm water system

As mentioned earlier, one could design a wet basin with a larger volume so that there will not be an overflow.

Advantages and Disadvantages of Retention Basins:

Advantages:
- Retention basins remove sediments and pollutants from the surface water.
- Provide a positive habitat for plants and animals
- Creates an aesthetic environment

Disadvantages:
- Wet basins must contain enough water to maintain aquatic plants and fish. This may not be achievable in dry areas.
- There may not have enough land in highly developed areas
- Discharge from ponds to local streams may have different chemical constituents and temperatures. This might affect the local flora and fauna.

Underground Detention Basins:

Perforated pipes are laid underground. The whole system is covered with gravel. Water would flow from perforated pipes to the gravel layer and store water during a rain event.

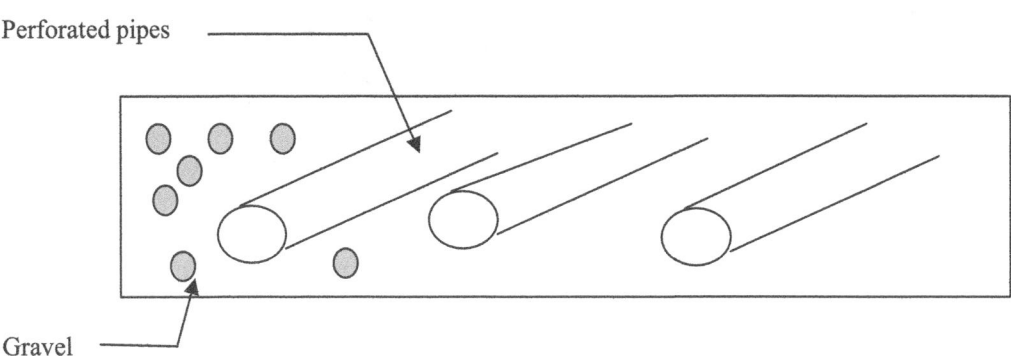

Perforated pipes

Gravel

8.4   Roadside Safety:

TTC: Temporary Traffic Control:
Temporary traffic control or TTC is required during construction activities. Prior to establishing TTC devices, a TTC plan needs to be prepared. The plan should consider passenger traffic, commercial traffic and pedestrians. Commercial traffic may not be able to go over certain bridges due to weight restrictions. Hence TTC planners need to be aware of special considerations for commercial traffic when diverting traffic.

When developing a TTC plan, one has to assume that drivers would reduce the speed only if they clearly perceive a need to do so.

Seven fundamental principles to follow when developing a TTC:
1) General plans or guidelines should be developed to ensure safety of motorists, bicyclists and pedestrians.
2) Road user movement should be inhibited as little as practical.
3) Motorists, bicyclists and pedestrians should be guided in a clear and positive manner while approaching and traversing TTC zones and incident sites.
Which means, signs should be visible and clear? Cones and other traffic guidance equipment should be properly used.
4) Routine day and night inspections of TTC elements should be performed:
Cones and signs could be misplaced due to wind, rain and kids playing around. Workers should routinely inspect cones and signs.
5) Attention should be given to maintenance of roadside safety during the life of the TTC zone
6) All personnel involved in developing and maintenance of a TTC program should be adequately trained.
7) Good public relations should be maintained. This can be done by providing advance notices of road closing and diversions etc. That way motorists can plan alternate routes.

Traffic Control Devices:

Traffic Signs: Traffic signs are a major part of any TTC plan. Traffic signs will let the motorists know what to expect.

Crash Attenuators: When workers are working in the side of the road, crash attenuator will protect the workers.

Cone Placing Trucks:

In this cone placing truck, the worker who places the cones has a place to stand up and place the cones. It is important that the truck provide a place to stand on.

Components of Temporary Traffic Control Zones:

TTC (Temporary Traffic Control) zones are divided into four areas:
A) The advance warning area
B) The transition area
C) The activity area,
D) The termination area.

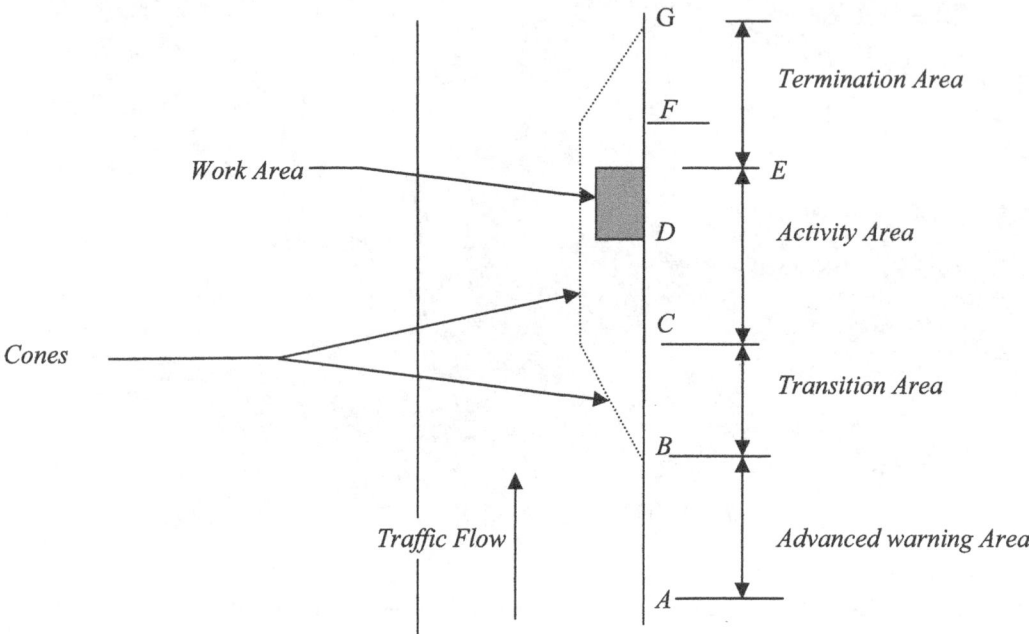

Above figure shows the four temporary traffic control zones. These four are further divided as shown below.

- **A to B: Advance Warning Area:**   Advance warning area consists of signs such as "Construction Ahead", "Your Tax Dollar at Work", or various signs indicating that construction zone is ahead.

- **B to C: Transition Area:**   In this zone, cones will be placed to guide the traffic.

- **C to D: Buffer Zone:**   Buffer zone is created to provide a safety area for workers. In case a runaway driver to come thru the cones, the workers will be able to see the vehicle that is coming towards them. In some cases, speed attenuators are placed in this zone.

- **D to E: Work Zone:**   Workers will be working in the work zone.

- <u>E to F: Buffer space:</u>   Another safe space for workers to move around.

- <u>F to G: Downstream Taper:</u>   Traffic is guided back.

Development of a TTC Plan:

TTC plan should be developed by qualified personnel. Depending upon the complexity of projects, TTC plans can be simple to very complicated. Generally, TTC plans are costly. Placing and maintaining cones, barriers, pillow trucks in a daily basis costs significant amount of money for contractors.

Flaggers:

One Lane Two-Way Traffic: (Conditions for No Flagmen)

As per MUTCD, if the workspace on a low-volume street or road is short and road users from both directions are able to see the traffic approaching from the opposite direction through and beyond the worksite, the movement of traffic through a one-lane, two-way constriction may be self-regulating. In other words, a flagman is not necessary.

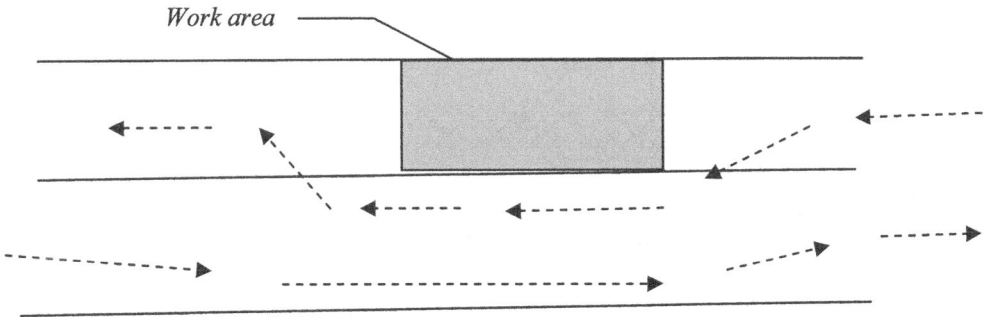

If the traffic is high or drivers cannot see each other due to a curve then flaggers are necessary.

One Lane Two-Way Traffic: (Conditions for One Flagman):
One flagman may be used only if the flagman can see one end to the other. When the road construction area is too long, the flagman may not be able to see the traffic coming from the other end.

One Lane Two-Way Traffic: (Conditions for Two Flagmen):
When one flagman cannot see the full length of the road, then two flagmen should be used.

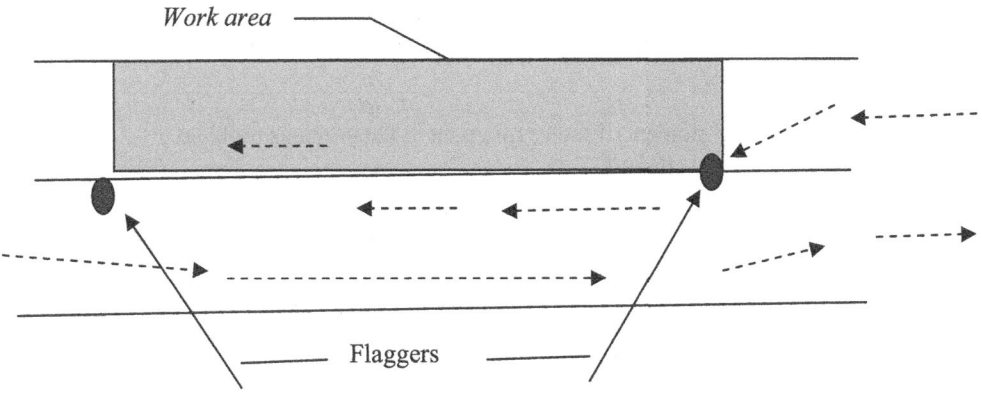

Problems from "Four Sample Exams for the Civil PE Exam"

Complete step by step solutions provided in "Four Sample Exams for the Civil PE Exam", Second Edition.

Problem 1)   A beam is loaded as shown. Find the maximum shear force of the beam

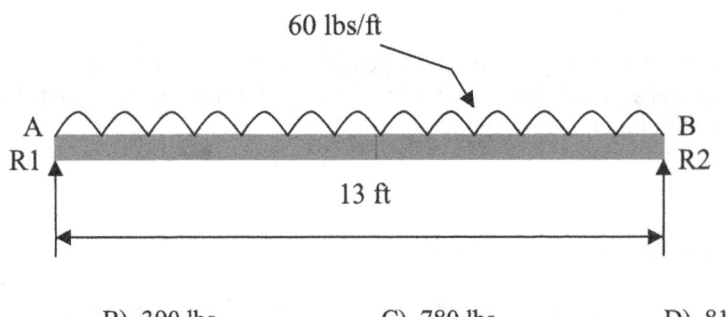

A) 450 lbs   B) 390 lbs   C) 780 lbs   D) 810 lbs
Ans B

Problem 3)   Find the maximum bending moment for the loaded beam shown in the previous example.

A) 2,367.5 lbs. ft   B) 1,345.9 lbs. ft   C) 1,267.5 lbs. ft   D) 2,340.8 lbs. ft

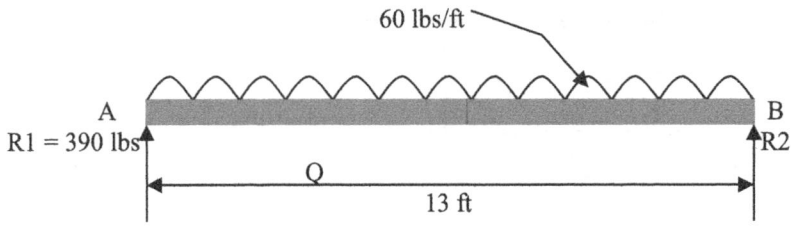

Ans C

Problem 2)   A beam is loaded with a uniform load of 5 kips/ft and a concentrated load of 30 kips. Find the shear force at point P.

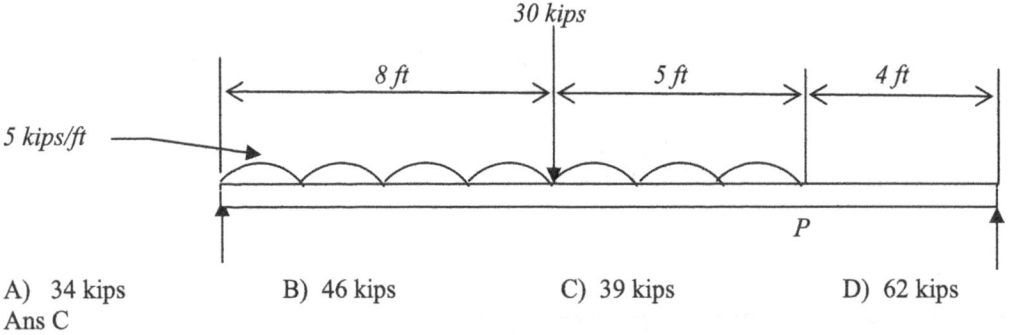

A) 34 kips   B) 46 kips   C) 39 kips   D) 62 kips
Ans C

Problem 4)   Find the maximum bending moment of the beam given in the previous problem.

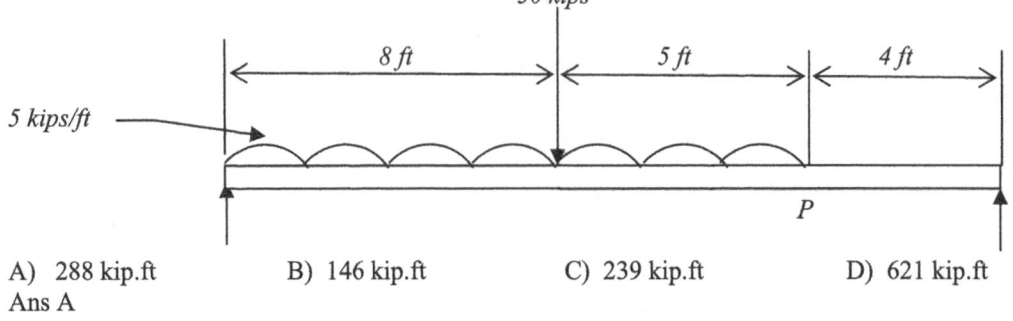

A) 288 kip.ft   B) 146 kip.ft   C) 239 kip.ft   D) 621 kip.ft
Ans A

Problem 5)   Find the deflection at the end of the hollow steel tube shown. The outside diameter of the steel tube is 8 inches and the wall thickness is ¼ in. Young's modulus of steel is 29 x 10^6 psi

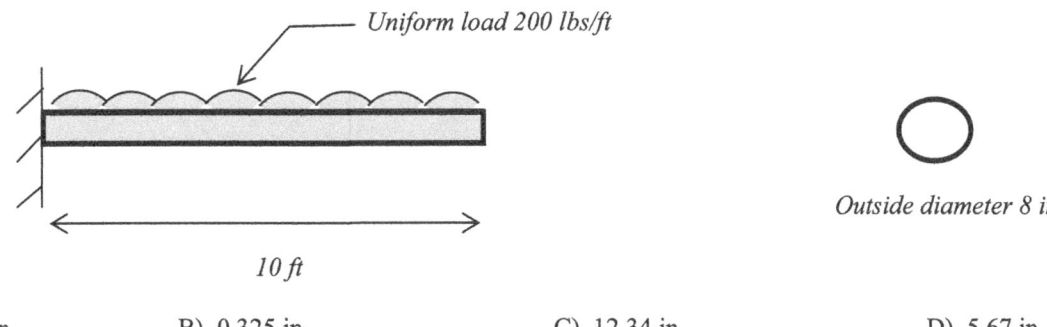

A) 3.61 in   B) 0.325 in   C) 12.34 in   D) 5.67 in

Problem 6) W12 x 96 column has a radius of gyration of 3.09 inches along weaker axis and 5.43 inches along stronger axis. Effective length factor (K) along both axis is equal to 1.0. Young's modulus of steel is 29 x 10^6 in^2. Area of the W section is 28.2 in^2. Unbraced length of the column is 15 ft. Find the elastic buckling load (Euler buckling load) of the column.

A) 1,245 kips   B) 2,368 kips   C) 1,765 kips   D) 876 kips

Problem 7) W12 x 96 column has a radius of gyration of 3.09 inches along weaker axis and 5.43 inches along stronger axis. Effective length factor (K) along both axis is equal to 1.0. Young's modulus of steel is 29 x 10^6 in^2. Area of the W section is 28.2 in^2. Unbraced length of the column is 15 ft. Find the elastic buckling load (Euler buckling load) of the column.

A) 1,245 kips   B) 2,368 kips   C) 1,765 kips   D) 876 kips

Ans B

Problem 8)   Find the deflection at the end of the hollow rectangular section shown. Outside dimensions of the section is 12 x 12 inches and the thickness is 0.25 inches. The Young's modulus of steel is 29 x 10^6 psi.

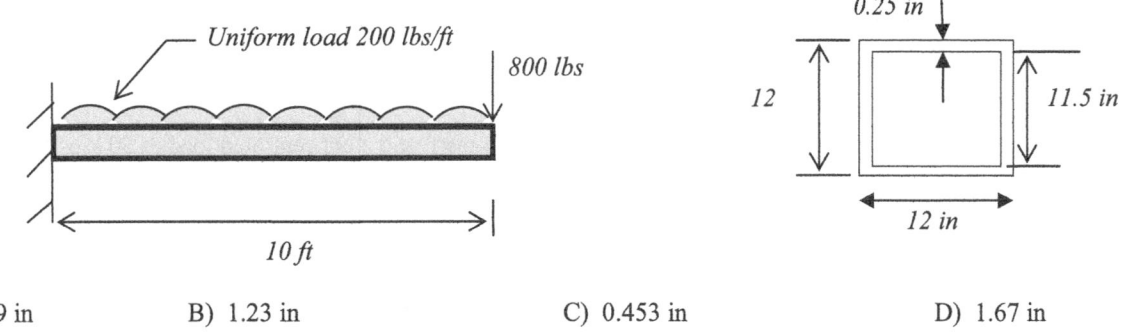

A) 0.149 in   B) 1.23 in   C) 0.453 in   D) 1.67 in

Ans A

Problem 9):   Find the nominal moment of the beam shown.

f_y= 60,000 psi
Concrete compression strength (f_c') = 4,000 psi
A_s = 4.3 sq. in

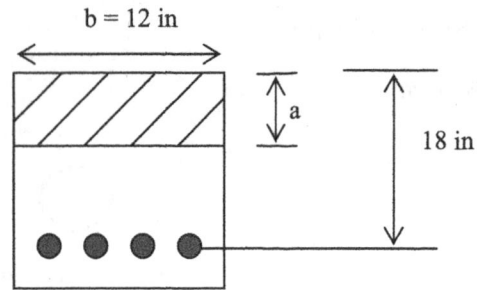

A) 213 kip. ft   B) 319 kip. ft   C) 154 kip. ft   D) 298 kip.ft

Ans (B)

Problem 10)   What is the maximum steel allowed by ACI 318 for the beam shown.
As per ACI 318, steel strain allowed = 0.004
Concrete crushing strain = 0.003
f_y= 60,000 psi
Concrete compression strength (f_c') = 4,000 psi

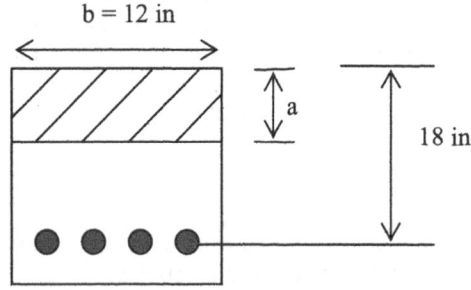

A) 3.45 sq. in   B) 4.46 sq. in   C) 5.93 sq. in   D) 6.12 sq. in

(Ans B)

Problem 11)   A beam is loaded with a uniform load of 5 kips/ft and a concentrated load of 30 kips. Find the shear force at point P.

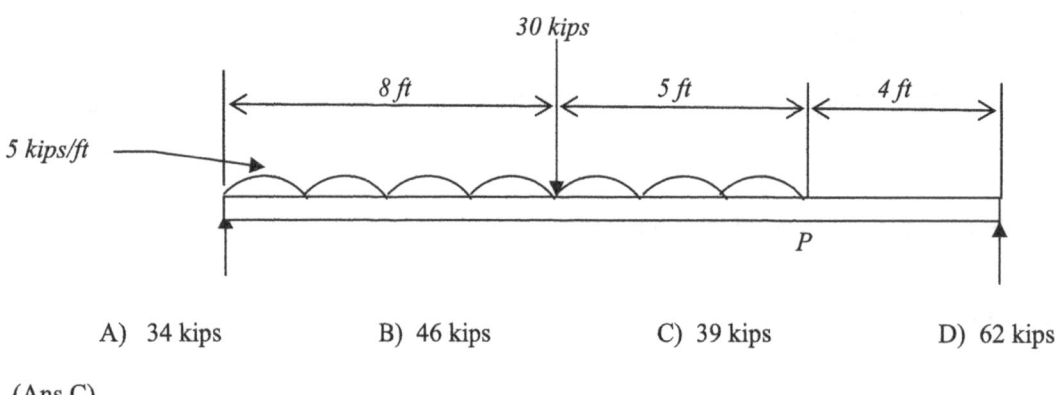

A) 34 kips   B) 46 kips   C) 39 kips   D) 62 kips

(Ans C)

Problem 12)   Find the maximum bending moment of the beam given in the previous problem.

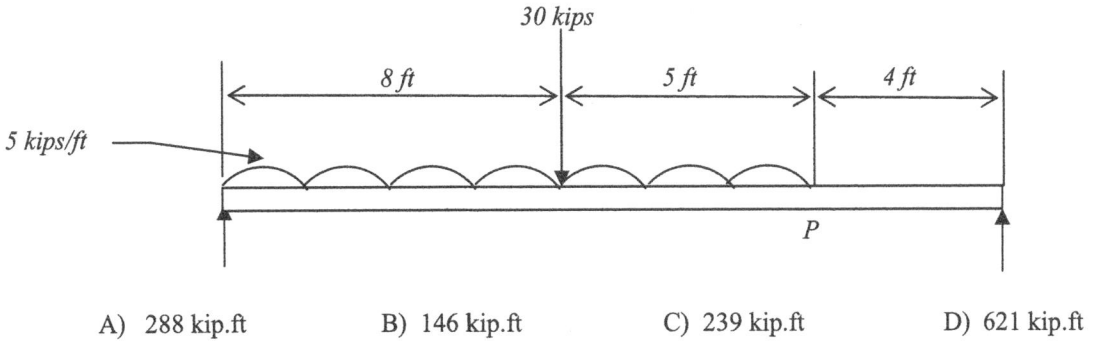

A) 288 kip.ft   B) 146 kip.ft   C) 239 kip.ft   D) 621 kip.ft

Ans A

Problem 13)   A beam is loaded as shown. A distributed load of 60 lbs/ft is applied and a concentrated load of 570 lbs applied at 5 ft from point A. Find the larger shear force at point M.

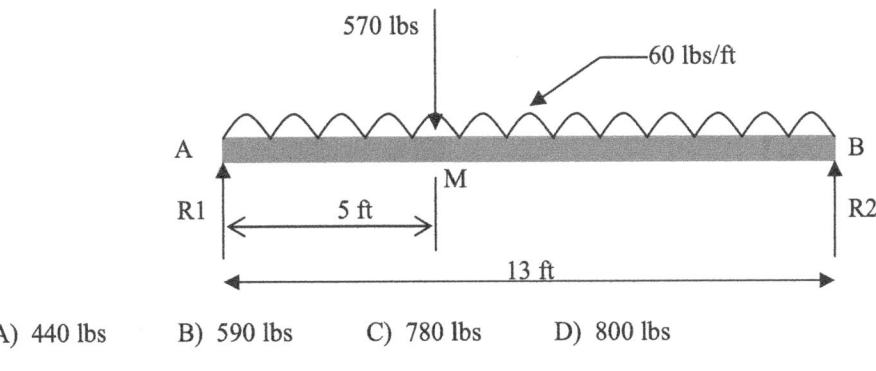

A) 440 lbs   B) 590 lbs   C) 780 lbs   D) 800 lbs

Ans (A)

Problem 14):   Find the maximum bending moment of the beam given in the previous problem.

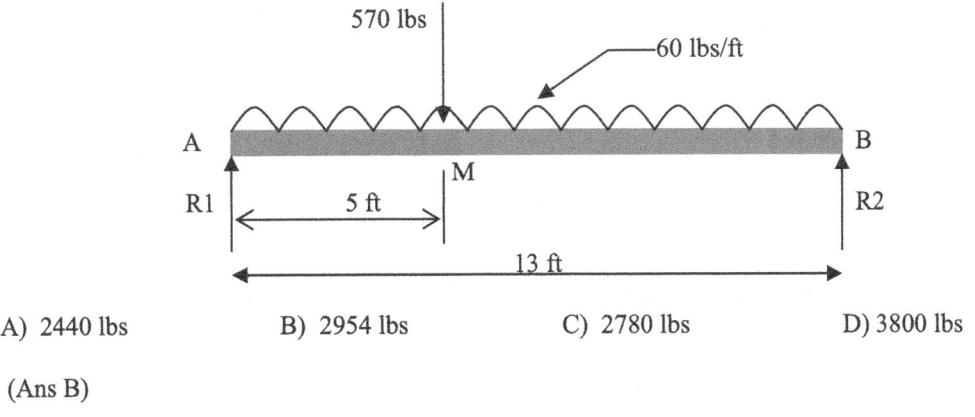

A) 2440 lbs   B) 2954 lbs   C) 2780 lbs   D) 3800 lbs

(Ans B)

Problem 15):   Find the distance (y) to centroid of the T-beam shown.

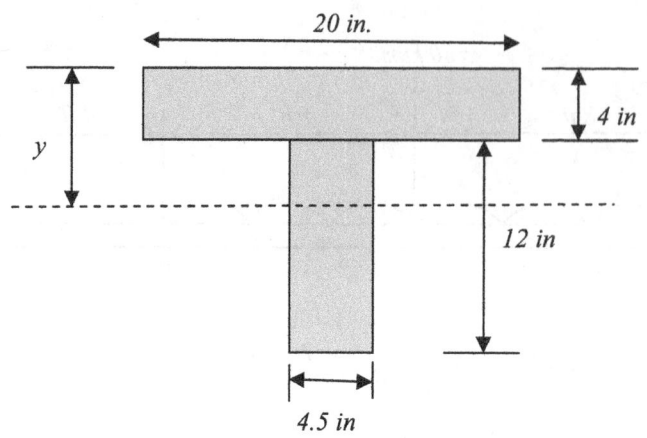

A) 5.22 in   B) 3.78 in   C) 4.12 in   D) 4.81 in
(Ans A)

Problem 16):   Find the nominal moment of the T-beam given.

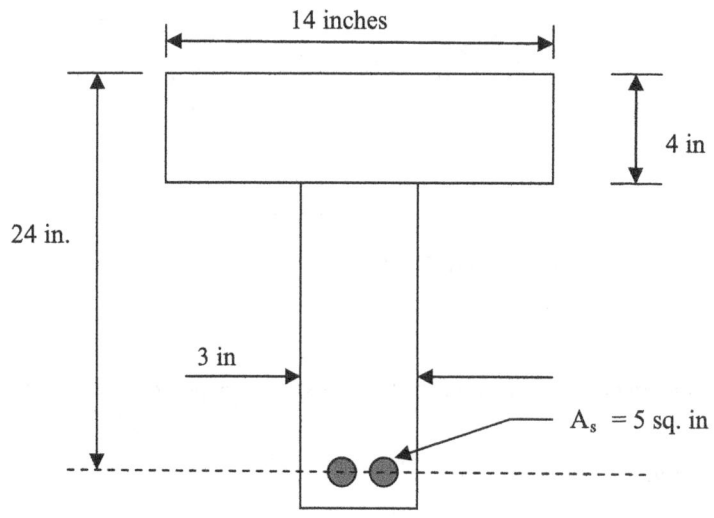

Steel yield stress = 60,000 psi
Concrete compressive strength = 4,000 psi
Steel area = 2 sq. in
A) 483 kip. ft   B) 227 kip. ft   C) 142 kip. ft   D) 331 kip. ft
Ans B)

Problem 17):   Find the total force in the reinforcements in the beam shown. f_y = 60,000 psi.

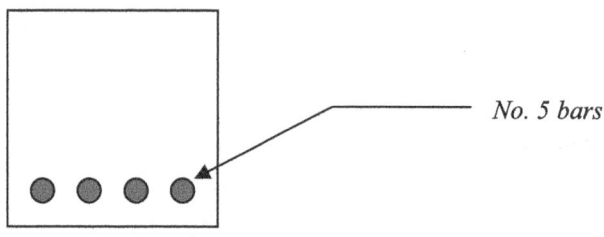

A) 23.8 tons   B) 45.7 tons   C) 36.8 tons   D) 76.9 tons

(Ans C)

Problem 18):   Find the area of steel required for a balanced section for the beam shown below.

Steel Young's modulus = 29 x 10^6 psi.
f_y= 60,000 psi
Concrete compression strength (f_c') = 4,000 psi
Concrete is assumed to crush when the strain in concrete is equal to 0.003.   (ACI 318)

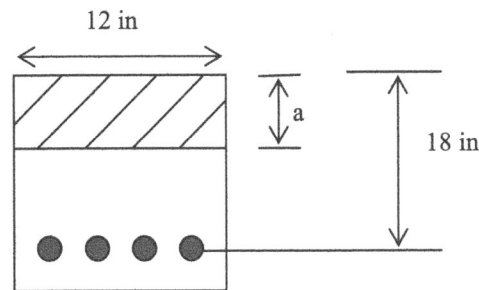

A)   7.08 sq. in   B) 6.12 sq. in   C) 1.24 sq. in   D) 5.44 sq. in

(Ans B)

Problem 19):   Find the nominal moment of the T-beam given.

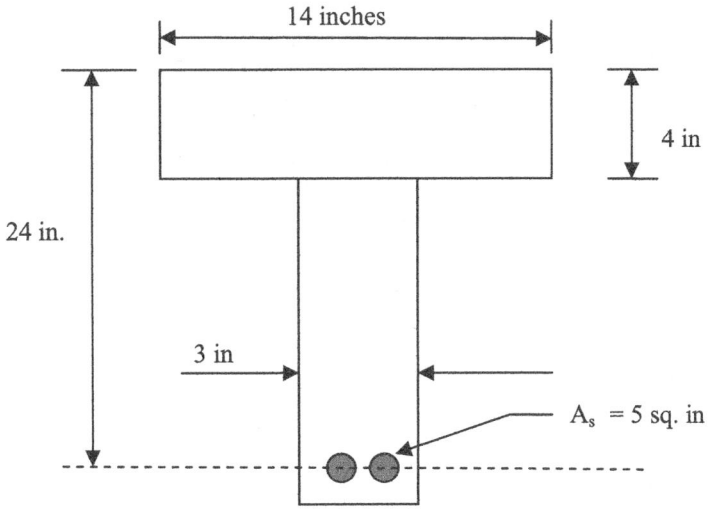

Steel yield stress = 60,000 psi
Concrete compressive strength = 4,000 psi
Steel area = 5 sq. in
A) 483 kip. ft   B) 271 kip. ft   C) 342 kip. ft   D) 381 kip. ft

(Ans A)

Problem 20):   Find the moment of inertia (I) thru the centroid of the T- beam shown.

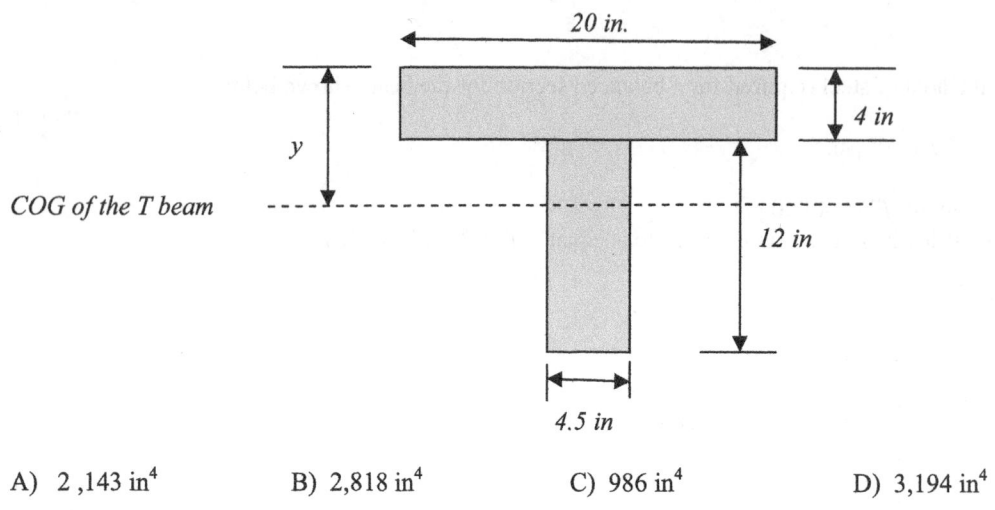

A) 2 ,143 in^4 B) 2,818 in^4 C) 986 in^4 D) 3,194 in^4

Ans (B)

Problem 21): W18 x 67 column has one end pinned and other end fixed. What is the nominal compression strength of the column in kips? Length of the column is 12 ft.

	Effective length factor (K)
Both ends pinned	1.0
One end pinned, other end fixed	0.8
Both ends fixed	0.65

Portion of AISC Table 4.1   (W 18 x 67)

K.L (with respect to least radius of gyration)	Axial compression strength (kips)
7	790
8	762
9	733
10	701
11	667
12	632
13	596

A) 596 kips   B) 632 kips   C) 667 kips   D) 701 kips

(Ans D)

Ruwan Books

Four Sample Exams for the Civil PE Exam
All Modules
Second Edition
With Step by Step Solutions

(WATER RESOURCES/ENVIRONMENTAL,
GEOTECHNICAL, TRANSPORTATION,
CONSTRUCTION AND STRUCTURAL)

Ruwan Rajapakse, PE, CCM, CCE, AVS

Practice for Exam Success!

I

L

M

N

O

P